SEMICONDUCTORS AND SEMIMETALS

VOLUME 17

CW Beam Processing of Silicon and Other Semiconductors

Semiconductors and Semimetals

A Treatise

Edited by R. K. WILLARDSON

WILLARDSON CONSULTING
SPOKANE, WASHINGTON

ALBERT C. BEER

BATTELLE COLUMBUS LABORATORIES
COLUMBUS, OHIO

SEMICONDUCTORS
AND SEMIMETALS

VOLUME 17

CW Beam Processing of Silicon and Other Semiconductors

Volume Editor
JAMES F. GIBBONS

DEPARTMENT OF ELECTRICAL ENGINEERING
STANFORD ELECTRONICS LABORATORIES
STANFORD UNIVERSITY
STANFORD, CALIFORNIA

1984

ACADEMIC PRESS, INC.
(Harcourt Brace Jovanovich, Publishers)

Orlando   San Diego   New York   London
Toronto   Montreal   Sydney   Tokyo

Academic Press Rapid Manuscript Reproduction

ACADEMIC PRESS, INC.
Orlando, Florida 32887

United Kingdom Edition published by
ACADEMIC PRESS, INC. (LONDON) LTD.
24/28 Oval Road, London NW1   7DX

Library of Congress Cataloging in Publication Data

Main entry under title:

Semiconductors and semimetals.

  Vol. 13 edited by K. Zanio; v. 17 edited by James F.
Gibbons.
    Includes bibliographical references and indexes.
    1. Semiconductors--Collected works. 2. Semimetals--
Collected works. I. Willardson, Robert K. II. Beer,
Albert C.
QC610.9.S47   537.6'22   65-26048
ISBN 0-12-752117-8 (v. 17)

Contents

Chapter 1 Beam Processing of Silicon

James F. Gibbons

Chapter 2 Temperature Distributions and Solid Phase Reaction Rates Produced by Scanning CW Beams

Arto Lietoila, Richard B. Gold, James F. Gibbons, and Lee A. Christel

Chapter 3 Applications of CW Beam Processing to Ion Implanted Crystalline Silicon

Arto Lietoila and James F. Gibbons

Chapter 4 Electronic Defects in CW Transient Thermal Processed Silicon

N. M. Johnson

Chapter 5 Beam Recrystallized Polycrystalline Silicon: Properties, Applications, and Techniques

K. F. Lee, T. J. Stultz, and James F. Gibbons

Chapter 6 Metal–Silicon Reactions and Silicide Formation

T. Shibata, A. Wakita, T. W. Sigmon, and James F. Gibbons

Chapter 7   CW Beam Processing of Gallium Arsenide

Yves I. Nissim and James F. Gibbons

List of Contributors

Numbers in parentheses indicate the pages on which the authors' contributions begin.

LEE A. CHRISTEL[1], *Stanford Electronics Laboratories, Stanford University, Stanford, California 94305* (71)

JAMES F. GIBBONS, *Department of Electrical Engineering, Stanford Electronics Laboratories, Stanford University, Stanford, California 94305* (1, 71, 107, 227, 341, 397)

RICHARD B. GOLD[2], *Stanford Electronics Laboratories, Stanford University, Stanford, California 94305* (71)

N. M. JOHNSON, *Xerox Corporation, Palo Alto Research Center, Palo Alto, California 94304* (177)

K. F. LEE[3], *Stanford Electronics Laboratories, Stanford University, Stanford, California 94305* (227)

ARTO LIETOILA[4], *Stanford Electronics Laboratories, Stanford University, Stanford, California 94305* (71, 107)

YVES I. NISSIM[5], *Stanford Electronics Laboratories, Stanford University, Stanford, California 94305* (397)

T. SHIBATA[6], *Stanford Electronics Laboratories, Stanford University, Stanford, California 94305* (341)

T. W. SIGMON, *Stanford Electronics Laboratories, Stanford University, Stanford, California 94305* (341)

T. J. STULTZ[7], *Stanford Electronics Laboratories, Stanford University, Stanford, California 94305* (227)

A. WAKITA, *Stanford Electronics Laboratories, Stanford University, Stanford, California 94305* (341)

[1]Present address: SERA Solar Corporation, Santa Clara, California 95054.
[2]Present address: Adams–Russell Company, Burlington, Massachusetts 01803.
[3]Present address: AT&T Bell Laboratories, Holmdel, New Jersey 07733.
[4]Present address: Micronas, Inc., 00101 Helsinki, Finland.
[5]Present address: CNET, 92220 Bagneux, France.
[6]Present address: VLSI Research Center, Toshiba Research and Development Center, Kawasaki City, Kanagawa, 201 Japan.
[7]Present address: TS Associates, San Jose, California 95128.

ix

Preface

This book had its beginnings at the 1979 winter meeting of the American Physical Society, at which I was asked to review the basic features of cw laser processing of ion implanted silicon. The review covered techniques for calculating laser-induced temperatures, experimental results on the annealing of ion implanted layers in silicon, and enough material on structural (TEM) and electronic (DLTS) defects to suggest that the process was reasonably well behaved and perhaps had a future in materials preparation and processing. Al Beer, who was attending the meeting, thought the subject might be suitable for a short research monograph that would be useful to those entering the field. We agreed that, as things seemed reasonably tidy, the project could easily be finished within a year. However, the field was just then entering a period of very rapid growth. The fabrication of MOSFETS on polysilicon films recrystallized by a cw scanning laser, reported in July 1979, opened up the field of silicon-on-insulators (SOI). There followed a succession of significant contributions to this technology at a number of laboratories, dealing with (1) such things as thermal profile management, beam shaping, seeding, and zone recrystallization on the materials side and (2) a number of new device configurations, including the general technique of three-dimensional integration, on the applications side.

Simultaneously, the use of a cw scanning laser beam to react metals with silicon to form single-phase silicide layers was being explored; and its potential consequences for materials, device, and integrated circuit technology required that it be included if the book was to offer a broad, balanced view of the field. Finally, as these areas began to settle down enough to be reviewed with some reasonable perspective, the rapid thermal processing technique was reported in Japan. Its natural relationship to the cw scanning laser process and its "throughput" advantage from a manufacturing point of view then required a further modification of many chapters of the book.

The result is that 5 years have passed since Al Beer and I shook hands on the project. The cw beam technology as it relates to semiconductor processing is now much more mature and better understood than it was in 1979–1980, though its applications in the semiconductor industry are very likely still in their infancy. It is on this hopeful note that my coauthors and I now offer this volume.

Chapter 1 provides an overview of the general field of beam processing, with a brief discussion of the main topics to be developed in depth in the succeeding chapters. Chapter 2 deals with the calculation of cw beam-induced temperature profiles in semiconductor substrates under a variety of conditions that are found in practical applications. A review of the use of cw beam processing for annealing ion implanted silicon is presented in Chapter 3, and the question of electronic defects remaining after various implantation and cw beam annealing cycles is reviewed in Chapter 4. In Chapter 5 we discuss the recrystallization of thin polysilicon films by cw beam systems and summarize its applications in device fabrication.

The use of a cw beam to react a thin metal film with a silicon (crystalline or polycrystalline) substrate to form a silicide is reviewed in Chapter 6, together with oxidation characteristics and other properties of the silicide layers that are important for practical applications. Finally, in Chapter 7 we review the current state of the art regarding the use of cw beams for processing of GaAs. It will be apparent from this chapter that much remains to be done in the beam processing of compound semiconductors.

To conclude this preface, I would like to express my most sincere appreciation to the authors of these chapters. We were all privileged to share the experience and excitement of working in a field during the early stages of its development, both as colleagues at Stanford and as members of the larger community of "beam annealers" whose pioneering work is described here in terms that we hope will reflect the respect we have for all of them. My coauthors and I would also like to acknowledge our enormous indebtedness to Ms. Mary Cloutier, both for her extraordinary skill in preparing this manuscript and for the patience and unfailing good humor that she maintained through its several revisions.

Finally, I want to thank Dick Reynolds and Sven Roosild at DARPA, Ben Wilcox at NSF, and Horst Wittman at ARO for the support that enabled the Stanford group to participate in the development of the field.

JAMES F. GIBBONS

CHAPTER 1

Beam Processing of Silicon

James F. Gibbons

STANFORD ELECTRONICS LABORATORIES
STANFORD UNIVERSITY
STANFORD, CALIFORNIA

High power lasers have been used in the semiconductor industry for many years, for applications such as wafer scribing, resistor trimming and the drilling of ceramics, where the clean removal of a specific amount of material is required; and for contact alloying, where permanent, relatively extensive chemical changes are sought [1.1]. Despite the mature development of these applications, however, lasers and electron beams have only recently been considered for the processing operations that are required during silicon device fabrication.

The earliest work with Q-switched lasers was performed by Russian workers [1.2-1.4], who showed that ion implanted silicon could be annealed successfully with laser pulses of sufficient energy and duration. The importance of this work was quickly appreciated by workers at several different laboratories in Europe [1.5-1.7] and the United States [1.8-1.10], and a large volume of published work has now appeared on the use of both Q-switched lasers and pulsed electron beams [1.11] for semiconductor processing. At the same time, the use of scanning cw laser and electron beams for annealing ion implanted silicon was being explored at Stanford [1.12-1.14], with a particular focus on its application to semiconductor device processing. More recently, both scanning and stationary incoherent light sources have been used to achieve annealing of ion implanted silicon with a wafer throughput that is compatible with the needs of the semiconductor industry.

As a result of these efforts, beam annealing of ion implanted silicon has now been rather thoroughly studied and shown to possess several potential advantages over the furnace annealing process. In addition, lasers, electron beams and arc sources have been used to recrystallize vapor deposited polysilicon, with substantial improvements in its electronic properties [1.15,1.16]; to facilitate the formation of metal silicides [1.17-1.20]; and to perform a number of other processing functions that are of increasing importance in the fabrication of fine geometry integrated circuits and high speed devices. In this chapter we will review the basic mechanisms by which beam annealing proceeds and summarize the basic metallurgical and electronic properties of beam-annealed material.

Two distinctly different beam annealing mechanisms have been identified, depending on the duration of the beam exposure. For Q-switched lasers or pulsed electron beams, exposure times are typically in the range of 5 ns - 500 ns and the annealing process then involves the formation of a thin surface layer of molten silicon that recrystallizes on the underlying substrate when the radiation is removed. If the irradiated sample is an ion implanted single crystal and the depth of the molten layer is sufficient to envelop the implantation damaged region, the molten layer regrows by a liquid phase epitaxial process on the crystalline substrate to produce material with a very high degree of structural perfection and very superior electronic properties.

For cw systems, on the other hand, the silicon surface is typically exposed to the beam for 0.1-10 ms and, in some cases, for durations of several tens of seconds. The annealing of ion implanted material can then proceed by a solid phase epitaxial regrowth process at temperatures that are well below the melting point. As in the pulsed beam case, a very high degree of crystalline perfection and very superior electronic properties can be obtained under appropriate annealing conditions. However, the absence of melting proves to be of some interest since no redistribution of the implanted impurity profile then occurs during the annealing process, whereas significant impurity redistribution occurs when annealing is effected by a pulsed laser or electron beam. These and other differences in the two annealing processes make it convenient to discuss them separately. In what follows we first consider the pulsed beam annealing process and then take up the cw alternative.

1.1 CENTRAL FEATURES OF PULSED BEAM ANNEALING OF ION IMPLANTED SILICON

We will begin with a brief review of experimental data that characterize the annealing of ion implanted silicon using a Q-switched laser and then turn to a brief mathematical analysis of the process.

Laser annealing has been demonstrated using a variety of sources and pulse durations. In most of the experiments reported so far, implanted samples have been annealed directly in the laboratory ambient; i.e., no special precautions have been taken to immerse the wafer in an inert environment during annealing. The area irradiated by the beam is, in most cases, large compared to both the diffusion length for heat in the solid and the thickness of the wafer, so the process can be treated as being basically one dimensional.

Experimental results from various laboratories are reasonably consistent and are summarized in Table 1.1. For convenience, the data are selected to illustrate the annealing of As^+-implanted silicon with the As^+ implantation conditions chosen to provide amorphous regions of differing thicknesses at the sample surface.

The principal features of the data are as follows. For pulses of 25-100 ns duration (typical of a Q-switched laser):

1. Surface melting is initiated at a threshold energy of 0.2-3.5 J/cm^2, depending on the laser wavelength and the thickness of the amorphous layer.

2. Full annealing of an ion implanted layer (as judged by crystal recovery and electrical activity) requires additional energy in an amount that also depends on both the laser wavelength and the thickness of the amorphous layer.

TABLE 1.1. Experimental Results for Pulsed Laser Annealing of As^+-Implanted Silicon

Laser	As^+ Imp. Par.	X_d	Melt Initiated	Fully Recovered	Ref.
Ruby, 50 ns	5×10^{15}, 400 keV	4300Å		2 J/cm^2	[5]
Ruby, 60 ns	1.4×10^{16}, 100	1700Å	0.64	1.4	[8]
Nd:YAG, 110 ns	8×10^{15}, 100 keV	~1500Å		6 J/cm^2	[7]
40 ns	10^{15}, 130 keV	468	3.5	4.5	[21]
Nd:YAG, Doubled 40 ns	10^{15}, 30 keV	468	0.2	0.8	[21]

Early evidence suggesting the surface melting theory of pulsed laser annealing was provided by Auston et al. [1.21], who measured the optical reflectivity of samples during the annealing cycle with a low power (2mW) continuous He-Ne laser. Figure 1.1 shows a typical time resolved surface reflectivity measurement when a Nd:YAG laser operated at a wavelength of 0.53 μm with a pulse width of 30 ns and a total energy density energy of 2.75 J/cm^2 is used to irradiate a $\langle 100 \rangle$ silicon sample implanted with As^+ at 30 keV to a dose of $10^{15}/cm^2$. As the annealing laser turns on, the initial reflectivity of the amorphous layer, R_a, rises with increasing sample temperature to a value R_a', where surface melting is initiated. The reflectivity then rises abruptly to the reflectivity of molten silicon and remains at this value as the liquid-solid interface penetrates more deeply into the sample. The penetration ceases at a time after the annealing laser pulse has terminated that depends on the total energy absorbed during the pulse.

As the liquid-solid interface reverses its direction and proceeds back toward the surface, the molten zone regrows on the underlying silicon crystal. The reflectivity ultimately drops to R_c', the reflectivity of the hot solid. The surface is melted for a total time τ. The fall time, τ_f, is the time required for the regrowth front to move through one optical skin depth and is a measure of the regrowth velocity. The reflectivity decreases to its room temperature value for crystalline silicon, R_c, as the surface cools.

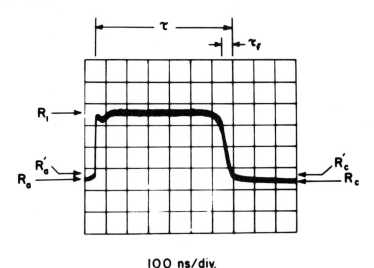

100 ns/div.

FIGURE 1.1.   Time resolved surface reflectivity trace (after Ref. 1.21).

1.1.1   Qualitative Analysis of the Melt Threshold

The initial interaction of a light beam with a material is always with its electrons. Energy absorbed by the electrons is ultimately shared with the atoms of the material as the electronic excitation is transformed into heat. Under most conditions the transfer of electronic excitation to heat is accomplished with relaxation times on the order of 1 ps or less and may therefore be considered to be instantaneous. The basic process of pulsed beam annealing then consists of the absorption of sufficient energy to melt a layer of silicon having a thickness at least equal to the thickness of the implantation-damaged region, followed by liquid phase epitaxial regrowth of the melted layer. The laser energy required to bring the surface of the sample to its melting temperature is therefore a critical parameter in the process.

To estimate the threshold energy required to produce surface melting, we consider the sample geometry shown in Fig. 1.2 [1.22]. The sample consists of an implantation-damaged layer of thickness x_d and optical absorption coefficient α_d resting on a crystalline solid with optical absorption coefficient α_c. A laser pulse of intensity I_0 and duration τ_p illuminates the front surface of the sample at normal incidence. The reflection coefficient at this surface is R. The laser intensity is assumed to be sufficiently low that ordinary one-photon optical absorption processes are dominant.

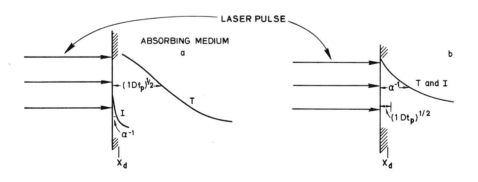

FIGURE 1.2.   Intensity and temperature profiles in laser-irradiated silicon. (a) Penetration depth of light small compared to $(D\tau_p)^{1/2}$. (b) Penetration depth of light large compared to $(D\tau_p)^{1/2}$ (after Ref. 1.22).

The relevant thermal properties of the material are its specific heat C_v, its mass density ρ and its thermal conductivity K. The heat diffusivity is then given by $D = K/\rho C_v$ and determines a characteristic length $(D\tau_p)^{1/2}$. This length gives an estimate of the distance over which the temperature profile is spread by heat diffusion during the laser pulse.

To estimate the surface temperature rise produced in a limiting case of practical interest, we suppose that the laser energy is absorbed in a distance small compared to the diffusion length for heat $[\alpha^{-1} \ll (D\tau_p)^{1/2}]$. Under these conditions [illustrated in Fig. 1.2(a)] the absorbed laser energy can be considered to heat a slab of thickness $(D\tau_p)^{1/2}$. The corresponding temperature rise ΔT is then obtained from

$$(I_0\tau_p)(1-R) = \rho C_v \Delta T (D\tau_p)^{1/2} \tag{1.1}$$

This formula suggests that the surface temperature is proportional to the <u>energy</u> in the laser pulse $(I_0 \tau_p)$ and predicts a <u>melt threshold energy</u> of

$$E_m' = \frac{\rho C_v (D\tau_p)^{1/2}}{(1-R)} (T_m - T_0) \qquad \alpha^{-1} \ll (D\tau_p)^{1/2} \tag{1.2}$$

where T_m is the melting temperature and T_0 is the ambient temperature.

The extreme temperature change envisaged ($\sim 1400°C$) of course makes it necessary to account for the temperature dependence of the thermal and optical parameters appearing in Eq. (1.2) in order to arrive at a meaningful estimate of E_m' for practical use. The parameters that are customarily used for process analysis are given in Table 1.2.

The thermal conductivity is an especially sensitive function of temperature and its temperature dependence must be taken into account to obtain accurate theoretical predictions. For numerical estimates we use the average value of K calculated from

$$\overline{K} = \frac{1}{(T-T_0)} \int_{T_0}^{T} K(T) \, dT \tag{1.3a}$$

with K(T) for silicon given by

$$K(T) = \frac{299}{T-99} \qquad (W/cm°K) \tag{1.3b}$$

TABLE 1.2. Thermal and Optical Parameters for Solid Silicon. The optical absorption coefficients for implantation-amorphized silicon have not been measured; values quoted for α_d are for sputtered silicon.

Parameter	Value at T = 300°K	Value at T = 1685°K	Ref.
C_V, J/gm-°K	0.65	0.95	[23]
ρ, gm/cm^3	2.33	2.33	[23]
K, W/cm-°K	1.45	0.25	[24]
R	0.35	0.7	[25]
α_d (λ = 0.69 μm)	3×10^4	–	[26]
α_d (λ = 1.06 μm)	$3-6\times10^3$	–	[26]
α_c (λ = 0.69 μm)	3×10^3	–	[27]
α_c (λ = 1.06 μm)	10	–	[27]

Substitution of numerical values from Table 1.2 leads to values of E_m' that are in satisfactory agreement with experiment for short wavelength irradiation such as that from ruby or frequency-doubled Nd:YAG lasers, where the basic assumption of heat transport made in the analysis is at least approximately valid [i.e., $\alpha^{-1} < (D\tau_p)^{1/2}$]. The validity of this assumption is improved if an implantation-amorphized sample is irradiated, of course, since the optical absorption coefficient is then increased.

For Q-switched Nd:YAG lasers operated at the fundamental frequency, however, the optical absorption depth (α^{-1}) in crystalline silicon is substantially greater than the diffusion length for heat. An order of magnitude estimate of the melt threshold energy E_m^* can be made for this case by assuming that the thickness of the slab that is heated by the radiation is (α^{-1}). Replacing $(D\tau_p)^{1/2}$ with α^{-1} in Eq. (1.2), we obtain

$$E_m^* = \frac{\rho C_v(T_m-T_o)}{\alpha(1-R)} \qquad \alpha^{-1} < (D\tau_p)^{1/2} \qquad (1.4)$$

In practical circumstances, however, the melt threshold predicted by Eq. 1.4 proves to be too large for a variety of reasons. For the particular case of ion implanted silicon with damage depths of 0.1 - 0.3 μm, free carrier absorption in the underlying crystalline material and bandgap narrowing during the heating cycle have a profound effect on the surface temperature. The possible effect of a dense free carrier plasma is also potentially important and needs to be included in the analysis. In the next section we describe a model that has been developed to calculate surface temperatures when these and other effects are included.

1.1.2   Quantitative   Analysis   of   the Melt Threshold Energy for
          Ion-Implanted Crystalline Si

     In this section we present the results of computer modeling
of the temperature and carrier concentration versus time in pulsed
laser irradiated Si before melting occurs [1.28,1.29]. The results
are obtained by simultaneously solving the differential equations
for temperature, carrier concentration (including diffusion) and
laser intensity within the sample.   Free carrier absorption,
Auger recombination and the temperature dependence of the thermal
conductivity are included.   In addition, the model accounts for
the temperature dependence of the bandgap and lattice absorption
coefficient; the electronic contribution to the thermal conductivity (this turns out to be insignificant); the fact that when
a photon creates an electron-hole pair, the photon energy is not
immediately transferred to the lattice; and electron-hole scattering when calculating the ambipolar diffusivity of carriers.
In addition, accurate values for the reflectivity and absorption
coefficient of the amorphous silicon are employed.

     With regard to the kinetics of the energy coupling, the
following assumptions are made:

     1.   Electron and hole concentrations are equal; carrier
diffusion is characterized by the ambipolar diffusion coefficient.

     2.   The dominant recombination mechanism is Auger recombina-
tion, where the gap energy is transferred to an already existing
free carrier; the reverse process, impact ionization, is not
included.

     3.   The release of energy from the hot carriers to the lattice
is characterized by an energy relaxation time, τ_e.

     4.   The diameter of the laser spot is large compared to the
thickness of the amorphous layer; i.e., the problem is one dimensional.

     Using these assumptions the differential equations for tem-
perature T, carrier concentration p, and laser intensity I, are
formulated and solved for the surface temperature as a function
of time prior to melting. The results of these calculations can
be summarized as follows.

1.1.2.1 Comparison of Calculated Melt Threshold with Experimental
          Data

     Auston et al. [1.21] have determined the threshold energy,
E_r, required to reach an enhanced reflectivity of ~70% with

a 30–40 ns pulse at .53 μm and 1.06 μm; both crystalline Si and samples with a 468 Å amorphous surface layer were studied.

The computer model was used to study these annealing cases. A comparison between experimental and calculated threshold energies is given in Table 1.3. As can be seen, the agreement is reasonable, and the shape of the pulse does not appear to be a crucial parameter. Note also that the melt thresholds for crystalline Si and implantation amorphized silicon are very similar at a laser wavelength of 1.06 μm. The greatest difference between calculation and experimental data is at a laser wavelength of .53 μm. The difference may be due to inaccuracy in the formula used to calculate optical absorption coefficient at a laser wavelength λ =.53 μm at high temperatures (see Ref. 1.28).

TABLE 1.3. A Comparison Between Experimentally Determined Thresholds, E_r, for Achieving the Enhanced Reflectivity (Ref. 1.21), and Calculated Energies, E_m, Required to Reach Silicon Melting Temperature for Nanosecond Pulses.

λ (μm)	Amorphous Layer (nm)	Pulse Length (ns)	E_r, Exp. (J/cm^2)	E_m, Calc. (J/cm^2) Constant Intensity	E_m, Calc. (J/cm^2) Gaussian Pulse
1.06	0	40	5.5	3.8	4.8
1.06	46.8	40	3.7	3.3	4.1
0.53	0	30	0.35	0.47	0.56
0.53	46.8	30	0.2	0.27	0.35

1.1.2.2 Free Carrier Absorption (FCA) and Bandgap Narrowing

The importance of free carrier absorption and the variation of the semiconductor bandgap with temperature in determining the melt threshold are shown in Fig. 1.3, which illustrates the surface temperature vs time of a crystalline Si sample being irradiated with a constant intensity of 160 MW/cm^2 at λ = 1.06μm. The figure shows that the predicted E_m is doubled if the temperature dependence of E_g is omitted. If free carier absorption is omitted, then the behavior changes drastically: after 70 ns of radiation, which corresponds to a total energy density of 11.2 J/cm^2, the surface temperature has risen only to about 70°C.

The evolution of the lattice and free carrier absorption coefficients (for 1.06 μm, 160 mW/cm^2) is shown in Fig. 1.4. Using the results presented in Fig. 1.4, the temperature behavior shown in Fig. 1.3 can now be explained as follows. At the outset

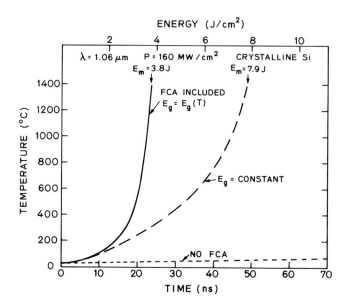

FIGURE 1.3.   The surface temperature vs. pulse duration in crystalline Si, when free carrier absorption and temperature dependence of bandgap are included (solid line) or ignored (dashed lines); λ = 1.06 μm and P = 160 mW/cm².

of the pulse, lattice absorption (~ 10 cm^{-1}) is sufficient to create some free carriers. Those carriers then absorb radiation which is converted to heat. This causes the bandgap to shrink and lattice absorption to increase. Thus more carriers are generated, which means stronger free carrier absorption, and so on. The process is thus self-regenerative, so that after about 20 ns the temperature rises rapidly to the melting point. The two absorption coefficients are of the same order of magnitude most of the time, but the self-regenerative process could never start without free carrier absorption.

In contrast to the behavior at 1.06 μm, the free carrier absorption process is found to be completely insignificant at a laser wavelength of .53 μm. Also, the effect of bandgap narrowing is much less profound, as is illustrated in Fig. 1.5. Notice that for this figure a higher constant intensity (50 mW/cm²) has been used than that used for comparison with the experimental data. The reason is that 50 MW/cm² is more typical of the range used for practical annealing. The consequence of this increase in the power density is a slight decrease in the melt threshold. Figure 1.5 also shows the rather minute effect of the electronic heat conduction, κ_e.

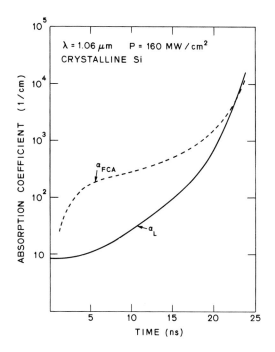

FIGURE 1.4. Lattice and free carrier absorption coefficients vs. time under same conditions as in Fig. 1.3 (temperature dependence of bandgap is included).

For the .53 μm radiation, the temperature rise with time does not exhibit the explosive character observed at 1.06 μm. The reason is that only lattice absorption is important, and it is already strong at room temperature. The time derivative of the surface temperature increases, however, during the pulse, since lattice absorption increases with temperature.

1.1.2.3 Energy Relaxation Time

A very imaginative proposal regarding the mechanism of pulsed laser annealing has been made by Van Vechten [1.30], who argues that intense laser irradiation may produce such large carrier concentrations that carrier screening effects would increase the energy relaxation time substantially. If this were the case, electron–hole plasma effects rather than simple surface melting would be required to explain the annealing phenomena observed under pulsed laser irradiation. To study this possibility, the energy relaxation time τ_e was varied in the computer program

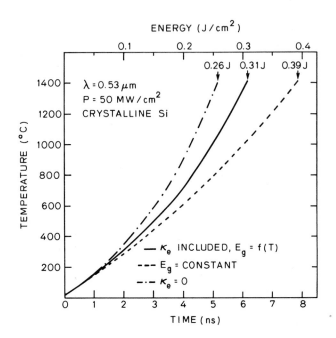

FIGURE 1.5.   The effect of bandgap narrowing and electronic heat conduction on the temperature evolution during annealing with 0.53 μm radiation. Free carrier absorption does not have any appreciation effect at this wavelength.

from 10 ps up to 10 ns for the .53 μm, 50 mW/cm^2 radiation. The The results are given in Fig. 1.6, where the surface temperature vs. time is plotted for several values of τ_e. For τ_e of less than 0.1 ns, there is <u>no</u> effect of τ_e on the results. If τ_e is increased to 1 ns, there is a slight increase in E_m. Increasing τ_e to 10 ns causes E_m to be about three times as much as for the low τ_e values, but melting temperature is still reached with less than 1 J/cm^2.

These results indicate that unless the energy relaxation time is at least comparable to the length of the laser pulse, the coupling between the carriers and the lattice is strong enough to assure that the sample surface reaches the melting point in typical pulsed laser processing cases. Of course, extremely large values of τ_e would delay lattice heating so much that carrier diffusion would cause the energy to be deposited in a large enough volume to prevent surface melting, as suggested by Van Vechten. Calculations show that, for the pulse used above (1 J/cm^2 in 20 ns), τ_e would have to be about 100 ns before melting could be avoided [1.28].

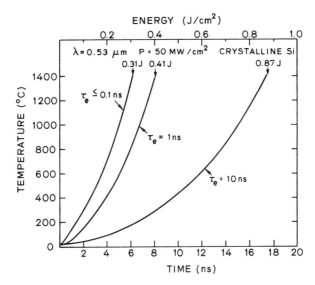

FIGURE 1.6. The effect of the energy relaxation time, τ_e, on the annealing results for 0.53 μm laser radiation.

1.1.3 Calculation of Carrier and Lattice Temperature Induced in Si by Picosecond Laser Pulses

While the energy relaxation time is a noncritical parameter for nanosecond pulse annealing, the situation may be different when picosecond laser pulses are used. To study this case, the previous equations were used together with the assumption that excited carriers instantaneously establish a Maxwell-Boltzmann distribution characterized by a carrier temperature T_e (equal for holes and electrons) [1.29].

The computer model was used to study annealing with laser pulses of 20 ps duration in the energy range 0.1 - 0.5 J/cm^2 at a laser wavelength of 0.53 μm. Both carrier-concentration-dependent and constant values of the energy relaxation time τ_e were used with little difference in the basic results. Table 1.4 shows that the choice of τ_e has a profound effect on both the carrier temperature and the melt threshold.

For comparison with experiment, Liu et al. [1.31] have determined that the melt threshold for 20 ps pulses at a wavelength of 0.53 μm is 0.2 J/cm^2 for <111> crystalline Si. Liu et al. also performed in situ electron emission measurements which indicated that the maximum electron temperature achieved is ~5000°K. As can be seen from Table 1.4, these data fit the calculations extremely well for a constant energy relaxation time of 1 ps.

TABLE 1.4.   Results of Calculations for the Constant Thermal-
ization Time τ_e.   The   melt   thresholds   are   indicated   on   the
bottom line.

Energy density (J/cm^2)	Maximum carrier temperature		
	$\tau_e = 0.1$ ps	1 ps	10 ps
0.1	950 K	2600 K	6 400 K
0.2	2200 K	5500 K	9 500 K
0.4	2600 K	7500 K	21 000 K
0.5	2900 K	8500 K	23 500 K
E_m	0.16 J/cm^2	0.19 J/cm^2	0.4 J/cm^2

1.1.4   Calculation of Melt Depth vs Pulse Energy

In practical annealing situations it is necessary to melt the
silicon to a depth that exceeds the depth of the damage produced by
the ion implantation.   This requires increasing the pulse energy
beyond the melt threshold value by an amount that is, very roughly,
$H \rho x_d/(1-R_\ell)$, where   H   is   the   heat   of   fusion   and R_ℓ is   the
reflection coefficient of liquid silicon.

To study   this   situation,   calculations   of   the   melt   depth
versus time have been carried out by a number of authors (Baeri et
al. [1.32] and Wang et al. [1.33]) with the   general   result   shown
in Fig. 1.7.   These calculations were performed [1.32] for irradiation with a ruby laser with a pulse duration of 50 ns and the
laser energies as shown in the figure.   A surface amorphous layer
of silicon having a thickness of 0.2 μm is assumed to approximate a typical damage depth produced by ion implantation.   As an
example we see that for a pulse energy of 2.5 J/cm^2, the calculations predict a maximum melt depth of about 6000 Å which occurs
about 100 ns after the laser pulse is initiated.   The propagation
rate of the the liquid-solid interface toward the surface can be
estimated by observing that the melt front moves from a depth of
~0.6 μm to   the   surface   in   about   300   ns,   which   leads   to   an
estimate of the regrowth rate of ~2 m/sec.   The fact that this
growth process   is   actually   epitaxial   recrystallization   is   of
course not predicted by the   model   and   must   be   established   by
transmission electron microscopy, ion channeling or other crystallographic measurement technique.

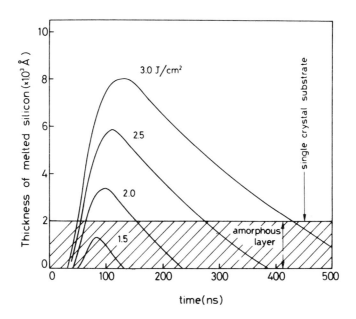

FIGURE 1.7.   Calculations of the penetration and retreat of the melt front in a-Si (200 nm)/Si structure for ruby laser pulses of various energies.

1.1.5   Impurity Profiles in Pulse Annealed Silicon

From a device applications viewpoint, one of the most important features of the process has to do with impurity redistribution effects which occur during the annealing process. Those effects have been thoroughly catalogued by the extensive early work of White and co-workers [1.9,1.34].

Figure 1.8 shows B profiles in as-implanted and laser annealed silicon samples taken from their work. The measurements were made by secondary ion mass spectroscopy (SIMS) and therefore yield the total concentration of boron without regard for lattice location. The samples were implanted at 35 keV to a dose of $1 \times 10^{16}/cm^2$. The as-implanted profile is approximately Gaussian as expected, and is only marginally changed by conventional furnace thermal annealing at 900°C for 30 minutes. In contrast, the profile measured after laser annealing (1.6 J/cm^2, 60 ns pulse duration) exhibits a substantial redistribution of boron. In particular, after the pulsed laser annealing cycle, the impurity profile becomes almost uniform from the surface down to a depth of approximately 0.2 µm in the crystal, and significant quantities of boron are observed at depths of approximately 0.5 µm.

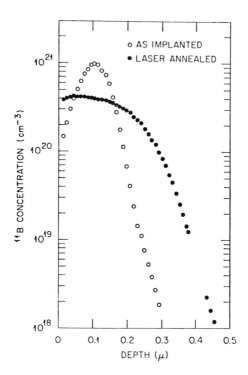

FIGURE 1.8.   Illustrating the redistribution of boron that occurs as a result of pulsed laser annealing.

The redistribution of the implanted boron is found to be both pulse energy density and pulse number dependent.   Figure 1.9 shows experimental results measured after laser annealing with different pulse energy densities in the range 0.64 J/cm^2 to 3.1 J/cm^2.   At a laser energy density of approximately 0.6 J/cm^2, only surface melting is initiated and the boron profile is indistinguishable from that of the as-implanted sample.   Hall effect measurements and transmission electron microscopy show about 30% of the ex- pected electrical activity and significant damage remaining in the form of dislocation loops under this pulse energy.   At energy densities 1.1 J/cm^2 and greater, the impurity profiles are almost flat-topped in the surface region, and the boron spreads deeper in the sample as the energy density is increased.

The substantial redistribution of boron induced by pulsed laser annealing cannot be explained by thermal diffusion in the solid phase because the time duration is too short.   However, the

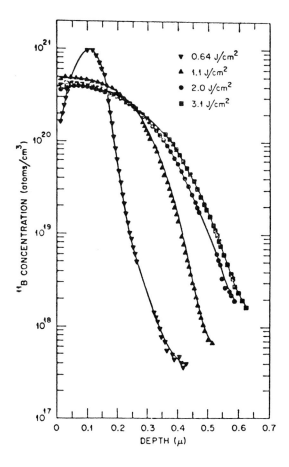

FIGURE 1.9.   Illustrating   the   redistribution   of boron in pulse annealed silicon using pulses of different energies.

melt depth calculations just discussed show that a region several thousand angstroms deep can be melted for pulse energies greater than about 1 J/cm². The dopant atoms have a very high diffusion coefficient in the melt and the implanted profile can diffuse rapidly through the liquid while the crystal is molten. For boron in silicon, the experimental profile can be fit by using a liquid diffusion coefficient of $2.4 \pm 0.7 \times 10^4$ cm²/sec [1.35] and a diffusion time of 180 ns. These parameters provide the calculated fit shown in Fig. 1.10 and adequately justify the conclusion that normal diffusion processes in the molten state account for the spreading of the implanted profile during pulsed laser annealing.

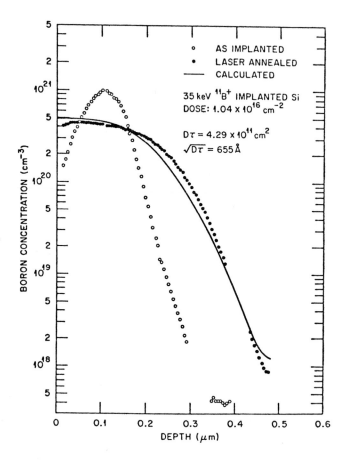

FIGURE 1.10.　Comparison of theory　and experiment for ruby pulse annealed ion implanted silicon.

　　Hall effect　and　transmission　electron　microscopy　measurements made on implanted material annealed in this way show 100% electrical activity and no defects to a resolution of at least 50 Å, provided that the amorphous layer is sufficiently thin. However, it should be mentioned that, since the reflection coefficient R at the surface of an irradiated sample changes from 0.35 to 0.7 when the surface becomes molten, only 30% of the pulse over and above the melting threshold is absorbed by the silicon. The result is that it is difficult to melt layers that are in excess of approximately 1 μm without heating the surface to the boiling point, which leads to very poor surface morphology. As a result, pulsed laser annealing is generally useful only for the annealing of relatively shallow implanted layers.

1.1.6   Supersaturation and Segregation

In normal crystal growth from the melt, which is carried out under near-equilibrium conditions (and therefore at very low growth rates), the solid which forms from the liquid at any moment has a concentration of dopant given by $C_s = k_0 C_L$, where k_0 is the equilibrium distribution coefficient and C_L is the dopant concentration in the liquid. It follows that, unless $k_0 = 1$, the dopant will segregate into the liquid as crystal growth proceeds under near-equilibrium conditions. However the recrystallization process for pulsed laser annealing is far from equilibrium; growth rates are very high (on the order of meters/sec) and there is then the possiblity that dopants can be incorporated into the solid at concentrations that far exceed the solid solubility limit for near-equilibrium growth processes, thus producing a substitu-tional, supersaturated alloy. Supersaturation will require, or be characterized by, a nonequilibrium distribution coefficient k' that is appropriate for the high speed, nonequilibrium recrystallization process.

If we assume as a limiting case that the diffusion coefficient of the dopant in the liquid phase is so large that the dopant is always uniformly distributed in the liquid that remains at any stage of the recrystallization process, then the impurity distribution following a pulsed laser annealing process will be controlled completely by a "nonequilibrium" distribution coefficient k'. In particular, if the maximum melt depth is denoted by x_m, then the concentration incorporated into the solid, $C_s(x)$, is given by

$$C_s(x) = k' \frac{\Phi}{x_m} \left[\frac{x_m}{x_m - x} \right]^{(1-k')} \tag{1.5}$$

where Φ is the dose of the implanted impurity and x is measured from x_m toward the surface.

Equation (1.5) shows that, if k' = 1, the implanted dopant will be distributed uniformly over a depth x_m. However if k is less than 1, dopant will be swept toward the surface as recrystal-lization proceeds, and the dopant concentration will approach infinity at the surface (for the simple model under consideration). Figure 1.11 shows profiles obtained from Eq. (1.5) for k' = 1, 0.1 and 0.01, respectively. Under realistic conditions, of course, it is necessary to include both diffusion and dopant segregation into the liquid to calculate the final impurity profile.

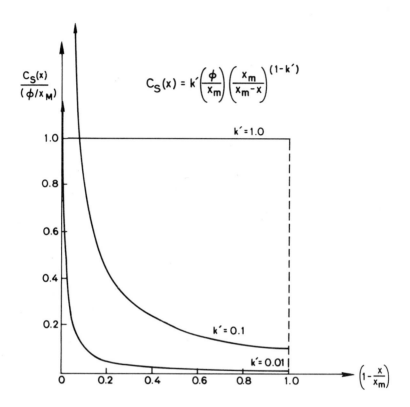

FIGURE 1.11.   Illustrating dopant segregation during resolidification.

White et al. [1.36] have studied pulsed laser annealing of
high dose implants in Si, using both diffusion and distribution
coefficient phenomena in the theoretical analysis of their experimental data.   An example of their analysis for a high dose As
implant into Si is shown in Fig. 1.12.   Rutherford backscattering
and channeling analysis shows that all of the As atoms were incorporated into substitutional sites by the pulsed laser anneal.
Furthermore, as indicated on Fig. 1.12,

(1) the shape of the impurity profile after annealing can be
be explained entirely by diffusion, so the effective value of k'
for [75]As in Si is unity for these conditions.   This is to be compared to an equilibrium value of k_o for As in Si of ~0.3.

(2) [75]As has been incorporated substitutionally at a concentration that is approximately four times the reported maximum
equilibrium solubility.

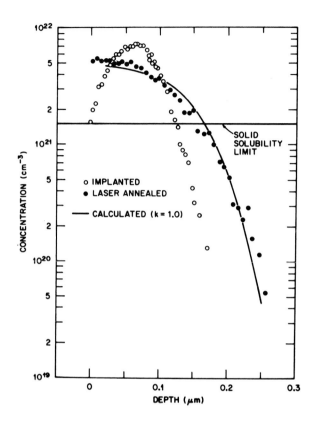

FIGURE 1.12. Dopant profiles for ^{75}As (100 keV, $6.4\times10^{16}/$ cm^2) in silicon compared to model calculations. The horizontal line indicates the equilibrium solubility limit (after Ref. 1.36).

Very substantial increases in k' over k_0 were observed by White et al. [1.36] for a very wide range of impurities, though k' was found to be unity for only B, P and As in Si. For impurities with low values of k', such as Cu in Si, for example, segregation of the impurity to the surface is the dominant process, as shown in Fig. 1.13. This process is clearly similar to the idealized process described by Eq. (1.5) and shown in Fig. 1.11 for a very low value of k.

White et al. [36] have also established maximum substitutional concentrations for dopant incorporation following pulsed laser annealing, for dopants that can be readily studied by backscattering and channeling analysis. The results of their study for both k' and $C_{s,max}$ are given in Table 1.5.

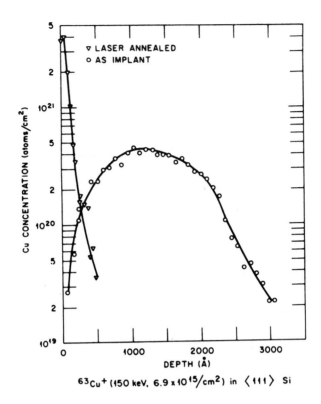

$^{63}Cu^+$ (150 keV, 6.9 x 10^{15}/cm^2) in $\langle 111 \rangle$ Si

FIGURE 1.13.   Profile of ^{63}Cu (150 keV, 6.9 x 10^{15}/cm^2) implanted into $\langle 111 \rangle$ Si before and after laser annealing (after Ref. 1.36).

TABLE 1.5   Distribution Coefficient and Maximum Substitutional Solubility for Dopants in Si Following Pulsed Laser Annealing.

Dopant	k'	k_o	c_s^o	c_s^{max}
As	1.0	0.3	1.5×10^{21}	6×10^{21}
Sb	0.7	0.023	7×10^{19}	1.3×10^{21}
Bi	0.4	0.0007	8×10^{17}	4×10^{20}
Ga	0.2	0.008	4.5×10^{19}	4.5×10^{20}
In	0.15	0.0004	8×10^{17}	1.5×10^{20}

With regard to device fabrication, these supersaturated
alloys may prove to be of some interest, though is should be mentioned that precipitation of these supersaturated dopants occurs
rapidly in subsequent thermal processing steps, with the development of dislocation loops, rods and precipitation networks that
can be expected to have deleterious effects on device properties.

1.1.7 Crystallographic Analysis of Pulsed Laser Annealed Ion Implanted Silicon

The damage remaining in ion implanted samples after an
annealing cycle is traditionally studied by two techniques.
Transmission electron microscopy (TEM) is used to identify crystallographic defects such as dislocation loops, rods and precipitate structures; and deep level transient spectroscopy (DLTS)
is used to measure the concentration and distribution of residual
point defects. In this section we summarize the TEM results for
pulse annealed samples obtained by Narayan and co-workers [1.37].

1.1.7.1 Ion Implanted Samples

Figure 1.14 shows transmission electron micrographs which
compare directly the damage that remains in the lattice after
laser annealing and conventional thermal annealing for silicon
crystals implanted by B, P, and As. After thermal annealing at
temperatures up to 1100°C, significant damage remains in the
implanted region in the form of dislocation loops as shown in the
right column of micrographs. By contrast, after laser annealing
($\sim$ .5 J/cm^2, 60 nsec) no damage (dislocation loops or stacking
faults) remains in the implanted region down to the resolution
of the microscope ($\sim$ 10 Å). Diffraction patterns from these
same crystals have no irregularities, and the orientation of the
recrystallized region is the same as that of the underlying
substrate, (001). Upon subsequent heating to 900°C, a low concentration ($<$ 5 x 10^{14}/cm^3) of small defects (average diameter
$\sim$ 30 Å develops) in the implanted region.

Figure 1.15 shows three different types of microstructure
and crystal structure in the implanted region which was observed
after annealing with different laser energy densities. These
results were obtained using silicon crystals implanted with ^{75}As
(100 keV), 1.4 x 10^{16}/cm^2). Both ion channeling and transmission
electron microscopy showed that a region $\sim$ 1600 Å deep was made
amorphous by implantation. Annealing with 0.63 J/cm^2 caused a
region $\sim$ 1000 Å thick to become polycrystalline as indicated by
the diffraction pattern. This suggests that the crystal was
melted to a depth of $\sim$ 1000 Å . The regrowth was not epitaxial

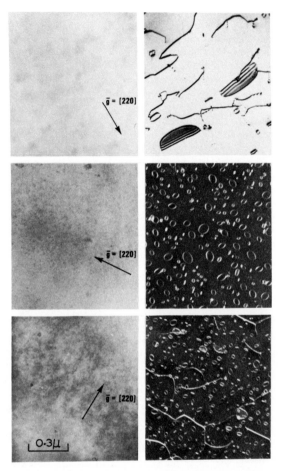

FIGURE 1.14.   Comparison of laser (left column) and thermal
(right column) annealing of ion implanted silicon, (001) orientation. Implanted species, energy and dose were: top row – ^{11}B(35
keV, 3×10^{15}/cm^2); middle row – ^{31}P(80 keV, 1×10^{15}/cm^2); bottom
row – ^{75}As(100 keV, 1×10^{16}/cm^2). Conditions for thermal annealing were B,P(1100°C – 30 mins); As(900°C – 30 mins). Conditions
for laser annealing were 1.5 J/cm^2, 60 nsec (after Ref. 1.37).

because the melted region did not extend into the undamaged substrate. After annealing with 1.07 J/cm^2, which is very near
the threshold for complete electrical activation, the electron
diffraction pattern showed the recrystallized region to be single
crystal, but a thin surface layer was observed containing residual
damage in the form of dislocation loops. Stereomicroscopy showed
that the residual damage was confined to a depth of ~ 200 Å. The
origin of this damaged layer at the surface is not well understood,

^{75}As(100 keV, 1.4×10^{16}/cm^2)

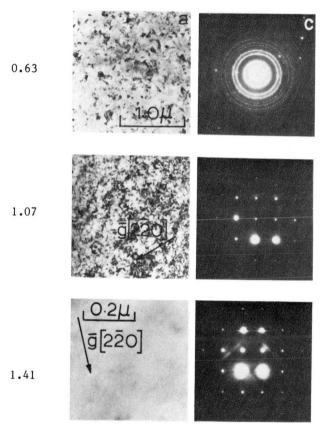

FIGURE 1.15. Microstructure developed in silicon by anneal-ing with different laser energy densities. The energy density used for annealing is indicated beside each micrograph and cor-responding diffraction pattern. Pulse duration was 25 nsec (after Ref. 1.37).

but it may be related to radiant heat losses in the surface region during the regrowth process. Residual damage at the surface has been observed for B, P and As implants after irradiation with laser energy densities very near the threshold for complete annealing (~ 1.1 J/cm^2 for our implant conditions).

1.1.7.2   Thermally-Diffused Samples

In thermally diffused specimens, Narayan and White [1.37] have shown that dislocation loops and precipitates formed during phosphorus or boron diffusion at 1100°C for 30 minutes can be dissolved by the use of pulsed laser irradiation. For a phosphorus-diffused sample, one pulse of energy density 1.5 J/cm^2 produced an annealed, defect-free region (i.e., containing no phosphorous precipitates and dislocation loops) that was 0.29 μm deep. The depth of the defect and precipitate free region was found to be 0.38 μm after two such pulses; and after seven pulses the depth was determined to be 0.58 μm. A saturation phenomenon in the depth of the annealed region was observed after approximately five pulses. The results of 1.5 J/cm^2, 50 n sec pulse are summarized in Table 1.6.

TABLE 1.6   Ruby Laser (λ = 0.694 μm) Annealing Parameters in Phosphorous Diffused Silicon.

Pulse Energy Density $E(Joule\ cm^2)$	Pulse Duration (Nano secs)	Number of Pulses	Depth of Annealing (μm)
1.5	50	1	0.29
1.5	50	2	0.38
1.5	50	5	0.56
1.5	50	7	0.58
2.0	20	1	0.57
2.0	20	3	0.70
2.0	20	5	0.75
2.0	20	14	0.77
3.0	20	1	0.95

The depths of the annealed regions were similarly determined for 1, 3, 5 and 14 laser pulses of energy density 2.0 J/cm^2 and pulse duration 20 n sec with the results summarized in Table 1.6. Again saturation in the annealing depth was observed after about five laser pulses. It should be mentioned that in the specimens irradiated with three or more laser pulses, a small amount of reprecipitation of phosphorous near the surface was observed. This is probably related to the segregation coefficient of phosphorous in silicon being slightly less than unity. For 2.0 and 3.0 J/cm^2 (20 n sec) laser pulses, the increase in the annealing depth (from 0.57 m to 0.95 m) is approximately proportional to the pulse energy density. A complete annealing of defects up to one micron was possible with one pulse of E = 3.3 J/cm^2 and τ = 20 n sec.

1.1.8   Deep Level Transient Spectroscopy

Capacitance transient spectroscopy provides a very powerful technique for measuring the energy levels and defect state densities in processed silicon. The principle of the method was first described by Lang [1.38] and recently summarized for laser annealing applications by Johnson et al. [1.39]. Briefly, carrier trapping centers surrounding a pn junction or a Schottky barrier are charged by application of an appropriate voltage pulse. Their discharge is then encouraged by scanning the sample over a temperature range from liquid nitrogen temperature to room temperature. Each trapping center is ideally characterized by an energy level and therefore a specific temperature at which carrier emission from the trap will be maximized. This carrier emission will change the charge density in the depletion region of a reverse biased pn junction or Schottky barrier, thus providing a capacitive transient signal that can be used to detect the trapping center.

Figure 1.16 shows a typical defect state spectrum observed in a pulse laser annealed, boron-doped, Czochralski grown Si substrate, after Benton et al. [1.40]. The dominant defect state is a hole trap [H(0.36)] located approximately at an energy level that is 0.36 eV above the valence band. By measuring the concentration of this defect vs. laser power and implantation dose, Benton et al. were able to establish that:

1.   The threshold power for residual minimum defect density after pulsed laser annealing decreases with increasing implant dose, due presumably to the enhanced coupling of the near-band-edge light from the Nd:YAG 1.06 μm laser.

2.   A higher defect state concentration was observed near n^+p junctions than in the substrate (the pn junctions were fabricated by As implanted followed by laser annealing at a wavelength of 1.06 μm).

3.   The average defect state concentration appears to show a minimum when the laser power is just sufficient to produce a melt depth that precisely matches the range of the ion damage. Deeper melting may leave quenched-in defects beyond the junction. The concentration of these defects can be reduced by a subsequent low temperature furnace anneal (800°C for 30 min).

1.1.9   Pulsed Electron Beam Annealing of Ion Implanted Silicon

The use of pulsed electron beams for annealing ion implanted silicon was first developed for the purpose of annealing ion implanted solar cells [1.41]. The details of the annealing mechanism were explored later by Kennedy et al. [1.42], using Rutherford

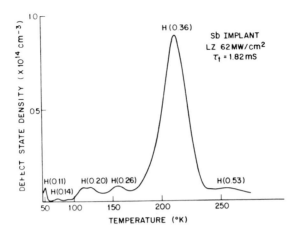

FIGURE 1.16.   Capacitance transient defect state spectra of laser annealed, boron doped Czochralski grown silicon (after Ref. 1.40).

backscattering/channeling spectroscopy and TEM to study the annealing process.

The silicon crystals used for the study were (100) and (111) in orientation, and were implanted with either $^{28}Si^+$ or $^{75}As^+$ or both. The implants were carried out at 25, 50, 100 and 200 keV. The conditions used for the pulsed electron beam were: an energy density of 0.8-1.2 J/cm^2 (over a 50 cm diameter beam) and a pulse length of 100 ns. The results were as follows.

1.1.9.1 $^{75}As^+$-implanted Samples

Pulsed electron beam annealing at an energy of 0.8 J/cm^2 succeeded in completely annealing the samples for all of the implant energies. Substantial redistribution of the ^{75}As was observed, leading to an impurity profile similar to that obtained with pulsed laser annealing. Kennedy et al. estimated the diffusion coefficient for As required to explain the As profile depth to be 10^4 cm^2/sec, which is typical of liquid state diffusivities. SIMS measurements of a high dose, ^{75}As implanted silicon sample annealed with a pulsed electron beam at an incident energy of 1.1 J/cm^2 were obtained by Wilson et al. [1.42] and are shown in Fig. 1.17. These data are very similar to the results for pulsed laser annealing. They support the melting hypothesis and lead to the conclusion that pulsed laser and electron beam annealing of an ion implanted sample proceed by the same basic mechanism, though of course the coupling of the beam to the solid is different for the two cases.

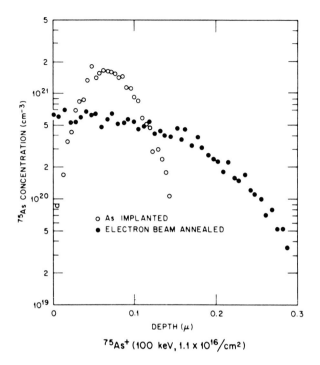

FIGURE 1.17. ^{75}As profile   in the as—implanted   and pulsed electron beam annealed condition.

1.1.9.2 ^{28}Silicon—implanted Samples

A similar experiment carried out by Kennedy et al. [1.41] on Si implanted samples (i.e., Si into Si) showed that full annealing was possible for the implants at 25 and 50 keV only. For higher implant energies the melt depth was insufficient to envelop the implantation damaged layer so that the recrystalli- zation did not start from a single crystal.

Such a result is also obtained for pulsed laser annealing when the pulse energy is insufficient to melt the sample to a depth that exceeds the implantation—damaged region.

The central conclusions from the studies of Wilson et al. [1.42] and Kennedy et al. [1.41] are that, when laser and electron beam systems are optimized, they give essentially the same re- sults. In particular, the dopant redistribution is the same; the fraction of dopant atoms (^{75}As) annealed into substitutional sites is the same; and TEM shows an absence of defects to a resolution of ~50 Å in both cases.

1.1.10   Summary of Results for Pulsed Beam Annealing

To summarize the foregoing, the mechanism of annealing for both Q-switched lasers and pulsed electron beams involves melting of a surface layer followed by a liquid phase epitaxial regrowth on the underlying substrate. The critical parameter for this process is the pulse energy density. The principal electrical characteristics of the annealed layer are:

1. 100% substitutionality of the implanted dopant, even for concentrations that exceed the solid solubility.

2. No residual defects in TEM to 50 Å resolution.

3. Redistribution of the implanted dopant via diffusion in a liquid layer during the recrystallization process.

4. Residual point defects that may require a subsequent low temperature (800°C, 30 min) anneal for their removal.

1.2 CENTRAL FEATURES OF CW BEAM ANNEALING OF ION IMPLANTED SILICON

In contrast to the pulsed annealing process, a scanning laser or electron beam provides an extremely convenient and highly controllable means for heating the surface of a semiconductor to a given temperature while holding the body of the material at a convenient low temperature (typically 350–500°C). Recognition of this fact leads naturally to a comparison of the annealing mechanism with the annealing process for furnace annealing. In what follows we first give a brief description of furnace annealing as it applies to implantation–amorphized layers, and then proceed to a discussion of the scanning beam alternative.

1.2.1   Furnace Annealing Process

Conventional furnace annealing of implantation–amorphized silicon proceeds by what is called a solid phase epitaxial (SPE) recrystallization process. Atoms at the amorphous–crystalline interface rearrange themselves into crystalline locations by a process that appears to involve vacancy diffusion in both the crystalline and the disordered layers. The process is described quantitatively by an activation energy E_a and proceeds at a rate

$$r(t) = r_0 \exp(-E_a/kT) \tag{1.6}$$

where r(t) is normally measured in Å/second. Measurements of r(t) made by Cspregi et al. [1.43] in the temperature range 400°–600°C undoped <100> silicon (amorphized by the implantation of Si$^+$) show an activation energy of E_a = 2.35 eV and a pre-exponential multiplier of 3.22 x 10^{14} Å/sec. Using these values at a temperature of 800°C gives a furnace regrowth rate of of about 0.3 µm/sec for <100> undoped Si. In other words, a 0.5 µm thick amorphized layer can recrystallize by the solid phase epitaxial process process on the underlying substrate in less than two seconds if the layer is held at a temperature of 800°C. The preexponential factor is found to be larger if dopants such as B, P and As are added, and smaller if oxygen is incorporated [1.43].

Using Rutherford backscattering/channeling analysis, the implanted dopants are found to be normally incorporated into substitional sites during this recrystallization process, leading to high electrical activity. However, point defects and trapping centers form during the recrystallization process (related to the vacancy motion by which the process proceeds), and these defects must be annealed to obtain good pn junction character-istics; i.e., high carrier mobilities and reasonable carrier lifetimes.

The motion and agglomeration of defects during the anneal-ing is especially troublesome if the implant dose and energy are not sufficient to produce an amorphous layer that envelops the implanted dopant. Such a condition is most often obtained when B is implanted into Si, a case which has been thoroughly studied. A 30 minute anneal in the temperature range 800°C–1000°C then produces a combination of dislocation loops, rods and precipi-tates that are nearly immune to further furnace annealing [1.44, 1.45].

Partly as a result of these problems it is customary to use a so-called "two-stage" annealing process [1.46]. The first stage is carried out at ~600°C for 1/2 hour and is intended only to recrystallize the implanted region. Point defects and small damage clusters remain in the regrown layer after this stage, which often lead to low electrical activity and poor carrier mobility and lifetime. The low temperature anneal is then typically followed by a 1000°C, 10 minute anneal to remove point defects, increase carrier mobility and lifetime, and provide for some impurity diffusion so that pn junctions formed by the process are located outside the region of residual damage. Such a process is especially important in the annealing of im-planted <111> Si; i.e., for most bipolar applications. In con-trast a single stage anneal can be sufficient for less demanding applications in <100> Si.

1.2.2   Basic CW Laser Annealing Systems

The basic cw laser annealing process is identical to the furnace annealing process just described, except that the sample is
maintained at the annealing temperature for such a short time
that only the fastest thermal processes, including solid phase
epitaxy, can be carried to completion. Dopant precipitation and
the formation of dislocation loops and rods are generally not
observed since there is insufficient time for them to form.

A basic system that can be used for scanned cw laser annealing is shown in Fig. 1.18. It consists of an Ar (or other) cw
laser that is passed through a series of lenses, beam-shaping optics and filters and then deflected onto a sample that is mounted
in the focal plane of the final lens [1.47]. The beam shaping
optics are used for the recrystallization studies described in
Chapter 5 of this volume. For the annealing experiments reported
here no beam shaping is required. The sample is mounted on a
computer driven, heated XY table which permits the sample to be
scanned underneath the beam. As an alternative, the beam can be
swept with X and Y mirrors. The X mirror can be mounted on a
galvonometer that is driven with a triangular waveform and the Y
mirror mounted on a galvonometer that is driven by a staircase
waveform. This arrangement permits the beam to be scanned across

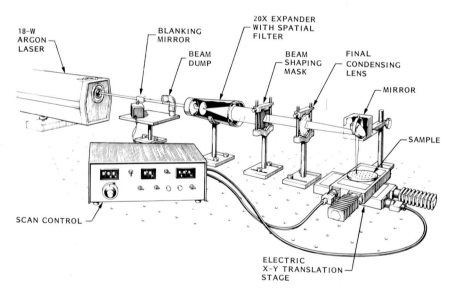

FIGURE 1.18. A general schematic of the annealing apparatus,
including Ar$^+$ laser, lens perpendicular X and Y mirrors, and
a vacuum sample holder.

the target in the X direction, stepped by a controlled Y increment and then scanned back across the target in reverse X direction. Individual scan lines can be overlapped or not by appropriate adjustments of the Y step.

The samples are mounted on a stage that can be heated to about 500°. Control of the annealing ambient can be obtained by placing a cylindrical quartz jacket around the sample holder and pumping appropriate gases into this jacket. A variety of lenses, laser powers, scan rates, and sample temperatures have been found to produce essentially perfect annealing of ion implanted semiconductors. If the samples are held at room temperature, a typical set of annealing conditions consists of a laser output (Ar, milti-line mode) of 7 W focused through a 79 mm lens into a 38 micron spot on the target. The spot is typically scanned across the target at a rate of approximately 2.5 cm per second. Adjacent scan lines are overlapped by approximately 30% to produce full annealing in the overlapped areas. The laser power for full annealing can be reduced (and the width of the annealed line increased) by increasing the sample temperature.

1.2.3   Surface Temperature Profiles

The beam geometry and scanning conditions listed above lead to a spot dwell time on the order of 1 ms. This is to be compared to a thermal time constant (for a 40 μm cube of Si) of approximately 10 μs, from which it follows that the semiconductor surface has adequate time to come to thermal equilibrium with the scanning heat source. It is therefore possible to calculate exactly the surface temperature that will be produced at the center of the spot by simply solving the steady state equation for heat flow. Calculations of this type have been carried out for both circular and elliptical scanning beams [1.48]. The details of the calculations are presented in Chapter 2 of this volume for a number of different sources and scanning geometries. The results for a beam of circular cross section irradiating a silicon substrate are shown in Fig. 1.19. The calculations leading to these results account for the temperature dependence of the thermal conductivity and specific heat, and have proven to give very accurate estimates of the surface temperatures achieved with the laser beam.

Two features of these curves are worth particular attention. First, the horizontal axis is in units of power per unit radius of the beam, this being due to the fact that, for a small beam diameter, the heat flow problem has essentially hemispherical symmetry. As a result the surface temperature is a function of power divided by spot radius rather than power per unit area of the beam. Secondly, the sizeable decrease of thermal conductivity

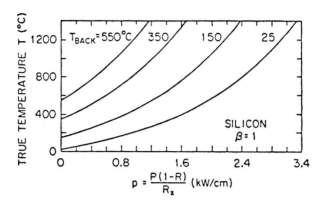

FIGURE 1.19.   The true   maximum   temperature   (X=Y=Z=V=0) in
Si is plotted versus the   normalized   power   p   (p=P(1-R)/R_x) for
different substrate backsurface temperatures.

of Si with temperature leads to a situation in which the temperature at the irradiated surface is a very sensitive function of
the backsurface temperature.   For example, a surface temperature
at the center of a 40 micron spot of 1000°C can be obtained with
a 5 watt laser output if the backsurface temperature (or thermal
bias) is 350°C.   If the backsurface temperature is reduced to
150°C, the temperature in the center of the irradiated area
drops by nearly 500°C.   Hence it is very important to control
the backsurface temperature accurately, and it is also possible
to use relatively low laser powers if substantial thermal bias
can be employed.

1.2.4   Basic Mechanism of CW Beam Annealing

As mentioned earlier, the basic mechanism of cw beam annealing is solid phase epitaxy.   A very convincing illustration of
this fact is provided by the work of Roth et al. [1.46] and illustrated in Fig. 1.20.   Here the regrowth rate of both implantation-
amorphized and UHV deposited Si layers on a <100> Si substrate
is plotted as a function of the surface temperature produced by
the laser.   The growth rate was determined from time-resolved
reflectivity measurements.   The data show exactly the form expected for a process described by Eq. (1.6), and in fact the growth
rate at 800°C is found to be 0.3 μm/sec, in excellent agreement
with the furnace annealing result.   Hence the regrowth process
is quantitatively identified as solid phase recrystallization.
Additional information on the growth mechanism is provided in
Chapter 3 of this volume.

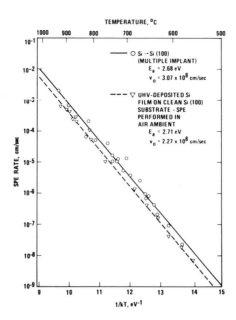

FIGURE 1.20. SPE growth rates for amorphous Si layers formed by ion-implantation and UHV evaporation, determined from time-resolved reflectivity measurements (after Ref. 1.46). ©1981 Elsevier Science Publishing Co.

1.2.5 Impurity Profile

Figure 1.21 shows the impurity profiles obtained by SIMS under as-implanted, laser annealed and thermally annealed conditions for As^+ implanted into ⟨100⟩ Si under typical conditions. The most striking feature of this figure is that the laser annealed profile is identical to the as-implanted profile. In other words, there has been <u>no diffusion</u> of the implanted species during the laser anneal. Furthermore, the as-implanted impurity distribution is matched exactly by the Pearson type IV distribution function using LSS moments [1.50], so the experimental and theoretical profiles are in excellent agreement. The thermal anneal shows the well-catalogued impurity redistribution given by the open triangles in Fig. 1.21.

Majority carrier profiles and carrier mobilities were obtained by sheet resistance and Hall effect measurements and are shown in Fig. 1.22. As can be seen the carrier concentration profile also fits the Pearson type IV distribution quite well. Under very high dose conditions (> $10^{16}/cm^2$), ^{75}As precipitation can occur near the peak of the profile, leading to a residual, nonsubstitutional As content of approximately 5% [1.48]. As in the case of pulsed beam annealing, dopant concentrations can

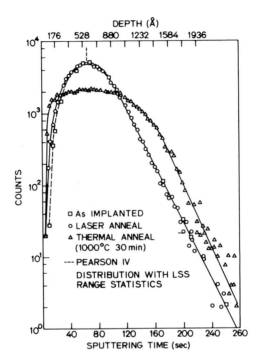

FIGURE 1.21. Carrier concentration and mobility profiles obtained on As-implanted samples annealed with a scanning Ar cw laser beam.

exceed the equilibrium solid solubility limits when implanted samples are annealed using a scanning beam, though precipitation of these dopants occurs when the samples are subjected to further high temperature processing steps. The subject of dopant incorporation and precipitation is thoroughly discussed in Chapter 3 of this volume.

1.2.6. Transmission Electron Microscopy

Typical results of TEM performed on thermally annealed and laser annealed samples are shown in Fig. 1.23. The thermally annealed sample (Fig. 1.23a) shows a single crystal diffraction pattern with a bright field micrograph containing the usual variety of ~200Å diameter defect clusters and dislocation loops. The laser annealed sample shown in Fig. (1.23b) is essentially free of any defect observable in TEM except near the boundary between crystallized and amorphous regions. It should be emphasized also that no defects are observed in TEM in the region where adjacent scan lines overlap.

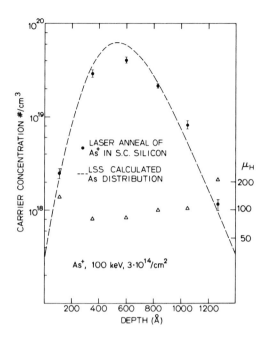

FIGURE 1.22. Carrier concentration and mobility profiles obtained on As-implanted samples annealed with a scanning Ar cw laser beam.

1.2.7 B-implanted Single Crystal Si

A similar set of experiments has been performed in B-implanted Si. The central results are identical to those described above for As-implanted Si. In particular, 100% electrical activity can be obtained with no diffusion of the implanted species from its as-implanted profile. Recrystallization is also perfect as judged by TEM to a resolution of ~ 20 Å. These results are independent of whether the B is implanted into pre-amorphized Si or directly into single crystal material.

1.2.8 Residual Defects Following CW Laser Annealing

As mentioned above, cw laser annealing of ion-implanted silicon is typically carried out under conditions that produce a surface temperature of approximately 1000-1100°C with a back-surface temperature of 350°C. This thermal gradient can produce a rapid diffusion of vacancies and other point defects from the surface into the cooler interior of the sample and a consequent reduction of minority carrier lifetime.

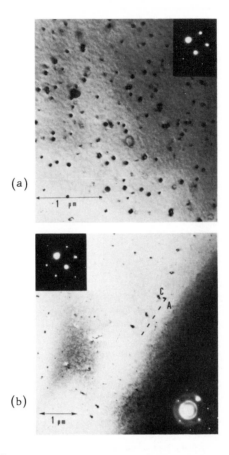

(a)

(b)

FIGURE 1.23.   Electron   micrographs of As−implanted silicon
subjected to thermal anneal of 1000°C 30 min (a) and laser annealing (b); inserts show diffraction patterns which are typical for
their regions.

Point defect density profiles in both cw laser annealed and
cw electron beam annealed ion−implanted silicon have been measured
by Johnson et al. [1.51].   These results and others of importance
for understanding and applying cw beam processing are presented
and thoroughly discussed in Chapter 4 of this volume.

As an example of the results, we show in Fig. 1.24 the defect
density signal obtained from a constant capacitance (CC)−DLTS measurement made on cw laser annealed silicon.   From this figure we
see that a significant density of point defects remains immediately after the annealing cycle.   The figure also shows that these
defects can be reduced to low levels (comparable to furnace annealing) by a subsequent heat treatment at 850°C for 1/2 hour.

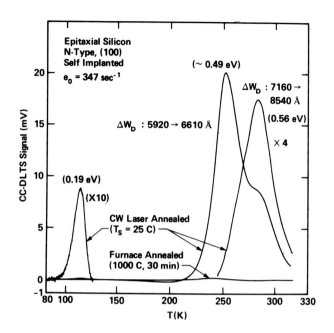

FIGURE 1.24.   Defect density signal obtained from DLTS measurements of cw laser annealed silicon.

The density profile for a particular defect can also be measured by a sophisticated application of the CC-DLTS technique. An example is shown in Fig. 1.25, where we see that a substantial density of defects can be produced at depths up to ~1 μm immediately following cw beam processing under typical conditions. As mentioned above, the defect density can be removed by subsequent thermal processing, sometimes at temperatures as low as 450°C (see Ref. 1.51 and Chapter 4 of this volume).

1.2.9   Summary

To summarize the foregoing, cw laser annealing proceeds by solid phase epitaxial recrystallization. For "small" irradiated areas (40 μm spot), the critical parameter is the beam power per unit radius. The principal results of the annealing are:

1.   100% substitutionality of the implanted dopant even for concentrations that exceed the solid solubility.

2.   No residual defects in TEM to 50 Å resolution.

3.   No dopant redistribution during annealing.

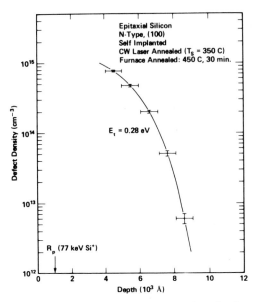

FIGURE 1.25. Defect density vs depth for a cw laser annealed silicon sample.

4. Residual point defects remain below the recrystallized layer that require a subsequent low temperature anneal for their removal (800°C for 10 minutes).

For VLSI applications, the absence of dopant redistribution may be a significant feature of this annealing process. In any case it is an ideal companion for the implantation process because the dopants are annealed into their as-implanted sites. Since the implanted impurity profile can be calculated with considerable accuracy, this process provides a means of assuring that computer simulations of impurity profiles based on the implantation process can be realized with precision.

1.3   ANNEALING WITH LARGE DIAMETER SCANNING CW SOURCES

A very attractive alternative for cw laser annealing is the use of cw arc sources. Arc sources can be configured to produce both large diameter annealing spots and ribbon-shaped beams. These beams are scanned at speeds that produce dwell times on the order of seconds, which substantially reduces the severity of the thermal shock that is often obtained with a cw laser. Such an annealing time is, however, still short enough to prevent excessive dopant redistribution during annealing of ion implanted semiconductors. Furthermore, the throughput of an arc lamp or ribbon electron beam annealing system can also be very high (hundreds of 4" wafers per hour) and, in addition, both of these annealing techniques are insensitive to anti-reflective dielectric coatings, unlike their laser counterpart.

Both arc lamps and electron beams can be used for annealing in two possible modes, an isothermal mode and a heat sunk mode. In the isothermal mode the entire wafer reaches the annealing temperature. The dominant heat loss mechanism in this case is radiation; the time required to reach a given temperature is typically on the order of a few seconds. This method has the advantage of requiring low beam power. However, in certain cases it is desirable confine the heating to the sample surface. The heat sink mode must then be employed in which the backsurface of the sample is kept at a specified temperature T_O. The principal heat loss mechanism under these conditions is conduction through the wafer. Calculations of sample temperature vs beam power for both the isothermal and heat sunk modes of operation, including the estimated spectral response of the arc source, are presented in Chapter 2. Here we discuss the basic experimental results obtained with such systems.

1.3.1   Rapid Annealing of Silicon With a Scanning cw Hg Lamp

A practical "large diameter" scanned annealing system has been built by Stultz et al. [1.52] using a 3" long mercury arc lamp with an elliptical reflector. The light from the lamp is focused into a narrow (<5mm wide) ribbon by a 4" long reflector with an elliptical cross section. By using this shape of reflector to collect and focus the light, an intense linear heat source can be created while still maintaining a reasonable working distance between the sample and the lamp. The reflector has major and minor axes of 10 cm and 8.6 cm, respectively. This configuration gives a magnification factor of about 3.2 and a working distance of about 26 mm between the edge of the reflector and the focal plane.

The wafers are placed on a heated sample stage, which can be translated beneath the lamp at speeds up to 10 cm/s. A schematic representation of the arc lamp annealing system is shown in Fig. 1.26. A variety of arc sources can be used, though a mercury lamp is recommended because the spectral distribution of this lamp is most heavily weighted in the uv range. Consequently, the mean absorption depth for this source is shallower than that for the other lamps. Specifically, the fraction of light absorbed versus depth is given by:

$$F = I(\lambda)[1-R(\lambda)](1 - \exp[-\alpha(\lambda)x]$$

(1.15)

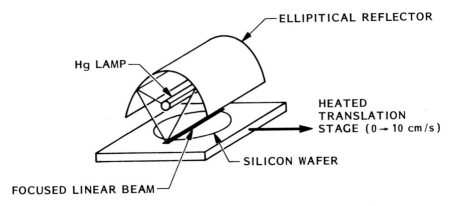

FIGURE 1.26.   Schematic representation of scanning cw arc lamp annealing system.

where

     I(λ) is the wavelength dependent radiation intensity;
     R(λ) is the wavelength dependent reflectively;
     α(λ) is the wavelength dependent absorption coefficient.

Using published absorption and reflectivity data for silicon and the spectral distributions for xenon, krypton and mercury lamps, the percent of incident light absorbed versus depth in silicon was calculated for each of the lamps. These data are given in Fig. 1.27. In thin film processes such as annealing and recrystallization, the region of interest is generally less than 1 μm thick. As shown in Fig. 1.27, the fraction of light absorbed from a mercury lamp in this region is more than twice that for the other lamps.

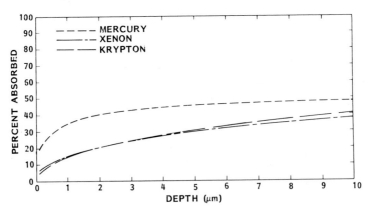

FIGURE 1.27.   Percent of incident light absorbed versus depth into silicon for three different sources.

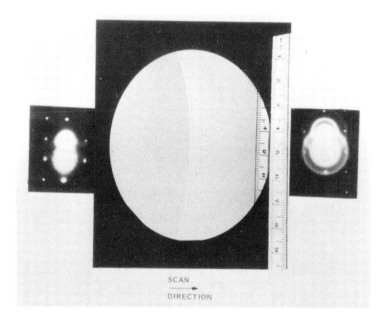

SCAN
→
DIRECTION

FIGURE 1.28. Photograph of a partially annealed silicon wafer, and TEM diffraction patterns from the annealed and unannealed regions. ©1983 Elsevier Science Publishing Co.

Annealing experiments have been carried out using 2 and 3 inch 1-8 Ω-cm p-type (100) silicon wafers which were implanted with $^{75}As^+$ at 100keV to 1 x $10^{15}cm^{-2}$. Several combinations of arc lamp power, scan rate and substrate temperature can be used to achieve a wafer temperature sufficient for good annealing, ~1000°C. In general, scan rates from 5mm/s to 1cm/s, substrate temperatures of 400°C to 600°C and arc lamp input power of 1Kw/inch to 1.5Kw/inch give good results.

Because of the length of the arc lamp, an entire wafer can be annealed in a single scan. Figure 1.28 is a photograph of a 3" wafer in which the scan was stopped at the center of the wafer. Because of the difference in reflectivities between single crystal and amorphous silicon, the demarcation between the annealed and unannealed regions is readily seen. Transmission electron microscope (TEM) diffraction patterns made from samples taken from the unannealed and annealed regions clearly indicate the amorphous and completely recrystallized nature of the regions, respectively. (The picture is somewhat distorted due to the angle at which the photo was taken.)

The recrystallized layer was also analyzed by Rutherford Backscattering Spectroscopy (RBS) and directly compared to a

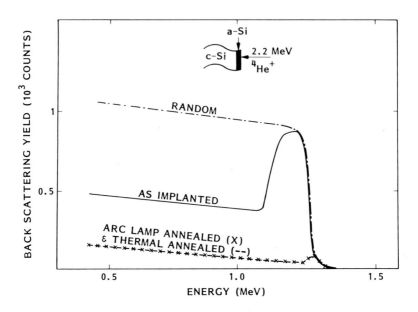

FIGURE 1.29. Rutherford backscattering spectra from as-implanted, arc lamp annealed and thermally annealed samples.

©1983 Elsevier Science Publishing Co.

sample which was thermally annealed at 850°C for 30 minutes followed by a 10 second 1000°C anneal. Figure 1.29 shows the backscattering spectra from an as-implanted, arc lamp annealed and thermally annealed sample. As shown, there is no detectable difference between the quality of the regrown material annealed by the scanned arc lamp and the thermally processed sample.

Carrier concentration and mobility of the scanning arc lamp annealed material as a function of depth were determined by means of differential sheet resistivity and Hall effect together with an anodic oxidation stripping technique. These results are shown in Fig. 1.30, along with the as-implanted profile calculated using LSS theory and published bulk silicon mobilities for the respective impurity concentrations. As shown, no measurable dopant redistribution occurred during the annealing process and the free carrier mobility in the annealed region is as good as found in bulk silicon.

Finally, to assess the uniformity of the anneal, four point probe sheet resistivity measurements were made on a wafer which was completely annealed in a single scan. The variation across the wafer from side to side and top to bottom (with respect to the scanning beam) showed no significant variation in either direction, indicating a very uniform annealing of the implanted wafer. These data are shown in Fig. 1.31.

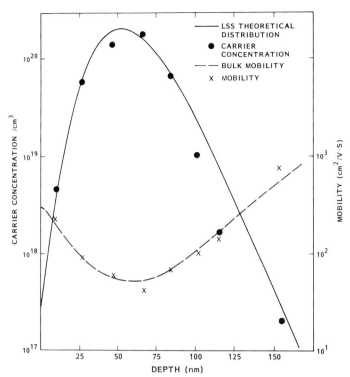

FIGURE 1.30.   Carrier concentration and mobility versus depth from an arc lamp annealed sample. [c]1983 Elsevier Science Publishing Co.

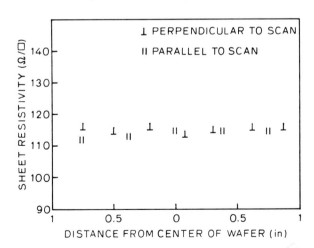

FIGURE 1.31.   Sheet resistivity profiles across an arc lamp annealed wafer. [c]1983 Elsevier Science Publishing Co.

The scanning mercury arc lamp thus provides a tool for annealing that is equivalent to the cw laser in results and has the added feature that the wafer throughput capability can be very high.

1.4   CW BEAM RECRYSTALLIZATION OF POLYSILICON FILMS

Heavily doped polycrystalline silicon (polysilicon) is a material that is widely used in present day silicon integrated circuit technology for gates and interconnection lines in MOS integrated circuits. Initial interest in the beam annealing of polysilicon arose because of its potential as a process that could be used to promote grain growth and hence obtain resistivity reduction in heavily doped films. Experiments undertaken to explore that potential are described in Chapter 5. These experiments showed that significant grain growth can be produced when a thin polysilicon film is irradiated by a scanning heat source under conditions that produce melting throughout the film. The majority carrier electronic properties of these recrystallized films closely approximate those of bulk silicon. This result has led to successful attempts to fabricate MOS transistors and integrated circuits directly in the beam recrystallized material. A substantial interest now exists in the potential of beam recrystallized silicon-on-insulators (SOI) as a substrate for integrated circuit fabrication.

Both the recrystallization process and its application to device fabrication are discussed at length in Chapter 5. We review here only the highlights of the subject.

1.4.1   Sample Preparation

The polysilicon films used in this work are typically 0.5-0.6μm thick films deposited by low pressure chemical vapor deposition (LPCVD) on an oxidized silicon substrate. The grain size in the as-deposited films is ~500Å. These films are then capped with an SiO_2 layer followed by an Si_3N_4 layer to provide good surface morphology in the recrystallized film. Details on encapsulation methods and results are provided in Sections 5.1 and 5.4 and the related references.

Beam processing can be performed with any scanning system that is capable of delivering sufficient energy to the film to melt it entirely. In the simplest type of laser processing, a cw argon laser is operated in the multiline mode with a laser beamwidth on the sample surface of ~40μm. The laser spot is scanned across the sample at a rate of 1-10 cm/sec. The sample

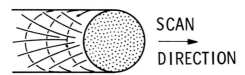

SCAN
→
DIRECTION

FIGURE 1.32. Illustrating the laser recrystallization of thin polysilicon films.

is usually held at 350°C during this operation, with the laser power adjusted to produce complete melting. A typical value of laser power under these conditions is 10-12 watts. Surface melting is initiated at 5-6 watts, after which there is an increase in reflectivity as noted previously, and a consequent reduction of power absorbed in the film. The result is that only a thin surface layer is melted until the laser power is increased to 10-12 watts, where sufficient energy is coupled into the film to produce melting throughout.

1.4.2 Basic Recrystallization Process

The basic recrystallization process is suggested in Fig. 1.32. First the laser beam melts a small area of the film that is approximately the size of the laser spot. This molten region cools as the spot is scanned to the right. Crystallites having various orientations grow as the melt cools, each serving as a potential seed crystal for the material that is being melted in the adjacent laser-irradiated region. The shape of the liquid-solid interface is clearly of critical importance in determining the number of crystallites that are formed and the likely crystalline orientation of the crystallites that are grown (see Sec. 5.4 for further details).

These recrystallized films are of course not normally single crystal, though single crystal islands can be grown by first etching the film into an array of islands [1.53], similar to the first step that would be taken in fabricating an SOS integrated circuit. The conditions for single crystal island growth are probably critical to the future of SOI, and a significant effort is now directed toward achieving this goal (cf Sec. 5.4).

1.4.3 Basic Device Results

Lee et al. [1.54] first showed that both enhancement and depletion mode MOSFETS could be made directly on the top surface of a laser-recrystallized polysilicon film with device properties that are surprisingly similar to devices made on single crystal substrates. Drain characteristics for both types of devices are shown in Fig. 1.33. Fabrication details are discussed in Sec. 5.3.

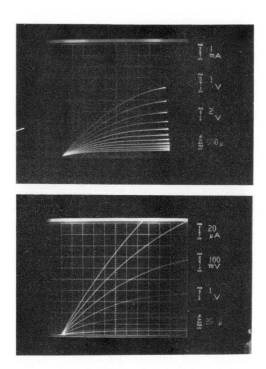

FIGURE 1.33.   (a) Source-drain I-V characteristics for a
deep depletion-mode device (V_G = 0 to -12 V).   (b) Source-drain
I-V characteristics for an enhancement-mode device.

The films used by Lee et al. were recrystallized on Si_3N_4/Si
substrates, however, which was ultimately found to be less satisfactory than SiO_2/Si substrates.   Lam et al. [1.55] obtained
improved characteristics by using a thermally grown SiO_2-on-Si
substrate rather than Si_3N_4, which led to less strain at the lower surface of the film; and they made 11 stage ring oscillators
with a minimum propagation delay of 44 ns per stage compared to
a propagation delay of 36 ns per stage for the same circuit made
on a single crystal substrate.

1.4.4   Improvements in the Recrystallization Process

These first results, together with the knowledge that polysilicon islands (rather than continuous films) could be recrystallized into single crystal silicon, led to a rapid development
of the field. Using a LOCOS technology, Lam and co-workers [1.56]
fabricated islands that were surrounded by an oxide "moat" that
helps to clamp the polysilicon island during its resolidification
from the melt.   This improves surface flatness and makes device

fabrication significantly easier. Biegelsen et al. [1.57] showed that recrystallization could be improved by shaping the island prior to recrystallization into a chevron-like configuration so that the point of the chevron melts first as the beam is scanned by. This point then serves as a seed for the subsequent recrystallization of the entire island. This, together with careful attention to the heat flow problem, led to significant improvement in the crystallinity of the laser processed films and corresponding improvements in devices. Later Stultz and Gibbons [1.58] showed that recrystallization could be improved by shaping the laser beam into a tilted, half moon configuration, obtaining large crystallites that were almost entirely <100> in their crystal orientation.

An alternative to the use of lasers or arc sources for both annealing and recrystallization, the graphite strip heater, was pioneered by Fan, Geis and ther co-workers [1.59,1.60]. For recrystallization studies (Geis, et al. [1.60]), a lower, large graphite strip serves as both the sample holder and a heater. It is heated to ~1200°-1300°C in typical recrystallization experiments with silicon films. A movable, linear strip heater is then mounted above the sample and scanned across it to produce the lateral thermal gradient necessary for recrystallization. The movable graphite strip is heated to a sufficient temperature to melt the silicon film as it is scanned over it.

A different approach was initiated when Kamins et al. [1.61] and Lam et al. [1.62] reported that single crystal films could be obtained by using a via through the SiO_2 film to permit the polysilicon to contact an underlying single crystal substrate, thus providing a single crystal seed for the subsequent laser-induced crystallization of the polysilicon that lies over the oxide (Fig. 1.34). Trimble et al. [1.63] then combined this seeded lateral epitaxy concept with the beam shaping technique of Stultz and Gibbons to produce very promising SOI films.

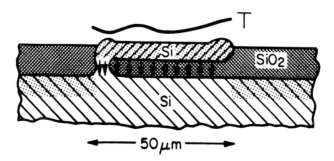

FIGURE 1.34. Geometry for laser-assisted seeded lateral epitaxy (after Trimble et al. [1.63]). ©1982 Elsevier Science Publishing Co.

Films recrystallized by many of these processes now have surface carrier mobilities that are essentially equal to those of bulk silicon, and superior to SOS. Furthermore the carrier mobility is found to be uniform through the film, in sharp contrast to SOS. Arrays of relatively large geometry MOSFETs have been made for use as large area LCD display decoders, an application that is based primarily on cost reduction and is relatively uncritical with regard to device performance. For smaller geometry (VLSI) applications, there remains one basic problem, however, related to the fact that none of the recrystallization processes has yet succeeded in fabricating single crystal films or islands every time. Instead, grain boundaries are usually present, and dopant diffusion along grain boundaries during conventional source and drain fabrication has been shown to lead to source-drain shorts when the gate length is too small (<3μm) [1.64].

This problem can be avoided by using a very short heat cycle to anneal the source and drain implants, such as is provided by any of the annealing techniques described above. Indeed, Ng et al. [1.65]; see also Sec. 5.4.4 of this volume] have built devices with a gate length of ~1.5μm and a 10 stage ring oscillator with a propagation delay of 118 ps/stage on non-seeded material that was recrystallized directly with a scanning circular laser beam; i.e., the least favorable recrystallization conditions. These results provide significant motivation for further work in this field.

1.4.5   Three Dimensional Integration

Further motivation for this field was provided by Gibbons and Lee [1.66], who demonstrated that SOI can be used to fabricate vertically integrated (i.e., three dimensional) devices. These authors used the <u>lower</u> surface of the recrystallized film to make an n channel device that is driven by the same gate that controls a p channel device in the underlying single crystal substrate. The basic structure is shown in Fig. 1.35. This single gate CMOS inverter is similar in principle to what one would obtain by conceptually folding the n channel device of a conventional CMOS inverter over the p channel device, thus obtaining immediately a nearly two-fold increase in inverter packing density. In addition, because the n and p channel devices are on physically different substrates, there is no possibility for latch up, a common problem with bulk CMOS circuits. Alternative processing sequences for stacked CMOS devices have been developed by several authors, [1.67, 1.68]. More recently, Kawamura et al. [1.69] have fabricated three dimensional CMOS integrated circuits with one type of device fabricated directly above a transistor of the opposite type with separate gates and an insulator between the devices. Seven

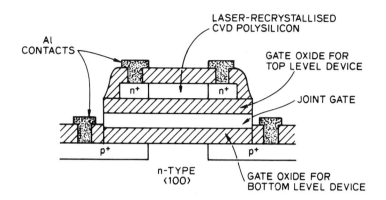

FIGURE 1.35.   One-gate-wide CMOS inverter with devices built in both the underlying single crystal silicon substrate and a laser recrystallized polysilicon film.

stage ring oscillators have been made with this form of three-dimensional integration having a propagation delay of 8.2 ns per gate.

1.5   SILICIDE FORMATION WITH SCANNING BEAMS

The reaction of thin metal films with both single crystal and amorphous silicon to form silicide compounds has been successfully demonstrated using both scanned cw and pulsed beam processing. A detailed summary of the work that has been done with cw beams is given in Chapter 6 of this volume. We give an overview of some of the principal results in this section.

Interest in the formation of silicide compounds arises in part from the fact that for given thickness, their sheet resistance is significantly lower than that of heavily doped poly-silicon. This is of importance in silicon integrated circuit technology since the drive toward higher device packing densities and decreased dimensions will require low resistance, thin film interconnects with high temperature properties that are compatible with other silicon processing operations.

Similar to the case for beam annealing of ion implanted sili-con, both the basic mechanisms of silicide formation and the experimental results are found to depend on whether pulsed or scanned beams are used to promote the reactions. For the pulsed beam (Q-switched laser) case, melting and rapid solidification occurs in the metal-silicon films. The principal result is that the reacted film is frozen in a mixed phase compound which

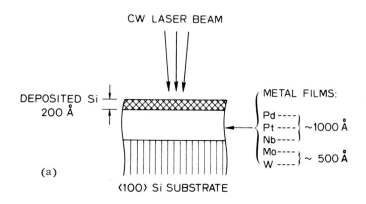

(a)

Pd₂Si (P/r = 1.12 kW/cm)

(b)

Pd₂Si 440x

FIGURE 1.36. (a) Example of sample structure used for scanned-cw laser silicide formation. (b) Nomarsky micrograph of cw laser formed Pd$_2$Si. ©1981 Elsevier Science Publishing Co.

normally includes several silicide compounds and silicon and metal precipitates. By contrast, the use of a scanned cw system to react the metal-silicon films results in the formation of a large area, uniform, single phase silicide with essentially no silicon or metal precipitation.

1.5.1   Sample Preparation

A schematic illustration of the samples that are used for this work is shown in Fig. 1.36. The metal film is electron-beam evaporated in a high vacuum, oil-free system. Sputter deposition of the metal film is generally avoided because of residual argon impurity incorporation and its affect on both growth mechanisms and film properties. For cw laser processing, a thin overlayer of silicon is also electron beam evaporated in the same pump down

as that used for depositing the metal film. This overlayer provides an anti-reflection coating that enables efficient coupling
of the laser into the film. If an electron beam is used to promote the reaction, this antireflection coating is not necessary.

We also show in Fig. 1.36 a Nomarski micrograph of a silicide
formed by cw laser processing. The surface of this material is
essentially featureless at the magnification indicated, suggesting
a very uniform reaction.

1.5.2   Growth Mechanisms

Measurements of the kinetics of silicide formation via
scanned laser processing have been carried out for several
metal/silicon systems. In Fig. 1.37 we show Rutherford backscattering data, taken from Shibata, et al. [1.70] obtained on cw
laser reacted Si/Nb/Si structures. The substrate temperature
during laser reaction for this example was 350°C. The advancement
of the interface between the silicon and niobium materials, indi-
cative of silicide formation, occurs for constant laser power as
a function of the number of laser scan frames. It can be seen
that the thickness of the silicide film ($NbSi_2$) increases with
the number of scan frames. The average composition of this
reacted layer was found to be $NbSi_2$, as determined from backscattering and verified by glancing angle X-ray diffraction analysis.

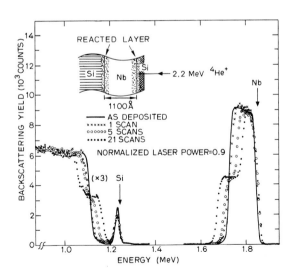

FIGURE 1.37.   Rutherford backscattering spectra taken on
laser reacted Nb/Si. The formation of $NbSi_2$ as a function of
number of laser scan frames is shown. ©1981 Elsevier Science Publishing Co.

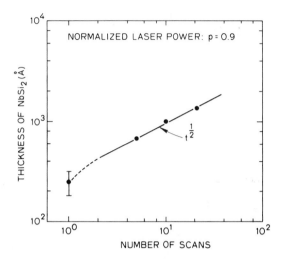

FIGURE 1.38.   Plot of the increase of NbSi$_2$ thickness versus
the number of laser scan frames. ©1981 Elsevier Science Publishing Co.
Reprinted by permission of the publisher,
the Electrochemical Society, Inc.

Using the near surface S factor, $[S_o]_{Nb}^{NbSi}2 = 83.7$ eV/Å, the
film thickness versus number of scan frames was determined and
plotted. This result is shown in Fig. 1.38. One can see that
a parabolic growth behavior exists for the NbSi$_2$ film since the
number of scans is proportional to the effective annealing time.
This result is in agreement with the general behavior of the
refractory metals which form disilicides by furnace annealing.
These facts thus suggest that the laser induced reaction for the
NbSi$_2$ formation process is via a solid phase diffusion mechanism.

Theoretical models presented in Chapter 2 of this volume
can be used to compute the effective surface temperature and
effective annealing time for given scanning conditions (laser
power, scan velocity, beam radius and sample reflectivity).
These data can be used together with measured film thickness
growth to study the growth mechanism and obtain kinetic data to
characterize the process.

We show such a development in Fig. 1.39. On the right hand
side of this figure the various normalized laser powers used to
obtain a certain thickness of growth of NbSi$_2$ are used to calculate
the maximum surface temperature for a substrate temperature of
350°C. The reciprocal of this temperature is then plotted vs the
square of the reacted layer thickness to determine the growth kinetics of the material. This plot is shown on the left side of Fig.
1.39. It can be seen that the data are well fit by a straight line
indicating that a first order process is occurring. A least
squares fit of the data results in an activation energy for the

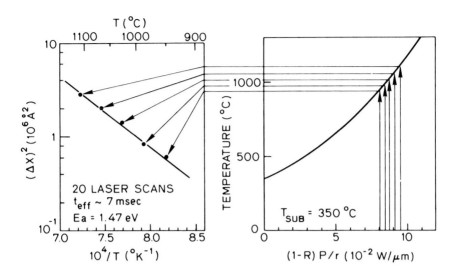

FIGURE 1.39   Comparison of theoretical calculations with experimental measurements for Pd_2Si formation by scanned cw laser.

process. By using an iterative procedure that takes into account that the effective time is dependent on both the activation energy and maximum surface temperature, Shibata, et al. were able to determine a value of activation energy of 1.47 eV. The rate constant for the process was found to be $A = 1.08 \times 10^{14}$ $\AA^2$/sec.

The formation and analysis of $MoSi_2$ and WSi_2, silicides of special importance for integrated circuit applications, are discussed in detail in Chapter 6 of this volume, along with attempts that have been made to form superconducting silicides by carrying out metal/silicon reactions at temperatures that are difficult to achieve without a laser system. Important results on silicide oxidation mechanisms and kinetics are also presented in that chapter.

1.6   RAPID THERMAL PROCESSING WITH STATIONARY INCOHERENT SOURCES

To conclude this chapter, we describe a simple form of cw beam processing in which annealing of an entire wafer may be achieved in seconds using a uniform source of incoherent radiation. As with the scanning Hg arc and graphite strip heater systems discussed previously, the high throughput capabilities of these systems make them attractive for application in practical cal semiconductor processing lines.

These systems are usually called "rapid thermal processing systems" because they are rapid compared to a conventional furnace processing. The basic physics of the process is very similar to furnace annealing, however.

Rapid thermal processing systems employ rapidly cycled banks of tungsten-halogen lamps, arc sources or electrically heated graphite strips to raise the temperature of a thermally isolated wafer to 1000°-1200°C in a few seconds. This temperature is maintained for a few seconds, after which the wafer is allowed to cool by radiation or forced air cooling. Calculations of the temperature-time profiles for such systems are given in Chapter 2.

The first system of this type used for annealing ion implant-ed silicon was reported by Nishiyama et al. [1.71], who used quartz-halogen lamps. A schematic illustration of a somewhat larger system, used by Lischner and Celler [1.83] for both anneal-ing several wafers at a time and recrystallizing thick films of silicon on insulating substrates, is shown in Fig. 1.40, along with a plot of lamp power and wafer temperature. The radiation field in this system is found to be uniform over an area of ~500cm². Wafers may be heated to 1000°-1200°C in a few seconds and cooled equally rapidly. The wafer peak temperature and also the various time intervals and ramps may be controlled with great precision using microprocessor-based control systems. The apparatus is also energy efficient.

For silicon integrated circuit technology, the movement of the field toward VLSI appears to provide an opportunity for applica-tions of rapid thermal processing that would probably not exist

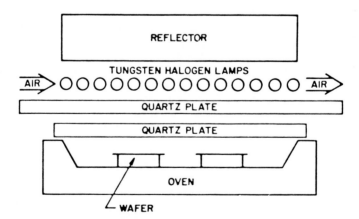

FIGURE 1.40   Schematic cross-section of the RTA apparatus. (after Lischner et al. [1.83]). ©1982 Elsevier Science Publishing Co.

otherwise. This opportunity can be fairly judged against specific technical criteria such as the uniformity of activation of implanted dopants, dopant redistribution, wafer warpage, minimization of residual defects, energy consumed during processing, and the throughput of large diameter (100 to 125 mm) wafers. We will review these points briefly below. Further details on specific aspects of the process will be found in succeeding chapters of the book.

1.6.1 Annealing Requirements for VLSI Applications on Large Diameter Silicon Wafers

Doping uniformity over a wafer and tight control of variations in routine production are key advantages of the implantation process in modern silicon technology. Current production ion implanters are capable of a throughput of at least 100 wafers per hour at doses up to about $2 \times 10^{15}/cm^2$, with doping uniformity better than 0.5% for both 100 and 125 mm wafers. Furnace annealing is capable of retaining this uniformity and, equally important, keeping wafers flat within the tolerances necessary for fine scale photolithographic steps. However the typical annealing conditions (600°C for 30 minutes followed by 1000°C - 1100°C for 20 minutes) produce dopant redistribution that makes it difficult to fabricate the shallow junctions and small device structures required for VLSI.

Figure 1.41 has been prepared to provide a comprehensive visualization of annealing processes and results. Here we show a set of time-temperature coordinates on which lines (---As---) have been drawn giving the time-temperature loci for given amounts of diffusive redistribution of As $(3\sqrt{Dt})$. Several experimental results that are helpful for judging the relevance of rapid thermal processing for VLSI applications are also given. It should be mentioned that the diffusion coefficient used for this construction is obtained from Tsai et al. [1.72] and is concentration dependent. We have assumed an As concentration of $5 \times 10^{19}/cm^3$ for the construction.

1.6.1.1 Uniformity

The uniformity of activation of implanted dopants over large wafers is of paramount importance for any process that is a candidate for application in silicon VLSI technology. Current and Yee [1.73], quoting a private communction from Perloff, state that full activation of moderate dose $(10^{15}/cm^2)$ implants of B, P and As carried out at low energies (~50 keV) into large wafers (100-125mm) can be successfully obtained at temperatures of 1000°C in ten seconds using a halogen lamp system (open circle in Fig.

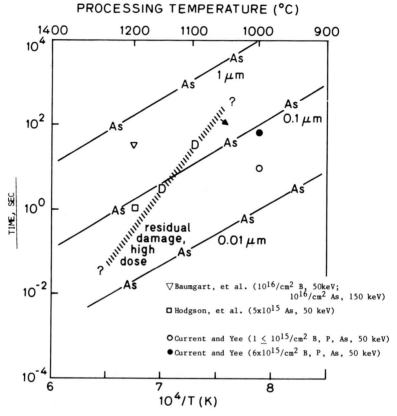

FIGURE 1.41.   Time-temperature coordinates with experimental annealing results and estimated damge annealing locus for heavy implants. ©1984 Elsevier Science Publishing Co.

1.41).   This confirms the earlier work of Burton, et al. [1.74] and Baumgart, et al. [1.75], who achieve high activity with little diffusive redistribution and who also show (on smaller samples) that residual defect densities as judged by TEM can be reduced to levels that are at or below those obtained in furnace annealing.   Figure 1.41 suggests that about 300Å of diffusive redistribution should be expected to take place under these conditions.   However, at higher doses, Current and Yee find that uniformity of activation is unacceptable until the annealing time is increased to 60 seconds at 1000°C (filled circle on Fig. 1.41), citing sensitivity of activation to temperature gradients across the wafer as the probable cause of this behavior.   The consequent redistribution is then estimated to be near 0.1 µm, which is approaching the limit of acceptability for MOS-VLSI and is already too large for really high speed bipolar devices.

1.6.1.2   Wafer Flatness

Wafer flatness is an absolute necessity for both mask regis-
tration and for high yield, since warped wafers are easily broken
in subsequent processing operations. Current and Yee [1.73] have
studied wafer flatness after implant and rapid anneal on both 100
and 125 mm wafers. Their results show that implantation improves
wafer flatness, probably by relieving surface stress; and that
halogen lamp annealed wafers have flatness that is equivalent to
that of furnace annealing. Table 1.7 gives a partial summary of
their data.

TABLE 1.7   Wafer flatness changes with implantation and anneal.

Wafer Diameter (mm)	Mount	Anneal	Peak to Valley in μm Before	After
100	Unclamped	Furnace	4.1	4.9
100	Unclamped	Rapid	5.5	5.1
100	Unclamped	None	4.1	3.6
100	Clamped	Furnace	4.4	4.9
100	Clamped	Rapid	4.6	4.6
100	Clamped	None	4.1	4.6
125	Unclamped	Furnace	6.5	4.8
125	Unclamped	Rapid	7.2	5.3
125	Unclamped	None	7.4	4.9
125	Clamped	Furnace	6.0	3.6
125	Clamped	Rapid	4.9	3.1
125	Clamped	None	4.9	1.9

1.6.1.3   Residual Damage (TEM)

A potentially troublesome feature of rapid thermal annealing not discussed by Current and Yee concerns the annealing
time/temperature combination necessary to achieve defect free
material, as judged by TEM, after high dose implants. Baumgart,
et al. [1.75] have shown that, for B implants of $10^{16}/cm^2$
at 50 keV and As implants of $10^{16}/cm^2$ at 150 keV, annealing
at 1200°C for 30 seconds is required to reduce the damage to
an acceptable level. Annealing at 1100°C for 30 seconds for
these implant conditions leaves a high density of irregularly
shaped dislocation loops that would almost certainly produce
unacceptable device properties.

Hodgson et al. [1.76], using an ultrahigh power, vortex cooled argon arc lamp obtain similar data for As implants of 5×10^{15}/cm^2 at 40 - 50 keV. To achieve no observable defects in TEM, temperatures in the vicinity of 1200°C are required with anneal times of a few seconds. Defects are observed when the anneal temperature is reduced to 1150°C, for ~3 second anneal-ing time. To call attention to these points, we have drawn a line in Fig. 1.41 (?'''''''D'''''''?) representing an estimate of the time-temperature locus that may be necessary to achieve high quality annealing for high dose implants.

Hill [1.77] has suggested, based on a plot similar to that shown in Fig. 1.41, that in the far future device dimensions might be reduced to the point where only 0.01 μm of impurity redistribution would be allowed, and that annealing times of 10 milliseconds and temperatures of 1200°C would then be re-quired, suggesting the use of a scanning cw laser. However, if the residual damage line has the shape suggested in Fig. 1 for high dose implants, higher temperatures and longer times than those suggested by Hill might be required for successful anneal-ing. In fact, it may be difficult to achieve a simple implant/anneal process with 100Å accuracy that is totally satisfactory. New device design procedures may simplify processing requirements, however.

1.6.1.4   Enhanced Redistribution

One of the advantages often cited for rapid cw processing is that little diffusive redistribution occurs if times and tempera-tures are managed properly. This conclusion has been based on work carried out primarily in the 1000°C temperature range. How-ever, Hodgson et al. [1.76] mention that enhanced diffusion may be occurring under the experimental conditions they employ; and the recent data of Narayan et al. [1.78] suggest that the diffu-sion coefficient for As at 1050°C is about two orders of magni-tude greater than the literature value. To illustrate this, we show in Fig. 1.42 As concentration profiles measured by Narayan et al. before and after a 1050°C, 10 sec annealing cycle. The implant was carried out at 100 keV to a dose of 10^{16}/cm^2.

The likelihood that the enhanced diffusion is a damage-enhanced phenomena is suggested by the work of Seidel [1.79], who implanted 2×10^{15}/cm^2 BF$_2$ into a preamorphized Si target and found full electrical activity following a rapid thermal anneal at 770°C for 10 seconds. SIMS profiles showed no motion of the annealed profile at temperatures up to 1000°C for 10 seconds. However, TEM studies show residual damage that may require higher temperature anneals for its removal. In any case, the possibility that enhanced diffusion is occurring during the rapid annealing

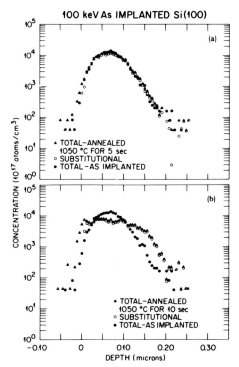

FIGURE 1.42   Arsenic concentration profiles before and after rapid thermal processing: (a) 1050°C/5 sec;   (b) 1050°C/10 sec.

of very shallow ion implanted layers seems to be worth attention. The future of rapid thermal annealing for very shallow structures and hence its appropriateness for VLSI applications will depend partly on the the answer to this question.

1.6.1.5   DLTS

A further point of interest regarding defect minimization arise from the work of Pensl et al. [1.80]. These authors find DLTS defect spectra and densities (Fig. 1.43) that appear to depend on both cooling rate and thermal gradients. If the cooling rate is indeed responsible for the excess defect density obtained, then more careful attention will have to be given to this feature of the annealing process.

There is also the possibility that, for high dose implants at least, the percolation of oxygen into the implanted layer prior to annealing (from direct exposure to the laboratory environment) may be responsible for some of the residual damage. These points may suggest that very well controlled annealing in situ (i.e., in

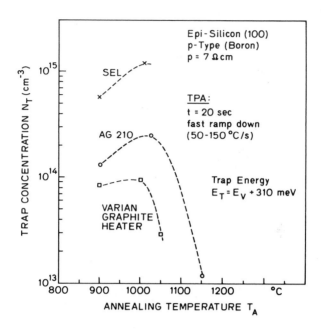

FIGURE 1.43.   DLTS spectra for various types of rapid thermal
annealing. ©1984 Elsevier Science Publishing Co.

the ion implantation chamber) may be required to achieve the
ultimate limits of the technology.

1.6.1.6 Summary

     To summarize, for doses of B, P and As below $10^{15}/cm^2$, any
of several forms of rapid thermal annealing (quartz-halogen
lamps, arc lamps, and hot graphite strips) appear to be satisfactory for application in silicon IC technology, measured against
reasonably exacting criteria. The adequacy of the technique
for very shallow structures and higher dose cases is less clear
if redistribution is to be strictly limited, though it is adequate
for the near future. In addition there are device structures
and alternative fabrication techniques to be discussed later
that may also create a role for rapid annealing.

1.6.2 Applications of Rapid Annealing for Si Microwave Bipolar
       Transistors

     It is clear from recent progress in the microwave semicon-
ductor industry that very shallow bipolar structures can be
made without putting the entire burden of profile control on low

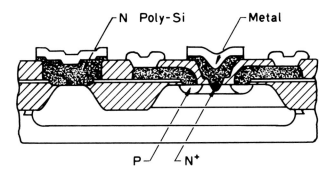

FIGURE 1.44. Super self-aligned bipolar transistor (Sakai et al. [1.81]. ©1984 Elsevier Science Publishing Co.

voltage ion implantation. As an example we show in Fig. 1.44 a bipolar transistor that uses a "double poly" process to form both low resistivity base contacts and a polysilicon "extended emitter" [1.81]. The active base region is formed by direct ion implantation and furnace annealing. Rapid thermal annealing could perhaps be used to some advantage here, though base widths much less than 0.1 μm are not as useful as might appear at first sight because of the undesirably large base resistance (r_b) that is then obtained. The advantage of rapid thermal processing would be primarily one of convenience.

The emitter in this structure is prepared by ion implantation of As into the polysilicon "extended emitter", followed by furnace annealing to drive the As into the underlying single crystal silicon. The final device has an f_T of ~12GHZ and has been used to make ECL-type counting circuits that operate at 5GHZ clock rates and dissipate ~0.5 mW per gate. These are among the fastest IC building blocks ever made.

Attempts to use rapid annealing to fabricate a similar bipolar structure have been made by Natsuaki, et al. [1.82], who have concentrated on using the process both to activate a 1.5×10^{16} As implant into the polysilicon extended emitter and to control As penetration into the underlying silicon. The "rapid anneal" was accomplished by simply mounting wafers on a quartz boat of low thermal mass, inserting them rapidly into a furnace, and then removing them quickly.

Grain growth in the polysilicon and As diffusion into the underlying crystalline silicon were measured as a function of time at an annealing temperature of 1150°C. Figure 1.45 shows the cut off frequency f_T obtained as a function of collector current. The data is similar to that of Sakai et al. [1.81] described earlier, who employ furnace annealing. We also show f_T vs collector current for a state-of-the-art double-implanted, furnace-annealed

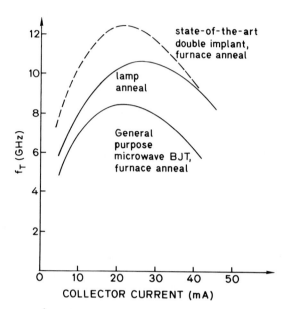

FIGURE 1.45. Dependence of f_T on collector current for microwave bipolar silicon transistors. ©1984 Elsevier Science Publishing Co.

bipolar transistor, and a similar curve for a general purpose microwave bipolar transistor, also double-implanted but with less attention given to ultimate frequency performance. The rapid anneal looks promising but there is need for improvement before it replaces existing technology.

It should also be mentioned that SIPOS polysilicon emitter structures are very likely to become the dominant technology for emitter fabrication in bipolar transistors intended for very high speed, high gain operation. The structure is shown in Fig. 1.46 and has an As-doped polysilicon extended emitter with a thin oxygen doped layer at the base-emitter junction. The absence of significant As penetration through this oxide layer gives very low emitter side wall capacitance. In addition, the asymmetry in electron and hole injection through this oxygen-doped layer results in very large emitter efficiency, leading to gain enhancements of more than an order of magnitude. As pointed out by Hill [1.77], this structure is such a step forward from conventional bipolar transistors that it will undoubtedly be the natural successor of the current microwave bipolar transistor.

From the processing viewpoint, rapid thermal annealing may be useful in redistributing the implanted As in the polysilicon layer and simultaneously activating the B base implant, though this structure will be more forgiving than a standard bipolar

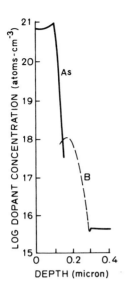

FIGURE 1.46   Bipolar transistor with SIPOS emitter (after Hill [1.77]).

©1983 Elsevier Science Publishing Co.

transistor of longer thermal cycles, so rapid annealing would be of importance primarily for convenience and possibly energy savings, both of which may be of considerable appeal.

1.6.3   Rapid Thermal Processing of Compound Semiconductors

Several rapid thermal processing techniques have been applied to GaAs with encouraging initial results and also important limitations. This subject has been reviewed by Williams [1.83] with the general conclusion that rapid thermal processing has not yet realized its initial promise of overcoming long standing problems with conventional furnace annealing for activation of implanted dopants in GaAs. Work done since the review of Williams has produced a somewhat more hopeful outlook, though problems still remain. This work will be reviewed in Chapter 7.

Apart from its use for standard annealing experiments, however, there is also now an increasing interest in more novel applications in which rapid thermal processing is used to study and control impurity diffusion in compound semiconductors and also to control surface properties without using ion implantation. We also review these applications in Chapter 7.

REFERENCES

1.1 Ready, J. F., "Effects of High Power Laser Radiation." Academic Press, 1971.

1.2 Khaibullin, I. B., Shtyrkov, E. I., Zaripov, M. M., Galyantdinov, M. F., and Zakirov. G. G., Sov. Phys. Semiconductors 11, 190 (1977).

1.3 Antonenko, A., Gerasimenko, N.N., Bvarechenskii, A., Smirnov, L. A., and Tseitlin, G. M., Sov. Phys. Semiconductors 10, 81 (1976).

1.4 Kachurin, G. A., Nadaev, E. V., Khodyachikh, A. V., and Kovaleva, L. A., Sov. Phys. Semiconductors 10, 1128 (1976).

1.5 Baeri, P., Campisano, S. U., Foti, G., and Rimini, E., Appl. Phys. Lett. 33, 2, 137 (1978).

1.6 Muller, J. C., Grob, A., Stuck, R., and Siffert, P., Appl. Phys. Lett. 33, 2, 137 (1978).

1.7 von Allmen, M., Luthy, W., and Affolter, K., Appl. Phys. Lett. 33, 9, 824 (1978).

1.8 Celler, G. K., Poate, M. J., and Kimerling, L. C., Appl. Phys. Lett. 32, 8, 464 (1978).

1.9 White, C. W., Christie, W. H., Appleton, B. R., Wilson, S. R., Pronko, P. P., and Magee, C. W., Appl. Phys. Lett. 33, 7, 662 (1978).

1.10 Auston, D. H., Surko, C.M., Venkatesan, T. N. C., Slusher, R. E., and Golovchenko, J. A., Appl. Phys. Lett. 33, 5, 437 (1978).

1.11 Kirkpatrick, A. R., Minnucci, J. A., and Greenwald, A. C., IEEE Trans. on Elec. Dev. ED-24, 429 (1977).

1.12 Gat, A., and Gibbons, J. F., Appl. Phys. Lett. 32, 3, 142 (1978).

1.13 Gat, A., Gibbons, J. F., Magee, T. J., Peng, J., Williams, P., Deline, V., and Evans, C. A. Jr., Appl. Phys. Lett. 33, 5, 389 (1978).

1.14 Regolini, J. L., Gibbons, J. F., Lietoila, A., Sigmon, T. W. Magee, T. J., Peng, J., Hong, J. D., Katz, W., and Evans, C. A., Jr., Appl. Phys. Lett. 50, 6, 4388 (1979).

1.15 Gat, A., Gerzberg, L., Gibbons, J. F., Magee, T. J., Peng, J., and Hong, J. D., Appl. Phys. Lett., 33, 8, 775 (1978).

1.16 Lee, K. F., Gibbons, J. F., Saraswat, S., Kamins, T. I., Appl. Phys. Lett. 35, 2, 173 (1979).

1.17 Poate, J. M., Leamy, H. J., Sheng, T. T., and Celler, G. K., Appl. Phys. Lett. 33, 11, 918 (1978).

1.18 Von Allmen, M. and Wittmer, M., Appl. Phys. Lett. 34, 3, 221 (1979).

1.19 Liau, Z. L., Tsaur, B. Y., and Magee, J. W., Appl. Phys. Lett. 34, 3, 221 (1979)

1.20 Shibata, T., Gibbons, J.F., and Sigmon, T. W., Appl. Phys. Lett. 36, 7, 566 (1980).

1.21   Auston, D. H., Galovchenko, J. A., Simons, A. L., Slusher, R. E., Smith, P. R. and Surko, C. M., "Laser-Solid Interactions and Laser Processing – 1978," (S. D. Ferris, H. J. Leamy, and J. M. Poate, eds.) pp. 11–16. AIP (1979).

1.22   Bloembergen, N., op cit, p. 1–9.

1.23   "Thermophysical Properties of Matter," Vol. 4, p. 204. IFI/Plenum, New York (1970).

1.24   Ho, C. Y., Powell, R. W. and Liley, P. E., J. Phys. Chem. Ref. Data 3, 1, I–588 (1974).

1.25   Phillip, H. R. and Taft, E. A., Phys. Rev. 120, 37 (1960).

1.26   Brodsky, M. H., Title, R.S., Weiser, K. and Pettit, G. D., Phys. Rev. 1, 6, 2632 (1970).

1.27   Dash, W.C. and Newman, R., Phys. Rev. 99, 4, 1151 (1951).

1.28   Leitoila, A. and Gibbons, J. F., J. Appl. Phys. 53, 4 (1982).

1.29   Leitoila, A. and Gibbons, J. F., Appl. Phys. Lett. 40, 7 (1982).

1.30   Van Vechten, J. A., Tsu, R. Saris, F. W. and Hoonbout, D., Phys. Lett. 74A, 417 (1979).

1.31   Liu, J. M., Yen, R., Kurz, H., and Bloembergen, N., Appl. Phys. Lett. 39, 755 (1981).

1.32   Baeri, P., Campisano, S. V., Foti, G. and Rimini, E., J. Appl. Phys. 50, 2, 788 (1979).

1.33   Wang, J. C., Wood, R. F. and Pronko, P. P., Appl. Phys. Lett. 33, 5, 455 (1978).

1.34   Young, R. T., White, C. W., Clark, G. J., Narayan, J., Christie, W. H., Murakami, M., King, P. W. and Kramer, S. D., Appl. Phys. Lett. 32, 3, 139 (1978).

1.35   Kodera, H., Jpn. J. Appl. Phys. 2, 212, (1965).

1.36   White, C. W., Wilson, S. R., Appleton, B. R. and Young, F. W., Jr., J. Appl. Phys. 51, 1, 738 (1980).

1.37   Narayan, J. and White, C. W., "Laser and Electron Beam Processing of Materials", p. 65–70, (C. W. White and P. S. Peercy, eds.). Academic Press (1980). See also Narayan, J., Young, R. T. and White, C. W., J. Appl. Phys. 49, 3912, (1978).

1.38   Lang, D. V., J. Appl. Phys. 45, 3023 (1974).

1.39   Johnson, N. M., Bartelink, D. J., Gold, R. B., and Gibbons, J. F., J. Appl. Phys. 50, 7, 4828 (1979).

1.40   Burton, J. L., Kimerling, L. C., Miller, G. L., Robinson, D. A. H. and Celler, G. K., "Laser-Solid Interactions and Laser Processing – 1978," (S. C. Ferris, H. J. Leamy, and J. M. Poate, eds.) p. 543. AIP (1979).

1.41   Kirkpatrick, A. R., Minnucci, J. A., and Greenwald, A. C., IEEE Trans. on Elec. Dev. ED-4, 429 (1977).

1.42   Kennedy, E. F., Lau, S. S., Golecki, I., Mayer, J. W., Tseng, W., Minnucci, J. A., and Kirkpatrick, A. R., "Laser-Solid Interactions and Laser Processing – 1978," (S. C. Ferris, H. J. Leamy, and J. M. Poate, eds.) p. 470. AIP (1979).

1.43   Cespregi, L., Mayer, J. W. and Sigmon, T. W., Phys. Lett. 54A, 157 (1975).

1.44   Bicknell, R. W. and Allen, R. M., Proceedings of the First International Conference on Ion Implantation in Semiconductors, Thousand Oaks, Calif. Gordon and Breach, New York (1970).

1.45   Chu, A. and Gibbons, J. F., Proceedings 1976 International Conference on Ion Implantation in Semiconductors, (F. Churno, J. Borders, and D. K. Price, eds.) p. 711. Plenum Press, New York (1977).

1.46   Gibbons, J. F., "Ion Implantation in Semiconductors – Part II: Damage Production and Annealing," Proc. IEEE 60, 9, 1062 (1972).

1.47   Stultz, T., Ph.D. Dissertation, Stanford University, Stanford, CA (1983).

1.48   Nissim, Y. I., Lietoila, A., Gold, R. B. and Gibbons, J. F., J. Appl. Phys. 51, 1 (1980).

1.49   Roth, J. A., Olson, G. L., Kokorowski, S. A. and Hess, L. D., "Laser and Electron-Beam Solid Interactions and Materials Processing," (J. F. Gibbons, L. D. Hess, and T. W. Sigmon, eds.) p. 413. North Holland (1981).

1.50   Gibbons, J. F., Johnson, W. S. and Mylroie, S. W., "Projected Range Statistics in Semiconductors," distributed by John Wiley & Sons. Dowden, Hutchinson and Ross (1975).

1.51   Johnson, N. M., Bartelink, D. J., Moyer, M. D., Gibbons, J. F., Lietoila, A., Taknakumar, K. N., Regolini, J. L., "Laser and Electron Beam Processing of Materials," (C. W. White and P. S. Peercy, eds.), p. 423. Academic Press (1980).

1.52   Stultz, T. J., Sturm, J. and Gibbons, J. F., "Laser-Solid Interactions and Transient Thermal Processing of Materials," (J. Narayan, W. L. Brown, and R. A. Lemons, eds.) p. 463. North Holland (1983).

1.53   Gibbons, J. F., Lee, K. F., Magee, T. J., Peng, J. and Ormond, R., Appl. Phys. Lett. 34, 831 (1979).

1.54   Lee, K. F., Gibbons, J. F., Saraswat, K. C. and Kamins, T. I., Appl. Phys. Lett. 35, 173 (1979).

1.55   Lam, H. W., Tasch, A. F., Jr., Holoway, T. C., Lee, K. F. and Gibbons, J. F., IEEE Elec. Dev. Lett. EDL-1, 6 (1980).

1.56   Lam, H. W., Tasch, A. F., Jr. and Holloway, T. C., IEEE Elec. Dev. Lett. EDL-1, 10 (1980).

1.57   Biegelsen, D. K., Johnson, N. M., Bartelink, D. J. and Moyer, M. D., "Laser and Electron-Beam Solid Interactions and Materials Processing," (J. F. Gibbons, L. D. Hess, and T. W. Sigmon, eds.) p.487. North-Holland (1981).

1.58   Stultz, T. J. and Gibbons, J. F., Appl. Phys. Lett. 39, 6 (1981).

1.59   Fan, J. C. C., Geis, M. W. and Tsaur, B.-Y., Appl. Phys. Lett. 38, 365 (1981).

1.60   Geis, M. W., Smith, H. I., Tsaur, B.-Y., Fan, J. C. C., Silversmith, D. J., Mountain, R. W. and Chapman, R. L., "Laser-Solid Interactions and Transient Thermal Processing of Materials (J. Narayan, W. L. Brown, and R. A. Lemons, eds.) p. 477. North Holland (1983).

1.61   Kamins, T. I., Cass, T. R., Dell'Oca, C. J., Lee, K. F., Pease, R. F. W. and Gibbons, J. F., J. Electrochem. Soc. 128, 5 (1981).

1.62   Lam, H. W., Pinizzotto, R. F. and Tasch, A. F., Jr., J. Electrochem. Soc. 128, 9 (1981).

1.63   Trimble, L. E., Celler, G. K., Ng, K. K., Baumgart, H. and Leamy, H. J., "Laser and Electron Beam Interactions with Solids", p. 505. North Holland (1982).

1.64   Ng, K. K., Celler, G. K., Povilonis, E. J., Frye, R. C., Leamy, H. J. and Sze, S. M., IEEE Trans. on Dev. Lett. EDL-2, 12 (1981).

1.65   Ng, K. K., Celler, G. K., Povilonis, E. I., Trimble, L. E. and Sze, S. M., "Laser-Solid Interactions and Transient Thermal Processing of Materials," (J. C. C. Fan and N. M. Johnson, eds.). North Holland, 1983, in press.

1.66   Gibbons, J. F. and Lee, K. F., IEEE Elec. Dev. Lett. EDL-1, 6 (1980).

1.67   Goeloe, G. T., Maby, D. J., Silversmith, D. J., Mountain, R.W. and Antoniadis, D. A., IEDM Conference Abstracts, p. 554 (1981).

1.68   Colinge, J. P. and Demoulin, E., IEDM Conference Abstracts, p. 557 (1981).

1.69   Kawamura, S., Sasaki, N., Iwai, T., Kakano, M. and Takagi, M., IEEE Elec. Dev. Lett. EDL-4, 10, 366 (1983).

1.70   Shibata, T., Gibbons, J. F. and Sigmon, T. W., Appl. Phys. Lett. 36, 566 (1980).

1.71   Nishiyama, K., Arai, M. and Watanabe, N., Jap. J. Appl. Phys. 19, 10 (1980).

1.72   Tsai, M. Y., Morehead, F. F., Baglin, J. E. E. and Michel, A. E., J. Appl. Phys. 51, 3230 (1981).

1.73   Current, M. and Yee, A., Solid State Tech., 197 (1983).

1.74   Burton, J. L., Celler, G. K., Jacobsen, D. C., Kemerling, L. C., Lischner, D. J., Miller, G. L. and Robinson, McD., "Laser and Electron Beam Interactions with Solids," (B. R. Appleton and G. K. Celler, eds.) p. 765. North Holland (1983).

1.75   Baumgart, H., Celler, G. K., Lischner, D. J., Robinson, McD. and Sheng, T. T., "Laser-Solid Interactions and Transient Thermal Processing of Materials," (J. Narayan, W. L. Brown, and R. A. Lemons, eds.) p. 349. North Holland (1983).

1.76   Hodgson, R. T., Baglin, J. E. E., Michel, A. E., Mader, S., Gelpy, J. C., "Laser-Solid Interactions and Transient Thermal Processing of Materials," (J. Narayan, W. L. Brown, and R. A. Lemons, eds.) p. 355. North Holland (1983).

1.77　　Hill, C., "Laser–Solid Interactions and Transient Thermal Processing of Materials," (J. Narayan, W. L. Brown, and R. A. Lemons, eds.) p 381. North Holland (1983).

1.78　　Narayan, J., Holland, O. W., and Eby, R. E., Appl. Phys. Lett. 43, 957 (1983).

1.79　　Seidel, T. E., IEEE Elec. Dev. Lett. EDL–4 (1983).

1.80　　Pensl, G., Schulz, M., Stolz, P., Johnson, N. M., Gibbons, J. F. and Hoyt, J., "Laser–Solid Interactions and Transient Thermal Processing," (J. C. C. Fan and N. M. Johnson, eds.) North Holland (1984) in press.

1.81　　Sakai, T., Konaka, S., Kabayashi, Y., Suzuki, M., and Kawai, Y., Elec. Lett., 284 (1982).

1.82　　Natsuaki, N., Tamura, M., Miyazaki, T. and Yanagi, Y., Extended Abstracts of 15th Conf. on Solid State Devices and Materials, Japanese Society for Appl. Phys., Tokyo (1983) p. 47–50.

1.83　　Lischner, D. J. and Celler, G. K., "Laser–Solid Interactions with Solids," (B. R. Appleton and G. K. Celler, eds.) p. 759. North Holland (1982).

CHAPTER 2

Temperature Distributions and Solid Phase Reaction Rates Produced by Scanning CW Beams

Arto Lietoila*, Richard B. Gold†, James F. Gibbons, and Lee A. Christel‡

STANFORD ELECTRONICS LABORATORIES
STANFORD UNIVERSITY
STANFORD, CALIFORNIA

2.1 INTRODUCTION

Quantitative studies of the effects of cw beam processing in various applications require accurate knowledge of the temperature produced by the scanning beam. By comparison to conventional furnace processing, a direct measurement of the temperature is difficult since at any one time only a small fraction of the sample is heated. As a result, theoretical calculations of the temperature produced by a scanning beam are important for choosing processing parameters which achieve a set of required experimental conditions. These temperature calculations are the subject of this chapter.

The temperature is calculated by solving the heat flow equation with appropriate source terms and boundary conditions. Since the thermal conductivity of semiconductors is a strong function of temperature, this equation is strongly nonlinear, and cannot be solved analytically in the general case. However, under steady state conditions, analytic solutions to the heat flow equation can be obtained using the Kirchhoff transform. We will review this method in Section 2.2 below.

*Present address: Micronas, Inc., 00101 Helsinki, Finland.
†Present address: Adams-Russell Company, Burlington, Massachusetts 01803.
‡Present address: SERA Solar Corporation, Santa Clara, California 95054.

With regard to the geometry of the calculations, several distinct regimes are important for practical applications. When lasers or circular electron beams are used, the dimensions of the focused beam are normally small compared to the semiconductor sample thickness. In this case, three-dimensional solutions for the heat equation need to be sought. At the other extreme, typical beam "diameters" in arc lamp and some ribbon e-beam annealing systems are at least a few millimeters. Since semiconductor samples are typically only a fraction of a millimeter thick, the lateral heat flow then becomes negligible and the problem can be treated as one-dimensional. In this case one also has a choice between two different boundary conditions at the sample back-surface. The sample may be left thermally isolated or it may be heat sunk to a fixed temperature. For very narrow ribbon beams, an intermediate regime obtains and the problem becomes two dimensional. Each of the above cases will be considered in more detail below.

In addition to the calculation of the temperature, it is also important to know how scanning speed, or more precisely the beam dwell time, affects various solid state reactions promoted by the beam. In principle this problem can be solved by computing the temperature vs time for a given point of the sample during the scan. However, the lateral intensity distribution of a laser or e-beam can often be well approximated by a Gaussian distribution, and the temperature profiles are similar in shape. This makes it feasible to model the effect of any scan by imagining it to produce at each point a constant temperature, equal to the maximum temperature achieved, with a reduced dwell time. This model, to be developed in Section 2.6, provides a quantitative framework for analyzing various beam-induced reactions such as solid phase epitaxial regrowth and silicide formation.

2.2   THE KIRCHHOFF TRANSFORM

The equation to be solved for the heat flow produced by the beam is the diffusion equation:

$$\rho C_p \, \frac{\partial T}{\partial t} = \vec{\nabla}(\kappa \vec{\nabla} T) + Q(\vec{r}, t) \tag{2.1}$$

where ρ is the density, C_p the specific heat and κ the thermal conductivity of the semiconductor; and Q is the rate of energy deposition.

Equation (2.1) is nonlinear, because all three coefficients are, in general, temperature dependent. The temperature dependences of ρ and C_p are usually weak, however, and will be neglected.

Equation (2.1) can be simplified by introducing the Kirchhoff transform [2.1],

$$\theta = \int_{T_0}^{T} \frac{\kappa(T')dT'}{\kappa(T_0)} \tag{2.2}$$

where θ will be called the <u>linear</u> <u>temperature</u> and T is the true temperature.

Using Eq. (2.2), the equation for θ becomes

$$\frac{\rho C_p}{\kappa(\theta)} \frac{\partial \theta}{\partial t} = \nabla^2 \theta + \frac{Q(\vec{r},t)}{\kappa(T_0)} \tag{2.3}$$

The nonlinearity is retained for the time-dependent case, but in the steady state, Eq. (2.3) further reduces to the <u>linear</u> heat flow equation

$$\nabla^2 \theta + \frac{Q(\vec{r},t)}{\kappa(T_0)} = 0 \tag{2.4}$$

In most semiconductors, the thermal conductivity can be accurately approximated by the following expression [2.2-2.6]:

$$\kappa(T) = \frac{A}{T-T_\kappa} \tag{2.5}$$

Values of the parameters A and T_κ for different semiconductors are given in Table 2.1. Using Eqs. (2.2) and (2.5), the Kirchhoff transform can now be expressed as

$$\theta = (T_0-T_\kappa) \ln \frac{T - T_\kappa}{T_0 - T_\kappa} \tag{2.6}$$

The inverse of the Kirchhoff transform is

$$T = T_\kappa + (T_0-T_\kappa) \exp \frac{\theta}{T_0-T_\kappa} \tag{2.7}$$

The steady state temperature distributions can now be readily obtained by first solving Eq. (2.4) for θ, and then applying the inverse Kirchhoff transform Eq. (2.7) to obtain the true temperature.

The normal boundary conditions for T are

$$\frac{\partial T}{\partial z} (z=0) = 0 \tag{2.8}$$

$$T \ (z=L) \ = \ T_o \qquad\qquad\qquad\qquad (2.9)$$

where z is the depth coordinate, and L is the sample thickness.

TABLE 2.1.   The Fitting Parameters, A and T_K, for the Thermal Conductivity of Various Semiconductors $[\kappa = A/(T-T_K]$.

Material	A(W/cm)	$T_K(^\circ K)$	Temperature Range	Refs.
Si	299	99	> 300°K	2.2,2.4,2.6
GaAs	91	90.9	> 250°K	2.2,2.3,2.6
InP	115.5	130.5	> 300°K	2.3, 2.6
SiC	430	320	> 500°K	2.5, 2.6

Condition (2.8) follows from the fact that the heat losses through the sample front surface (radiation, convection) are assumed to be negligible.   Condition (2.9) means that the sample is in perfect thermal contact with a heat sink having a temperature T_o. Those conditions imply for θ:

$$\frac{\partial \theta}{\partial z} \ (z=0) \ = \ 0 \qquad\qquad\qquad\qquad (2.10)$$

$$\theta \ (z=L) \ = \ 0 \qquad\qquad\qquad\qquad\quad (2.11)$$

In the case of a thermally isolated sample, radiation losses dominate and these boundary conditions become invalid.   This case will be discussed in Section 2.4.   In Section 2.3 we develop solutions to Eq. (2.4) using the boundary conditions Eqs. (2.10) and (2.11).

2.3   TEMPERATURE CALCULATIONS FOR CW LASERS

The majority of scanning beam annealing experiments to date have been performed using cw lasers.   We therefore begin by presenting, in this section, temperature calculations for cw laser processing.

As was mentioned in Section 2.1, cw laser (and circular electron) beams are, in annealing applications, typically so tightly focussed that the beam dimensions are clearly much less than the thickness of the semiconductor wafers.   Therefore, both the vertical and lateral heat flows must be considered, and the temperature calculations have to be three-dimensional.   On the other hand the actual layers which are subject to the beam processing (e.g., implanted layers or deposited thin films) are usually less than a micron thick, and are therefore thin compared to the typical beam diameters (20-100 μm). Therefore, the vertical temperature drop across the layer to be processed is negligible, and the surface temperature can be assumed to prevail across the entire layer. (An exception to this is the use of a modified SEM for beam annealing, in which case the beam dimensions can be comparable to the thickness of the layer.)

We will begin by showing how the maximum surface temperatures are obtained, and we will demonstrate that in practical applications the temperature can be assumed to be uniform in depth throughout the the layers to be processed. The effects of the wafer thickness, absorption coefficient, beam shape (elliptical, circular), and scanning speed are then studied. Finally, we will present a calibration procedure to facilitate the determination of the processing temperatures in practical annealing experiments.

2.3.1   The Maximum Temperature Obtained with a Scanning CW Laser

The maximum temperature is achieved at the sample surface at the center of the beam under the following conditions:

1.   The beam is circular.

2.   The beam is stationary.

3.   The wafer is semi-infinite.

4.   The absorption depth can be assumed to be zero.

In most annealing experiments, these conditions are practically met because: spherical lasers are used, the scanning speeds are slow, the wafer thickness is usually several times the beam diameter, and e.g., for the Ar+ laser, the absorption depth is only about 0.7 μm. In what follows we present the temperature calculations assuming that the above conditions 1-4 prevail. The effects of deviations from these conditions are studied in later sections.

The high power cw lasers used for beam annealing are usually designed to operate in the TEM_{oo} mode, in which case the lateral beam intensity distribution can adequately be approximated by a Gaussian distribution:

$$I = I_o \exp(-r^2/w^2) \tag{2.12}$$

where r is the distance from the beam center and w is defined as
the beam radius. We will follow this definition of w (1/e-points
in intensity) throughout this volume. However, other definitions
have also been used in the literature. Cline and Anthony [2.7],
and Nissim et al. [2.2] define the beam radius by the $1/\sqrt{e}$-points
in intensity, which gives a radius a factor of $\sqrt{2}$ smaller than
ours. On the other hand, those doing research on lasers per se,
sometimes define the beam radius by the 1/e point in the electric
field strength, which corresponds to the $1/e^2$ - points in intensity. This gives a beam radius which is a factor of $\sqrt{2}$ larger
than that determined according to the definition in Eq. (2.12).
The different definitions are compared in Table 2.2.

TABLE 2.2. Different Definitions of the Beam Radius for a
Gaussian Laser Beam (TEM$_{oo}$ Mode)

Defining value of intensity	Symbols used	Relative value for a given beam	Refs.
$1/\sqrt{e}$	R, r	$1/\sqrt{2}$	2.2, 2.6, 2.7
1/e	w	1	2.8, 2.9
$1/e^2$	w	$\sqrt{2}$	2.17

Assuming a Gaussian beam shape and conditions 1-4, above,
Lax [2.8] has shown that the maximum <u>linear</u> surface temperature
at the beam center is given by

$$\theta_{max} = \frac{1}{2\sqrt{\pi}\,\kappa(T_0)}\left[\frac{P(1-R)}{w}\right] \qquad (2.13)$$

We define the quantity within brackets in Eq. (2.13) as the
absorbed power per unit radius, P_a:

$$P_a = \frac{P(1-R)}{w} \qquad (2.14)$$

The maximum true temperature vs. P_a for different semiconductors can now be obtained using Eqs. (2.13-2.14), the inverse
Kirchhoff transform, and the expressions for the thermal conductivity, Eq. (2.5). The results are given in Fig. 2.1 for Si, SiC,
GaAs and InP using different substrate backsurface temperatures
T_0. That figure shows that use of higher values of T_0 results
in a substantial reduction of laser power for a given surface temperature. This is due to the fact that the thermal conductivity
for all of these materials decreases with increasing temperature.
It is for the same reason that the T vs. P curves are nonlinear
and concave up.

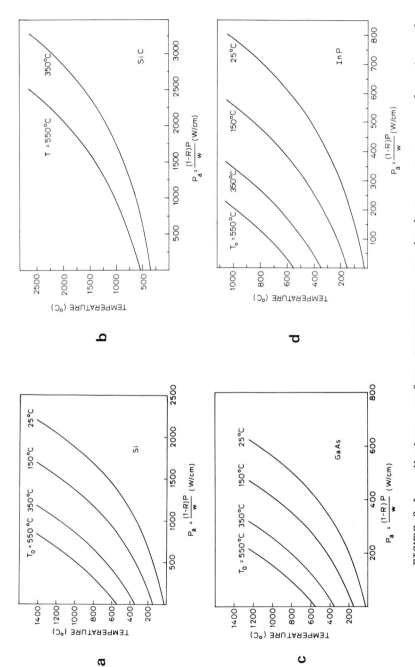

FIGURE 2.1. Maximum surface temperature at the beam center as a function of the absorbed power per unit radius for (a) silicon, (b) silicon carbide, (c) gallium arsenide and (d) indium phosphide.

2.3.2   Vertical and Lateral Temperature Distributions

If the maximum temperature vs. depth must be calculated, the following formula, derived by Lax [2.8], can be used for the linear temperature at the beam center:

$$\Theta(Z) = \Theta_{max} \; (Z{=}0) \; \exp(Z^2) \; \text{erfc}(Z) \qquad\qquad (2.15)$$

where $\Theta_{max}(Z{=}0)$ is given by Eq. (2.13) and $Z = z/w$. We plot in Fig. 2.2 the ratio $\Theta(z)/\Theta_{max}$, which we call the normalized linear temperature η_z, as a function of the dimensionless depth coordinate $Z = z/w$. From Fig. 2.2(a), it is evident that in most cases the surface temperature can be used throughout the layers of interest. For example, using a typical beam radius of 40 µm, the temperature within the first 1 µm of the sample does not vary by more than 4%.

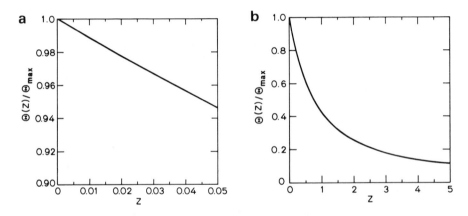

FIGURE 2.2.   Normalized linear temperature $\Theta(Z)/\Theta_{max}$ as a function of normalized depth $Z = z/w$ at the beam center (x=y=0) for (a) small and (b) large values of Z.

The linear temperature at the sample surface at a distance r from the beam center can be expressed [2.8] as a function of the dimensionless distance $R = r/w$:

$$\Theta(R) = \Theta_{max} \; (R{=}0) \; \exp\left(-\frac{R^2}{2}\right) I_0\left(\frac{R^2}{2}\right) \qquad\qquad (2.16)$$

where here I_0 is the modified Bessel function of order zero. We plot in Fig. 2.3 the ratio $\Theta(R)/\Theta_{max}(R{=}0)$ as a function of R and compare this to the normalized laser intensity distribution $I(R)/I_0$. That figure or Eq. (2.16) together with Eqs. (2.7) and (2.13), can be used to obtain the true surface temperature as a function of distance from the beam center.

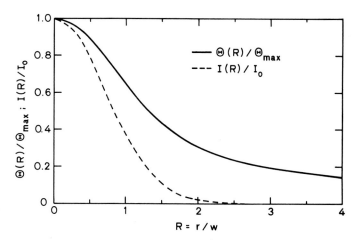

FIGURE 2.3. Normalized linear temperature $\Theta(R)/\Theta_{max}$ as a function of normalized lateral distance $R = r/w$ from the beam center at the sample surface ($z=0$; $r^2=x^2+y^2$). For comparison, the dashed line shows the profile of laser intensity.

2.3.3   Effect of Sample Thickness

The formulas given above for Θ, Eqs. (2.13–2.15), were derived assuming that the sample is infinitely thick compared to the beam dimensions. This is not always the case: use of very powerful lasers and long focal length lenses may result in beam diameters in excess of 100 μm and the sample thickness must then be taken into account.

A quick check of the validity of the infinite sample assumption can be performed by using Eq. (2.15) to calculate what the linear temperature rise would be at a depth equal to the actual sample thickness in an infinitely thick wafer. If the result is close to zero, no correction is required. Otherwise, the formulas given above, and Fig. 2.1, give temperatures that are too high, and an appropriate correction must be carried out. We will now present this correction for the linear temperature Θ. Equation (2.7) can then be applied to obtain the true surface temperatures in wafers of finite thickness.

It has been shown using the method of images [2.9] that the linear surface temperature, Θ_L, in a sample whose thickness is L can be expressed as the following series:

$$\Theta_L = \Theta_{max} \left[1 + \sum_{k=1}^{\infty} (-1)^k 2\exp(4k^2L^2/w^2) \; \text{erfc} \; (2kL/w) \right] \quad (2.17)$$

Figure 2.4 shows the ratio of $\Theta_L/\Theta_{max} = \eta_L$ as a function of w/L. We can see that for w/L < 0.3, the infinite sample approximation is good. On the other hand, for w/L > 2, the temperature approaches the values obtained from one-dimensional solutions indicated by the dashed line in Fig. 2.4.

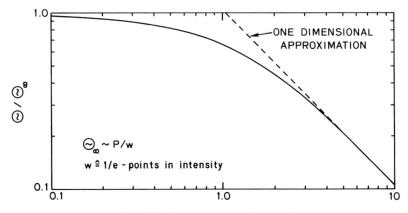

FIGURE 2.4. Dependence of the linear temperature Θ on the ratio w/L of the beam radius to the sample thickness. The value of Θ is normalized to the maximum value, Θ_∞, obtained for a semi-infinite sample. The dashed line is valid for the one-dimensional approximation. (After Ref. 2.9) ©1982 Elsevier Science Publishing Co.

For intermediate wafer thickness, the correct temperatures can be obtained in two ways. If the deviation of the infinite sample approximation is significant (i.e., w/L ≈ 1), one has to first calculate Θ_{max} using Eq. (2.13). Θ_L is then obtained using Eq. (2.17) or Fig. 2.4, and the true temperature is calculated using Eq. (2.7).

However, if the change in Θ due to the wafer thickness is small, a much simpler procedure can be used. Differentiation of Eqs. (2.6,2.7) leads to the following expression for the change in the real temperature:

$$\Delta T = (T_{max} - T_K) \ln\left[\frac{T_{max} - T_K}{T_0 - T_K}\right] \cdot \frac{\Delta\Theta}{\Theta_{max}} \qquad (2.18)$$

where T_{max} is the maximum real temperature achieved for infinite samples (see Fig. 2.1). [Note that temperatures in Eq. (2.18) have to be expressed in Kelvin.] The relative change in Θ, $\Delta\Theta/\Theta_{max}$, is obtained from Fig. 2.4.

2.3.4   Effect of Absorption Coefficient

Lax [2.8] has shown that for a finite absorption coefficient α, the maximum linear temperature is reduced. To account for this, we calculate the linear normalized temperature η_α which depends on the dimensionless parameter $W = \alpha w$:

$$\eta_\alpha(W) = \frac{\Theta_{max}(W)}{\Theta_{max}(W=\infty)} \tag{2.19}$$

where $\Theta_{max}(W=\infty)$ is given by Eq. (2.13). Lax [2.8] has shown that η_α can be expressed as

$$\eta_\alpha = \frac{2}{\sqrt{\pi}} \int_0^\infty \frac{e^{-u^2}}{1 + u/W}\, du \tag{2.20}$$

Figure 2.5 gives the normalized linear temperature η_α as a function of W. The figure shows that the assumption of an infinite absorption coefficient is generally good. For example, for the Ar^+ laser, α is 1.2×10^4 cm^{-1} at $300°K$ for Si, and a typical value for w would be 20 μm. This gives $W = 24$, and $\eta_\alpha = 0.97$. However, if a Kr laser is used, the absorption coefficient is substantially lower, namely about 4000 cm^{-1} at $300°K$. The lower power output of that laser may necessitate the use of a smaller spot to obtain high enough temperatures, and the effect of the absorption coefficient may then have to be taken into account. The situation will be further complicated by the fact that the absorption coefficient also depends on temperature.

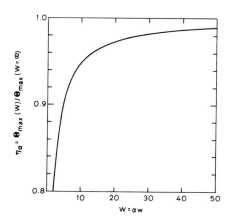

FIGURE 2.5.   Normalized   linear   surface   temperature $\eta_\alpha =$ $\Theta_{max}(W)/ \Theta_{max}(W=\infty)$ at the beam center (x=y=z=0) as a function of the product, W, of the absorption coefficient, α, and the beam radius, w. [After Ref. 2.9.]

2.3.5   Effect of Scanning Speed

Cline and Anthony [2.7] have shown that the effect of the scanning speed on the linear temperature depends on the dimensionless parameter V:

$$V = \frac{\rho \, C_p \, w \, v}{2 \, \sqrt{2} \, \kappa(T_o)} \qquad (2.21)$$

where v is the scanning speed. Again, the maximum linear temperature is reduced by a factor, η_v. Nissim et al. [2.2] give values of η_v for a wide range of scanning speeds and for various aspect ratios β for elliptical beams (see Section 2.3.7). For low scanning speeds, η_v can be approximated by the following expression:

$$\eta_v = 1.0 - 0.72 \, V \qquad (2.22)$$

Figure 2.6 gives η_v for large values of V.

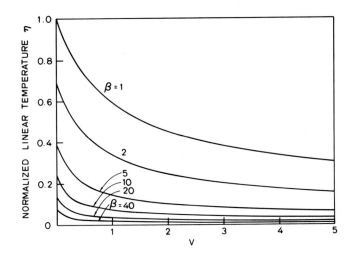

FIGURE 2.6.   Normalized linear surface temperature $\eta = \Theta(V)/\Theta_{max}$ at the beam center (x=y=z=0) as a function of the normalized scan speed V [see Eq. (2.21)] for a circular beam (β=1) and elliptical beams of various aspect ratios β (see Section 2.3.7). It should be emphasized that normally, V $\ll$ 1, in which case Eq. 2.22 can be used for η.   (After Ref. 2.2.)

In contrast to corrections due to the wafer thickness and absorption coefficient, the inverse Kirchhoff transform cannot be applied to the corrected linear temperature to obtain true temperatures for the scanning beam since the Kirchhoff transform is valid only in the steady state.   Upper and lower limits for the true temperature can, however, be obtained using the following procedure. Values for the parameter V in Eq. (2.21) are calculated using both the substrate backsurface temperature and the temperature which would be obtained if the beam were steady. Corresponding correction factors η_V for Θ are then determined. If η_V is close to one, Eq. (2.18) can be used to obtain the lower and higher limits for ΔT due to the scanning. For large deviations of η_V from unity, corrected values of Θ should be calculated and the inverse Kirchhoff transform should be applied to obtain the upper and lower limits for the true temperature.

2.3.6   Calibration Procedure for Temperature Calculations

The determination of the annealing temperature in a practical experiment is greatly simplified if the system is calibrated in terms of the melting point of the semiconductor to be processed. This can be achieved by noting the reading of the laser power meter at which melting of the material just occurs at the beam center (all other conditions, T_0, etc., should be the same as will be used for the actual annealing). This power level will be called P_{melt}. The annealing power is then expressed as a fraction, P/P_{melt}, of the melting power, and Fig. 2.7 can be used to determine the annealing temperature.

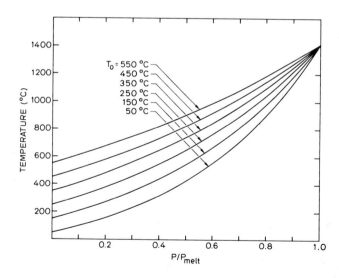

FIGURE 2.7.   Maximum true surface temperature at the beam center in silicon as a function of normalized power P/P_{melt} for various values of substrate backsuface temperature T_0. (After Ref. 2.9.)

This procedure is especially useful for Si, because no decomposition of the material occurs at submelting temperatures and once melting takes place, it leaves clearly visible rough traces on the otherwise mirror-like sample surface. The observation of melting is made easy if an interference contrast microscope is used. However, compound semiconductors, such as GaAs and InP, may decompose before melting occurs, and that will make the use of the calibration procedure more difficult.

The calibration procedure described above will completely remove the uncertainty in temperature calculations due to the inaccuracy of the laser power meter (providing it still is <u>linear</u>), the reflectivity of the material and the spot size of the laser beam. If the same wafer thickness is used for calibration and annealing, possible errors due to finite sample thickness are also eliminated. However, if the product of the absorption coefficient and beam radius is so low that it affects the maximum temperature, a slight error will result due to the fact that the absorption coefficient depends on temperature.

Finally, this calibration procedure will give the most accurate estimate for the temperature for scanning beams. However, some error will still occur, since the effect of the the scanning speed depends on the sample temperature as described above. Calculated temperatures which this procedure yields will be <u>lower</u> than the real temperatures, because the thermal conductivity in semiconductors decreases when temperature increases and the lowering of temperature due to scanning is more profound for low conductivity material (see Section 2.3.5 above). Thus, the effect of the scanning speed is more significant at P_{melt} than at the power used for annealing.

2.3.7   The Elliptical Beam

In certain applications it is desirable to use an elliptical laser beam rather than a circular one. For example, if wide scan lines are desired, one could use an elliptic beam with the long axis perpendicular to the scan direction. On the other hand, scanning in the direction parallel to the long axis would yield a long dwell time without reducing the scanning speed. Elliptic beam shapes are readily obtained by using cylindrical lenses, either alone or in combination with spherical lenses.

The incident intensity distribution of an elliptical beam can be described as

$$I = I_0 \exp - \left(\frac{x^2}{w_x^2} + \frac{y^2}{w_y^2} \right)$$

$$(2.23)$$

where the intensity I_0 is related to the total incident power as follows:

$$I_0 = \frac{P}{\pi \, w_x \, w_y} \qquad (2.24)$$

Nissim et al. [2.2], who have treated the case of elliptical beams in great detail, use the convention that w_x is the _short_ radius of the elliptic beam. The aspect ratio is then

$$\beta = \frac{w_y}{w_x} > 1 \qquad (2.25)$$

Figure 2.8 shows the effect of increasing β for the linear temperature Θ. In that figure, we plot the normalized linear temperature,* $\Theta/\Theta_{max}^{circ}$, as a function of position at the sample surface for different values of β. We can see that the increase of β results in a substantial decrease of the maximum temperature while the temperature distribution spreads out in the direction of the larger radius.

Nissim et al. [2.2] have also shown that the normalized linear temperature, $\eta = \Theta/\Theta_{max}^{circ}$, at the center of the beam at the sample surface can be expressed as:

$$\eta \, (x=y=z=0) = \frac{2}{\pi\beta} \, K \left(\sqrt{\frac{\beta^2 - 1}{\beta^2}} \right) \qquad (2.26)$$

where K is the complete elliptic integral of the first kind. We plot Eq. (2.26) as a function of β in Fig. 2.9. This figure, together with Eqs. (2.7) and (2.13), can be used to calculate the maximum annealing temperature for any elliptical beam.

We would like to emphasize that Eq. (2.26), and Figs. 2.8 and 2.9, have been obtained by assuming that the sample is infinite. This assumption is not valid, if even the larger radius of the elliptical beam is comparable to or larger than the wafer thickness. In that case, the best way to determine the actual annealing temperature is to use the calibration procedure described in Section 2.3.6 above.

---

*Θ_{max}^{circ} is the maximum linear temperature achieved with a circular beam of radius $w = w_x$, [see Eq. (2.13)].

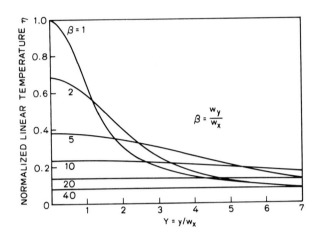

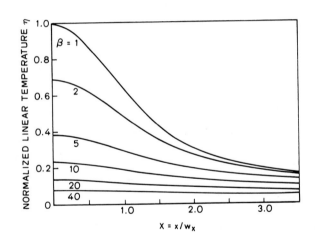

FIGURE 2.8.   Normalized linear surface temperature $\eta = \Theta/\Theta_{max}^{circ}$ for various aspect ratios $\beta = w_y/w_x$ as a function of normalized distance from the beam center along (a) the long radius $(Y = y/w_x)$ and (b) the short radius $(X = x/w_x)$ of the elliptical beam. $V = 0$ for all cases.

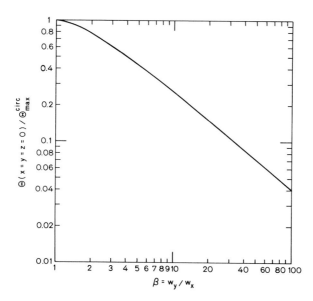

FIGURE 2.9.   Normalized linear surface temperature $\Theta/\Theta_{max}^{circ}$ at the beam center as a function of the aspect ratio β of the elliptical beam.

2.3.8   Thin Layers on the Semiconductor Sample: Effects of Reflectivity

In many beam processing applications the semiconductor sample is covered by thin metal or dielectric layers. These layers have two effects on the temperature calculations. First, the reflectivity of the layered structure can be significantly different from that of the bare semiconductor. Secondly, the thermal properties of the structure may also change somewhat. Fortunately the latter effect is negligible in most practical beam processing applications, since the layer thickness (typically less than 0.5 μm) is very small compared to both the beam diameter and the thickness of the underlying substrate. For this reason we consider in this section the effects of a change in the reflectivity only.

If the sample is covered by a thin nonabsorbing dielectric layer (e.g., SiO_2 or Si_3N_4), that layer will act as an antireflection coating. The reflectivity of such a structure is given by [2.10]:

$$R = \frac{n_f^2(1-n_s)^2\cos^2\phi + (n_s-n_f^2)^2\sin^2\phi}{n_f^2(1+n_s)^2\cos^2\phi + (n_s+n_f^2)^2\sin^2\phi} \qquad (2.27)$$

where n_f and n_s are the indexes of refraction of the film and the semiconductor substrate, respectively. The phase difference ϕ in the film is

$$\phi = \frac{n_f \, d}{\lambda_o} \, 2\pi \qquad\qquad (2.28)$$

where d is the film thickness and λ_o is the wavelength of the laser radiation in free space. The above formula is valid for thick substrates. We plot in Fig. 2.10 the reflectivity of Si coated with SiO_2 or Si_3N_4 for the Ar^+ laser (mean wavelength = 5000 Å). Note that the maximum reflectivity is that of an uncoated semiconductor.

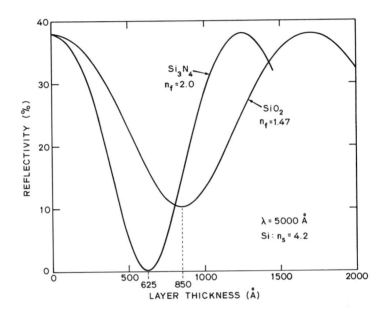

FIGURE 2.10. Reflectivity of Si covered by an AR coating of SiO_2 or Si_3N_4 for the Ar ion laser (mean wavelength = 5000 Å) as a function of coating thickness.

If the coating film is of strongly absorbing material, e.g., metal, the entire radiation is absorbed within the layer and the reflectivity of the structure is simply the reflectivity of the coating material.

2.4   TEMPERATURE CALCULATIONS FOR WIDE AREA OPTICAL SOURCES: THE ARC LAMP

One of the primary reasons for using arc radiation sources for annealing purposes, instead of lasers or tightly focused electron beams, is the need to achieve large spot sizes and thus large throughput for industrial applications. Large spot size of course implies that lateral heat flow in the sample is negligible, so the temperature calculations then become one-dimensional.

Arc lamp annealing has been performed in two ways. When the isothermal [2.11] mode is utilized, the sample is thermally isolated, and will be uniformly heated. In the heat sink [2.12] mode, the backsurface is kept at a constant temperature, as in the case of laser annealing described in Section 2.3 above. We will first consider the isothermal mode and then present calculations for the heat sink mode using a xenon arc lamp as the radiation source.

2.4.1   The Isothermal Mode

When this mode is used, the wafer to be processed is thermally floating, and the whole wafer is illuminated at the same time. The temperature of the wafer will be practically uniform (hence the term isothermal), and the temperature achieved is governed mainly by radiation of the sample. This method has the advantage of requiring very low power densities. However, the time constant for heating the sample is of the order of 10s [2.11], making very rapid anneals impossible. Also, if the heat is to be confined to a thin surface layer of the sample, this method is not practical.

A convenient geometry for this form of annealing is obtained by imagining the arc source to be located near the pole of a spherical volume having walls of reflectivity $R_s \approx 1$ with the sample mounted on thermal insulators in the center of this volume. Under these conditions, the wafer will receive radiation on both surfaces from all angles and will be cooled by radiation only. The total power Q (Watts) of the source (as well as radiation from the cavity walls) is coupled to the wafer with an efficiency η which is given by

$$\eta = \frac{2A_w(1-R_w)}{2A_w(1-R_w) + A_s(1-R_s)} \tag{2.29}$$

where A is area (cm^2) and R the reflection coefficient. The subscripts w and s refer to the wafer and sphere respectively. Since the radiation losses from the wafer are similarly recoupled back

to the wafer, these losses are reduced by the factor 1-η as compared with the losses which would occur in vacuum. Assuming that
the walls of the sphere remain at temperature T_0 (by external
cooling or high thermal mass), the equation describing the temperature rise of the wafer is given by

$$\rho C_p L A_w \frac{dT}{dt} = \eta(Q + \varepsilon_s \sigma A_s T_0^4) - (1-\eta)\sigma \varepsilon_w A_w T^4 \tag{2.30}$$

where σ is the Stefan–Boltzmann constant $(5.67 \cdot 10^{-12} Wcm^{-2} {}^{\circ}K^{-4})$
and the emissivity of silicon is taken to be $\varepsilon_w = 0.5$. (Note
we assume $\varepsilon = 1-R$ for both wafer and sphere.) The steady state
temperature T_f is

$$T_f = \left(T_0^4 + \frac{Q}{(1-R_s)A_s \sigma} \right)^{1/4} \tag{2.31}$$

By substituting for Q, Eq. (2.30) can be written

$$\frac{dT}{dt} = \frac{(1-\eta)2\sigma \varepsilon_w}{L\rho C_p} (T_f^4 - T^4) \tag{2.32}$$

As an experimental guide it is useful to note that, because of
the T^4 dependence, (dT/dt) is within 90% of its initial value
until $T = 0.6 \, T_f$. Hence a first order estimate of the time
required for T to reach its final value is obtained by extra-
polating the initial slope to the final temperature. This gives

$$\Delta t = \frac{(T_f - T_0)L\rho C_p}{2\varepsilon_w \sigma (1-\eta) T_f^4} \tag{2.33}$$

For a cavity of radius 15 cm and reflectivity of 99% containing a
300 μm thick 4" diameter wafer we obtain a coupling efficiency
η of 0.741. For $T_f = 1000°C$, such a wafer reaches steady state
in a time [from Eq. (2.34)] of $\Delta t = 16.3$ sec.

Although Eq. (2.33) predicts that the time to reach a given
steady state temperature increases as η increases, it should be
noted that for constant <u>power</u> Q, Δt will decrease rapidly as η
approaches unity.

Powell et al. [2.11] have calculated the temperature evolution of a Si wafer subjected to xenon arc radiation of 22 W/cm^2
in the absence of a reflecting cavity. The results are shown in
Fig. 2.11. In this case, one needs only to equate the blackbody
radiation of the sample with the incident intensity to obtain
the steady state temperature.

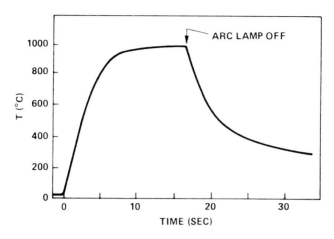

FIGURE 2.11.   Experimental temperature vs. time of a 2 inch diameter Si wafer exposed to constant xenon arc lamp radiation of 22 W/cm^2 in the isothermal mode. (After Ref. 2.11)

2.4.2   The Heat Sink Mode

This mode, where the backsurface of the sample is kept at a constant temperature, was first described by Gibbons [2.12]. The power densities required are substantially higher than in the isothermal mode, since the high thermal conductivity of silicon causes a large heat flow through the sample. On the other hand, the time constant describing the approach to equilibrium is substantially shorter (because of the high intensity used), and very rapid anneals are possible. Also, this method is preferred for applications where melting of the sample surface is desired; the melt depth is easier to control because there is a temperature gradient across the wafer.

The temperature calculations in this mode are complicated by the broad wavelength spectrum of arc radiation sources, which makes a single absorption coefficient inapplicable. This problem can be overcome by utilizing the superposability of the linear temperature Θ. To do this, the wavelength spectrum of the radiation source is divided into small intervals, each of which is assumed to have a constant absorption coefficient. The linear temperatures produced by these intervals are calculated and summed. The inverse Kirchhoff transform is then applied to the sum to obtain the real temperature.

Lietoila et al., [2.13] have applied this procedure to calculate the temperatures achieved in silicon irradiated by a xenon arc lamp. It was shown that <u>steady state</u> solutions to the

temperature can be used, if the dwell-time of the annealing spot
is about 30 ms or more. The results are given in Fig. 2.12 for
two different backsurface temperatures and wafer thicknesses. We
can see that use of a thicker sample and higher backsurface temperature results in a saving of arc lamp power. However, even in
the best case shown in Fig. 2.12, the absorbed intensity required
to reach useful annealing temperatures is about 5 kW/cm^2, which
may be difficult to achieve in practice.

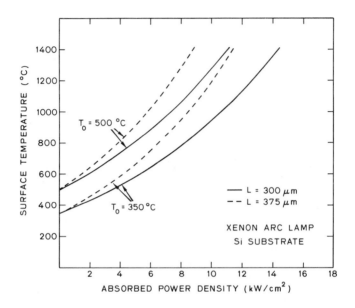

FIGURE 2.12.   Calculated steady state surface temperature
of Si in the heat sink mode as a function of absorbed xenon arc
lamp intensity for different values of heat sink temperature T_0
and sample thickness L.   (After Ref. 2.9) ©1982 Elsevier Science Publishing Co.

An easy way to achieve the temperature needed for annealing,
while still maintaining a well-controlled temperature gradient
across the wafer, is to place a poorly conducting layer between
the semiconductor sample and the heat sink.   A wafer of quartz
serves well as such a layer.   Lietoila et al., [2.13] have calculated temperatures achieved under the same conditions as in
the optimum case of Fig. 2.12 (375 μm thick wafer, T_0 = 500°C),
using quartz wafers of different thicknesses.   The temperature
dependent thermal conductivity of quartz given in Ref. [2.14]
was used in these calculations.   The results are given in Fig.
2.13, and show that the required intensities are in a range much
more easily achievable than if the wafer were placed directly on
the heat sink.

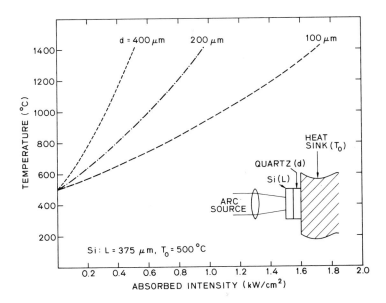

FIGURE 2.13.   Calculated effect of a quartz layer placed between the heat sink and the Si sample being annealed with a xenon arc lamp. The figure shows the surface temperature for the optimum case of Fig. 2.12 (T_0 = 500°C, L = 375 μm) when different quartz layer thicknesses are used. ©1982 Elsevier Science Publishing Co.

Finally, we would like to emphasize that the above calculations have been carried out assuming that the thermal contact between the wafer and the heat sink is perfect. This is not necessarily the case in practice, unless a good thermal conductor, e.g., liquid tin or gallium, is used to improve the contacts. Therefore, if accurate information about the annealing temperature is needed in a particular experiment, it would be advisable to perform the melt temperature calibration described in Section 2.3.6 above.

2.5   TEMPERATURE CALCULATIONS FOR SCANNING CW ELECTRON BEAMS

Processing of semiconductors by scanning cw electron beams is broadly analogous to cw laser processing. There are, however, a few differences which have to be considered in making temperature calculations. The principal dissimilarities are as follows.

a.   The reflectivity is essentially zero for any material.
b.   The lateral intensity distribution may not be Gaussian.

c. The energy deposition profile is not exponential in depth, but approximately Gaussian, with the peak normally inside the material; this "absorption" profile does not depend on temperature.

d. The beam will spread out when entering the material; this effect may be very significant if narrow beams are used.

The lack of reflectivity is a simplifying factor, since dielectric layers will not change the absorbed power. This feature is especially useful for processing of device structures, which have different layer thicknesses.

The other three points above, b-d, make the formulas presented in Sections 2.3.1-2.3.4 above inapplicable for e-beam annealing. However, the temperature calibration procedure given in Section 2.3.6 is valid for (stationary) e-beams, no matter what the intensity profile is laterally or in depth. Moreover, since the energy deposition profile vs. depth does not change with temperature as does the optical absorption coefficient, the procedure is accurate in this respect for the e-beam. However, if the beam is scanning fast, the determination of the temperature will become inaccurate in the same fashion as for scanning cw lasers.

2.5.1 Temperature Calculations for Finely Focused Electron Beams

Some applications, e.g., the direct writing of patterns during lithography, require an electron beam with extremely small lateral dimensions, on the order of 1 μm. In such cases the beam dimensions are not only small compared with the sample thickness (necessitating a three-dimensional analysis) but are in addition small compared with the range of the electrons themselves. In analogy with the absorption depth for laser radiation, the relevant parameter in this case is called the Grun range [2.18] and is denoted R_g. In Fig. 2.14 the Grun range R_g is plotted as a function of electron energy E_b for electrons in silicon. More generally, R_g is given approximately by

$$R_g = 4.57 \ E_b^{1.75} \ \mu g/cm^2 \qquad 5 \ keV < E_b < 25 \ keV \qquad (2.34)$$

where E_b is expressed in keV.

To calculate temperatures in this case, one must first obtain a realistic representation of the heat source. This has been accomplished using Monte-Carlo techniques to determine a universal energy dissipation function [2.19, 2.20]. Analytic techniques based on Green's functions were then used to calculate the temperature from the known source function.

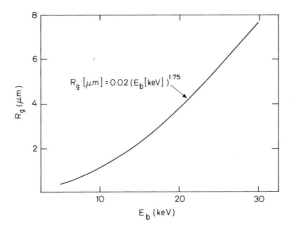

FIGURE 2.14.   Grün range R_g as a function of energy E_b for electrons in silicon.

In Fig. 2.15, we plot $\Theta_n = \Theta/\Theta_{max}$, the normalized linear temperature at the center of a circular Gaussian e-beam as a function of the dimensionless parameter R_g/w where w is the beam radius. Note that for R_g/w small, the results agree with the cw laser case in which the absorption coefficient $\alpha = 0$. An analytic approximation to the curve in Fig. 2.15 is given by

$$\Theta_n \approx 0.981 - 0.359 \left(\frac{R_g}{w}\right) + 0.0628 \left(\frac{R_g}{w}\right)$$

$$- 3.96 \cdot 10^{-3} \left(\frac{R_g}{w}\right)^3 \qquad (2.35)$$

The true maximum temperature is once again obtained via Eqs. (2.13) and (2.7) with the power P given by the beam voltage-current product and the reflectivity R = 0.

Similar calculations have been performed for a "ribbon" electron beam, i.e., a beam very long in the y direction but narrow in the x direction. In this case, instead of Eq. (2.13), the linear temperature at the center of the beam is given by the modified one-dimensional expression

$$\Theta \approx \frac{P_Y L}{\sqrt{\pi}\, w\, \kappa(T_0)}\; \Theta_n \qquad (2.36)$$

where P_Y is now the power per unit length of the beam. Θ_n is plotted in Fig. 2.16 as a function of w/L where here w is the half-width of the Gaussian ribbon beam and L is again the sample thickness. The results have been found to be insensitive to the value of R_g. Note that for w/L greater than about 10, the one-dimensional solution applies.

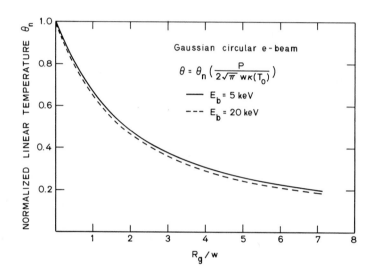

FIGURE 2.15. Normalized linear surface temperature at the center of a circular Gaussian e-beam as a function of the dimensionless parameter R_g/w.

The effect of scan speed on the lateral temperature distribution for a ribbon e-beam of Gaussian half-width 1.4 μm and energy 20 keV incident on a 250 μm thick silicon wafer is shown in Fig. 2.17. It can be seen from the figure that, for this case, distortion in the lateral temperature profile is significant only for velocities greater than about 20 cm/sec.

2.5.2 Temperature Calculations for Wide Area Ribbon Electron Beams

We now wish to present temperature calculations for the surface of a thin sample which is irradiated with a ribbon electron beam which has a width of order 1 mm. For these calculations it is assumed that the electrons deposit their energy in a very narrow region near the surface. For typical electron energies of 5 keV, this is a good approximation since the range of such an electron is only about 0.5 μm. We also assume a constant thermal conductivity κ and apply the Kirchhoff transform to obtain the true temperature when such a transformation is valid.

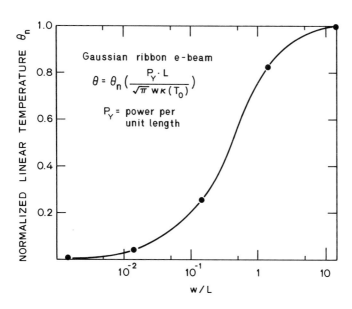

FIGURE 2.16.   Normalized linear surface temperature at the center of a Gaussian ribbon e-beam as a function of the ratio of beam half-width w to the sample thickness L.

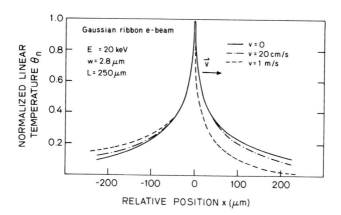

FIGURE 2.17.   Normalized linear surface temperature as a function of lateral position for a Guassian ribbon e-beam showing the asymmetry induced at high scan speeds.

Consider a beam which deposits energy at the rate Q per unit area per unit time in a region of width W = 2b at the surface of a semi-infinite solid. Furthermore, assume that this beam is scanned with velocity V in the −x direction. It has been shown [2.21] that the linear temperature rise due to such a beam in the steady state is given by

$$\Theta_\infty(x,z) = \frac{2Q}{\pi \rho C_p V} \int_{X-B}^{X+B} e^u K_0 \left(\sqrt{u^2 + z^2} \right) du \qquad (2.37)$$

where $X = \eta x$, $Z = \eta z$ and $B = \eta b$, $\eta = \rho C_p V / 2\kappa$ and K_0 and K_1 are modified Bessel functions of the second kind which satisfy

$$\int_0^a e^{\pm u} K_0(u) \, du = ae^{\pm a} \left\{ K_0(a) \pm K_1(a) \right\} \mp 1 . \qquad (2.38)$$

For a sample of finite thickness L and a back surface which is either thermally isolated or thermally sunk, we once again utilize the method of images to obtain the surface temperature

$$\Theta_L(x,0) = \hat{\Theta}_\infty (x,0) + 2 \sum_{n=1}^{\infty} (\pm 1)^n \hat{\Theta}_\infty (x,2nL) \qquad (2.39)$$

where the upper sign corresponds to the case of thermal isolation and the lower sign to the case of a heat sunk back surface. Equations (2.37) and (2.39) have been evaluated by numerical integration with the following results.

2.5.2.1   Thermally Isolated Wafer

When the scan speed is less than about 1 cm/sec, the front and back surfaces are very nearly at the same temperature for a given lateral position x and a first estimate of the temperature rise is simply given by the total energy deposited per unit volume divided by the specific heat.

The first approximation is independent of the thermal conductivity and scan speed (for slow enough scan speeds V). For very slow scan speeds, however, one must account for the cooling effect of radiation from both surfaces. In our calculations the effects of radiative cooling are accounted for by correcting the temperature rise at each lateral position (starting from the leading edge of the beam) by an amount equal to the cooling which occurs during a time $\tau = dx/V$ where dx is the distance between lateral points considered. This is just the dwell time corresponding to each lateral point.

When the scan speed exceeds about 1 cm/sec, the sample has little time to cool so radiative losses can be neglected. However, in this case another effect becomes important; i.e., the thermal conductivity limits the rate at which the back surface can come to equilibrium with the front surface. Thus the maximum surface temperature is higher than would be obtained assuming uniform heating. Furthermore, this maximum is dependent on the thermal conductivity. Since this maximum is also scan-speed-dependent, the Kirchhoff transform cannot be used to remove the temperature dependence of the thermal conductivity. For these cases we have performed calculations using both the thermal conductivity at the initial temperature (0°C), and the thermal conductivity at the final temperature. The true temperature must therefore lie between these two values.

The results of these calculations are presented in Fig. 2.18. The abscissa is now the deposited energy density and thus has units of J/cm^3. The dashed line gives the resulting temperature assuming uniform heating of the wafer with no losses due to radiation.

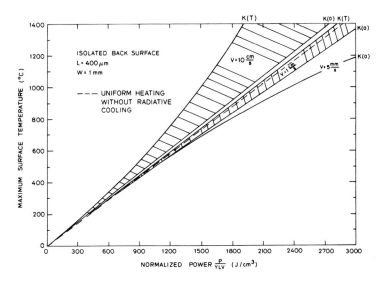

FIGURE 2.18. Maximum silicon surface temperature as a function of normalized power for a scanning ribbon e-beam with rectangular intensity distribution of width W. The sample is assumed to be thermally isolated. ©1982 Elsevier Science Publishing Co.

For slower scan speeds, radiative losses result in a lower maximum temperature while for higher scan speeds, the thermal gradient across the wafer thickness results in a higher maximum surface temperature. The resulting temperatures obtained for a scan speed of 5 mm/sec show that in this case radiative cooling is significant for temperatures above about 800°C. When the scan speed is 1 cm/sec, both cooling effects and thermal gradient effects are evident. The range of power values obtained using the maximum and minimum thermal conductivities is indicated in Fig. 2.18 by the width of the shaded region. For a scan speed of 10 cm/sec, radiative cooling is negligible, but the maximum surface temperature is strongly dependent on the thermal conductivity.

2.5.2.2 Heat Sunk Back Surface

When the back surface of the sample is heat sunk, much more power is necessary to reach a given temperature. As a result, losses due to radiation are negligible in all cases. In addition, for scan speeds less than about 5 cm/sec, the results are independent of scan speed. Thus in these cases the Kirchhoff transform is valid and the temperature dependence of the thermal conductivity can be properly accounted for.

In Fig. 2.19 we show the results for back surface temperatures of 0 and 350°C for a 1mm wide beam. Also shown are calculations at a scan speed of 1 cm/sec for beams of width 100 µm and 500 µm. For these narrow beams it is evident that lateral diffusion is severe enough to cause considerable reduction in the

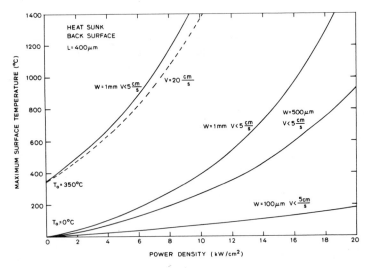

FIGURE 2.19. Maximum silicon surface temperature as a function of power density for a scanning ribbon e-beam with rectangular intensity distribution of width W incident on a sample which is heat sunk. ©1982 Elsevier Science Publishing Co.

maximum surface temperature achieved, even at constant power density. It has been found, however, that there is little dependence on the beam width for widths greater than about 1mm. Thus in such cases the heat problem is essentially one-dimensional.

2.6 REACTION RATE CALCULATIONS

A variety of physical processes induced by scanning cw energy beams have been interpreted in terms of solid-phase reaction mechanisms. These processes include the epitaxial regrowth of ion implanted silicon as well as the formation of certain metal-silicides, and are discussed in detail in subsequent chapters of this book. The behavior of these systems during low-temperature furnace processing can be described by a single activation energy, and the reacting interface can be considered to be at a constant temperature throughout the anneal. The effect of cw beam irradiation is to produce a high temperature and an accelerated reaction rate, but in this case the temperature is a rapidly changing function of time. We discuss here the analytical formulation for solid-phase reaction kinetics induced by a scanning cw beam, including the dependence of the temperature on time. The results of such calculations permit a comparison of observed beam processing results with extrapolated furnace-processing data.

For a rate-limited process such as the regrowth of amorphous implanted Si, the reacted-layer (or regrown-layer) thickness resulting from a constant-temperature furnace anneal can be expressed as

$$d = R_0 t \, \exp(-E_a/kT) \tag{2.40}$$

where t and T are the anneal time and temperature, respectively, and E_a is the activation energy. For diffusion-controlled reactions, such as the formation of certain metal-silicides, the d in Eq. (2.40) is replaced by d^2, but all other features of the following discussion are still applicable.

The transient nature of the temperature rise produced by a scanning cw beam means that the reacted-layer thickness should be expressed by

$$d = R_0 \int_{-\infty}^{\infty} \exp[-E_a/kT(t)]dt \tag{2.41}$$

If the dependence of temperature on time is known, Eq. (2.41) can be reduced to Eq. (2.40) by determination of an "effective annealing time," t_{eff}. The first step is the calculation of the spatial temperature profile. Then, using knowledge of experimental scan conditions, this spatial profile is converted to a temporal one.

We consider first the case of a single scan through the point of interest $(x,y) = (0,0)$. The beam is assumed to have a Gaussian intensity distribution given by Eq. (2.12), with an absorption depth α^{-1} much less than the beam radius w. We stipulate further that the beam is moving with a velocity v in the x direction, and that its center passes over $(x,y) = (0,0)$ at $t = 0$. Using the results of Eq. (2.16) the linear temperature rise $\Theta(t)$ at the surface can be shown to be

$$\Theta(t) = \Theta_{max} \exp(-v^2t^2/2w^2)\, I_0(v^2t^2/2w^2) \tag{2.42}$$

where $\Theta_{max} = \Theta(0)$ is given by Eq. (2.13), and I_0 is the modified Bessel function of order Θ. As pointed out earlier, we assume that the reacting-layer thickness is much less than the beam radius and that, consequently, the solution [Eq. (2.42)] is also valid at the reacting interface. Since the growth rate is an extremely strong function of temperature, very little accuracy is lost [2.15] by a first-order expansion of $\Theta(t)$ near Θ_{max}. Hence we express $\Theta(t)$ as

$$\Theta(t) \cong \Theta_{max} - \Theta_{max}\, \frac{v^2t^2}{2w^2} \tag{2.43}$$

The true temperature $T(t)$ is then approximately

$$T(t) \cong T_{max} - \Theta_{max}\, \frac{v^2t^2}{2w^2}\, \left.\frac{dT}{d\Theta}\right|_{\Theta_{max}} \tag{2.44}$$

Equation (2.6) can be used to eliminate Θ, with the result

$$T(t) \cong T_{max} \left[1 - \frac{v^2t^2}{2w^2}\, \frac{(T_0-T_\kappa)\ln[(T_{max}-T_\kappa)/(T_0-T_\kappa)]}{T_{max}}\, \frac{(T_{max}-T_\kappa)}{(T_0 - T_\kappa)} \right]$$

where
$$\tag{2.45}$$

$$T_{max} = T_\kappa + (T_0-T_\kappa) \exp \frac{P(1-R)}{2\pi^{1/2}wA} \tag{2.46}$$

can be derived using Eqs. (2.7) and (2.13), and A is the parameter listed in Table 2.1. The expression for $T(t)$ is now inserted in Eq. (2.41) to calculate the total reaction produced by a scan through the point of interest. This integration yields the result

$$d \cong R_0 t_{eff} \exp(-E_a/kT_{eff})$$

where

$$T_{eff} = T_{max}$$

and

$$t_{eff} = \frac{2w}{v} \left(\frac{\pi k\, T_{max}^2}{2E_a(T_{max}-T_K)\, \ln[(T_{max}-T_K)/(T_o-T_K)]} \right)^{1/2} \equiv \frac{2w}{v}\, f$$

$$(2.47)$$

The result has been expressed in terms of an "effective temperature," T_{eff}, and an "effective time," t_{eff}. The effective temperature is the peak temperature, while the effective time is equal to the beam dwell-time, $2w/v$, multiplied by a "dwell-time reduction factor," f. The calculated dependence of f on T_s and T_{max} is shown in Fig. (2.20). It can be seen that the effective reaction time for a thermally-activated process is typically one-third of the actual beam dwell-time. Higher activation energies and/or lower substrate temperatures lead to smaller values of t_{eff} for a given T_{max}, as would be expected intuitively.

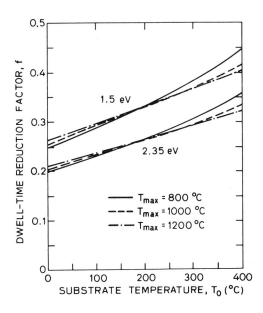

FIGURE 2.20. Calculated dependence on T_o and T_{max} of the dwell time reduction factor f. Two values (1.5 and 2.35 eV) were used for the activation energy of the reaction process.

This single-scan calculation can be extended to the case of multiply-scanned, uniformly-reacted samples, such as those which would be processed for large-area device structures. The reacted layer thickness produced at a given point is now the result of many different beam scans, not all of which pass directly through that point. The solution can be written as [2.16]

$$d \cong R_0 t_{eff} \exp(-E_a/kT_{eff})$$

where

$$T_{eff} = T_{max}$$

and

$$t_{eff} = t \frac{(2wf)^2}{A} \tag{2.48}$$

The variables t and A represent the total scan time and total scan area, respectively, and the dwell-time reduction factor f is identical to that defined in Eq. (2.48).

Examples of the use of these formulas will be given in Chapters 3 and 6, where beam-induced solid phase epitaxial regrowth of implantation-amorphized silicon and silicide formation are described, respectively.

REFERENCES

2.1   Carslaw, H. S., and Jaeger, J. C., "Conduction of Heat in Solids", 2nd ed., p. 11. Clarendon Press, 1959.

2.2   Nissim, Y. I., Lietoila, A., Gold, R. B., and Gibbons, J. F., J. Appl. Phys. 51, 274 (1980).

2.3   Maycock, P. D., Solid St. Electron. 10, 161 (1967).

2.4   Ho, C. Y., Powell, R. W., and Liley, P.E., J. Phys. Chem. Ref. Data 3, Suppl. 1, I-588 (1974).

2.5   Slack, G. A., J. Phys. Chem. Solids 34, 321 (1973).

2.6   Nissim, Y. I., Ph.D. Thesis, Stanford University, Department of Electrical Engineering (1981).

2.7   Cline, H. E., and Anthony, T. R., J. Appl. Phys. 48, 3895 (1977).

2.8   Lax, M., J. Appl. Phys. 48, 3919 (1977).

2.9   Lietoila, A., Ph.D. Thesis, Stanford University, Department of Applied Physics (1981).

2.10 Hecht, E., and Zajac, A., "Optics", p. 314. Addison-Wesley (1974).

2.11 Powell, R. A., Yep, T. O., and Fulks, R. T., Appl. Phys. Lett. 39, 150 (1981).

2.12 Gibbons, J. F., in "Laser and Electron-Beam Solid Interactions and Materials Processing" (J. F. Gibbons, L. D. Hess and T. W. Sigmon, eds.), p. 449. North-Holland, (1981).

2.13 Lietoila, A., Gold, R. B., and Gibbons, J. F., to be published.

2.14 Touloukian, Y. S., Powell, R. W., Ho, C. Y., and Klemens, D. G., "Thermophysical Properties of Matter," Vol. 2, p. 182. IFI/Plenum (1970).

2.15 Gold, R. B., and Gibbons, J. F., J. Appl. Phys. 51, 1256 (1980).

2.16 Gold, R. B., and Gibbons, J. F., in "Laser and Electron Beam Processing of Materials" (C. W. White and P. S. Peercy, eds.) p. 77. Academic Press (1980).

2.17 Siegman, A. E., "An Introduction to Lasers and Masers," p. 306. McGraw-Hill (1971).

2.18 Grun, A. E., Z. Naturfosch. 12a, 89 (1957).

2.19 Hawryluk, R. J., Hawryluk, A. M., and Smith, H. I., J. Appl. Phys. 45, 2551 (1974).

2.20 Iranmanesh, A. A., SEL Technical Report, Stanford University (1980).

2.21 Ref. 2.1, Chapter X.

CHAPTER 3

Applications of CW Beam Processing to Ion Implanted Crystalline Silicon

Arto Lietoila* and James F. Gibbons

STANFORD ELECTRONICS LABORATORIES
STANFORD UNIVERSITY
STANFORD, CALIFORNIA

3.1 INTRODUCTION

Historically, cw laser processing was first applied to the annealing of As-implanted crystalline Si [3.1]. The dose and energy chosen for these experiments was sufficient to amorphize the silicon to a depth of about 0.1 μm. It was found that with appropriate laser processing, the amorphized silicon could be completely recrystallized with 100% electrical activity and no residual defects as judged by TEM. Furthermore, the recrystallization was observed to produce no redistribution of the As from the as-implanted profile. The lack of redistribution itself strongly suggests that the recrystallization process is closely akin if not identical to solid phase epitaxy.

Subsequent to these first results, it was shown directly by several workers [3.2–3.4] that the annealing process for amorphized samples is a solid phase epitaxial regrowth identical to that obtained in conventional furnace annealing. It was also shown [3.5] that B-implanted silicon can be satisfactorily annealed (though with less ideal results than for amorphized samples); and it was further demonstrated [3.6–3.8] that 100% electrical activity can be obtained in As-implanted silicon even for impurity concentrations that exceed the equilibrium solubility. Metastable concentrations are successfully produced

---

*Present address: Micronas, Inc., 00101 Helsinki, Finland.

also by pulsed laser annealing, but since this process inevitably involves melting [3.9], the as-implanted profile cannot be preserved.

The crystalline quality, as judged by TEM, is good in cw laser annealed silicon layers. However, DLTS measurements have revealed production of high densities of point defects well beyond the range of the original implantation, a topic which is discussed in detail in Chapter 4.

Results obtained using a scanning e-beam, both in an electron beam welder and in a scanning electron microscope, are largely similar to those for cw laser scanning. The most significant difference between these methods and the use of a laser occurs if the silicon sample to be annealed is covered by an oxide or other dielectric layer. If a laser is used, the dielectric layer will act as an anti-reflective coating, and the laser power must then be adjusted accordingly (see Fig. 2.10 for reflectivity values). This may cause problems in device structures, where the oxide thickness is not uniform throughout the scanned area. For the e-beam, however, this problem does not exist, since no reflection takes place from either the silicon or from silicon dioxide. On the other hand, charging of the oxide may be important for certain choices of the beam current and voltage.

In this chapter, we first discuss the analytical techniques used in evaluation of cw laser and e-beam annealing of ion implanted single crystal silicon. We then give a brief discussion of solid phase epitaxy, after which we explain in detail the results of cw beam processing in amorphous and non-amorphous Si.

3.2   EXPERIMENTAL METHODS

When the results of cw beam annealing of ion implanted silicon are evaluated the experimentalist is usually most concerned about the behavior of the impurity atoms. It is important to know to what extent the doping atoms are electrically active and how the distribution of them has changed during the annealing cycle. These questions can be answered by Secondary Ion Mass Spectroscopy (SIMS) and electrical Van der Pauw (VdP) measurements. The former gives the distribution of all the impurity atoms, without making a distinction between electrically active and inactive species, whereas the Van der Pauw measurements yield the distribution of the electrically active atoms. The VdP measurements give also the Hall mobilities of charge carriers.

If the mass of the impurity atoms is larger than that of Si, the total number of impurity atoms (per unit area) can be accurately determined by Rutherford Backscattering (RBS) techniques; this provides a valuable calibration method for the implanted dose. The distribution of the impurity atoms can also be determined, though the accuracy is not very good unless a glancing angle method is used. When RBS measurements are performed in the channeling mode, information is obtained about the lattice location (interstitial/substitutional) of impurity atoms. A usual assumption is that shallow impurities which look substitutional by channeling are electrically active. However, we will show in Section 3.5.3 that this is not always true.

Since the practical goal of annealing of implanted Si is application in device processes, the crystalline quality of Si is also of major importance. RBS channeling techniques provide one powerful method for judging the crystalline quality of the annealed layers. Quantitative information about the structure of large defects (> 5 nm) can be obtained by Transmission Electron Microscopy (TEM). The depth distribution and electronic properties of point defects are detected by Deep Level Transient Spectroscopy (DLTS).

In this section, very brief reviews of VdP, SIMS and RBS measurement techniques are given. The DLTS technique and its applications are treated in Chapter 4. With regard to TEM, we refer to standard textbooks; e.g., Ref. [3.10].

3.2.1 Van der Pauw Measurements

Van der Pauw measurements of a doped layer can be performed in two ways. The sheet measurements give the true sheet resistivity of the layer and an estimate, N_s, for the active sheet carrier concentration; a weighed average mobility μ, can then be calculated. Differential Van der Pauw measurements, in turn, give the carrier concentration and mobility as a function of depth. In the following, the two types of measurements are explained. The details of the anodic sectioning for stripping measurements are then given. Some of the error sources are also discussed.

3.2.1.1 Sheet Van der Pauw Measurements

The geometrical configuration for the measurements has been presented by Van der Pauw [3.11]. These measurements give directly the sheet resistivity R_s and an estimated sheet carrier concentration N_s of the measured layer. The following formulas are valid for R_s and N_s [3.12]:

$$R_s = \cfrac{1}{q \int_o^d p(x)\ \mu(x)\ dx} \qquad (3.1)$$

$$N_s = \cfrac{[\ \int_o^d p(x)\ \mu(x)\ dx]^2}{\int_o^d p(x)\ \mu^2(x)\ r_H(x)\ dx} \qquad (3.2)$$

where d is the layer thickness, p is carrier concentration, and μ mobility. The Hall scattering factor r_H is defined as

$$r_H = \frac{\langle \tau^2 \rangle}{\langle \tau \rangle^2} \qquad (3.3)$$

where τ is the carrier momentum relaxation time. For degenerate electrons, $r_H = 1$ [3.13].

The weighed average mobility, μ, can be calculated from R_s and N_s:

$$\mu = \frac{1}{q \, R_s \, N_s} \qquad (3.4)$$

Use of formulas (3.1) and (3.2) yields

$$\mu = \frac{\int_o^d p(x) \, \mu^2(x) \, r_H(x) dx}{\int_o^d p(x) \, \mu(x) \, dx} \qquad (3.5)$$

If the mobility were constant throughout the doped layer, the sheet measurements would give correct values for the active sheet concentration and mobility. However, the carrier mobility is a monotonically decreasing function of carrier concentration. In this case, the measured N_s will be <u>lower</u> than the real active dose. The difference can be up to 30% for a heavy n-type dose, since μ varies rapidly when the electron concentration is greater than 10^{19} cm^{-3}. If the distribution of impurities is known, the above formulas (3.1, 3.2, 3.5) can be used to calculate the <u>expected</u> values for R_s, μ, and N_s. The measured values are then compared with the calculations, and if the agreement is good, the annealing is considered to be successful. This method is especially suitable for cw beam annealing, since the impurity redistribution is negligible during annealing, and the doping profiles can be accurately predicted.

3.2.1.2   Differential Van der Pauw Measurements

If information is needed about the carrier concentration and mobility vs. depth in the material, <u>differential</u> Van der Pauw measurements must be performed. In this method, thin sections of material are removed and the sheet measurements are performed after removal of each successive layer. This then allows the cal-culation of the sheet resistivity and sheet carrier concentration (per unit area) in each layer. In other words, the limits of integrals in Eqs. (3.1, 3.2) are replaced by the limits of each layer. The weighed average mobility in each layer can also be calculated: the expression for it is Eq. (3.5) with appropriately modified integration limits. Providing that the thickness of each layer is known, the mean carrier concentration in it can be calculated, and concentration and mobility can be plotted as a function of depth.

3.2.1.3   Anodic Sectioning

The most widely used method to perform sectioning for differential Van der Pauw measurements employs anodic oxidation and oxide stripping. This is an elaborate procedure, but it is far superior to etching, since the thickness of the grown oxide can be measured after each step, allowing accurate determination of of the layer thicknesses.

The standard anodization solution is prepared from [3.14] 1 gal. ethylene glycol, 15.4 g KNO_3 (0.04 N solution), 90 ml H_2O (2.5 vol - %), 6 g $Al(NO_3)_3 \cdot 9H_2O$.

It is customary to illuminate the sample through the front surface during the oxidation. This is not essential for n-type layers on p-substrate. However, for p on n, the light is necessary to get enough current through the reverse-biased pn-junction (as the name of the method suggests, the Si substrate is positively biased in anodic oxidation).

The thickness of the anodic oxide layer depends on the voltage prevailing across the oxide when this process is stopped, and calibration curves exist in literature [3.14]. However, the main reason for using the anodization method is a desire to obtain accurate values for layer thicknesses. Therefore, the oxide thicknesses should be measured by ellipsometry or some other suitable method.

Once the oxide thickness is known, it then has to be converted to the thickness of removed silicon. For thick oxide layers (> 1000 Å), the ratio of silicon thickness to oxide thickness is expected to be about the same as for thick thermal oxide, namely 0.44. However, this ratio is larger when the oxide gets thinner. The ratio for about 45 nm thick oxide has been measured to be 0.56 ± 0.01 [3.15].

3.2.1.4   Discussion of Error Sources

As was pointed out in Section 3.2.1.1., sheet Van der Pauw measurements give values that are systematically too low for the sheet carrier concentration, unless the mobility is constant throughout the whole layer. This is true also for differential Van der Pauw measurements: this measurement yields the sheet concentration in each layer and the total active dose is obtained by summing up the sheet concentrations of each layer. However, since the individual layers are thin, the variation of mobility is minute within each of them. Therefore, the differential measurements yield a much more accurate value for the active dose.

The inaccuracy in the determination of the thickness of the removed layers does not cause any error in the value of the total active dose, since the differential measurements give the total number of carriers in each layer. However, when the volume concentration versus depth is calculated, a possible error in the thickness measurement causes errors in the shape of the carrier profile.

The Van der Pauw method gives correct results only if the four contacts are point contacts and are at the edge of the sample. To achieve this, various etched "Van der Pauw structures" have been proposed [3.16]. Such structures are unnecessary, however, if sufficiently large square samples can be used. The contacts are formed by placing tungsten probes directly onto the sample surface. These probes can be regarded as point contacts, but of course they cannot be placed exactly at the sample edge. Probably the probes can be placed within 0.2 mm of the sample edge. Van der Pauw [3.11] has calculated that the relative error caused by this in sheet carrier concentration is about that fraction of the sample dimension by which the probes are separated from the edge. Thus if 5x5 mm samples are used, the error should be less than ~5%.

Additional uncertainty in differential Van der Pauw measurements results if only a small fraction (<5%) of the active carriers is removed with each step. [In this case, two numbers which are very close to each other will be subtracted.] This problem usually occurs for the first point of measurement because the surface concentration of typical implants is quite low unless a very low energy is used. However, from the second layer on, a sufficient fraction of the remaining dopants is usually removed at each step to avoid this type of error.

3.2.2. SIMS Technique

The popularity of secondary ion mass spectrometry (SIMS) as a technique for detecting impurities in semiconductors stems from the fact that, under appropriate conditions, it has the sensitivity to measure depth profiles of impurities down to concentrations on the order of 1×10^{15} cm^{-3} [3.17-3.19]. In this section we shall give a brief overview of the general technique and will show how SIMS may be applied to measure depth profiles of common impurities in semiconductors.

3.2.2.1   General Description

The traditional method of secondary ion mass spectrometry is schematically illustrated in Fig. 3.1 [3.20]. A primary source of low energy ions is generated, which collide with the atoms of the surface of a specimen. This induces a beam of sputtered secondary ions which may then be mass analyzed. Monitoring the concentration of constituents in this secondary ion beam as a function of time gives a depth profile of each of the host and impurity atoms.

ION MICROPROBE

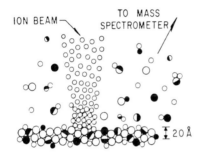

FIGURE 3.1.   Artist's   conception of the secondary ion production process (after Ref. 3.20).

The efficiency of ionization of the secondary ions determines the sensitivity of SIMS to these elements. There are two processes by which these atoms become ionized, (a) kinetic ionization and (b) chemical ionization. In the first of these, the incident ion must impart enough kinetic energy through electronic stopping losses to actually ionize the secondary sputtered ion. Chemical ionization is the process whereby the primary ions chemically ionize the surface atoms and is statistically the most favorable. It is therefore desirable to use a chemically reactive ion such as O_2^+, O^-, or Cs^+ as the primary beam. Other elements such as Ar^+ are used, but they tend to produce lower yields of either positive or negative secondary ions because they induce only kinetic ionization.

3.2.2.2 Depth Profiling of Impurities

Ion implanted impurities in Si are usually distributed at depths less than 300 nm, requiring a depth profiling technique with a depth resolution ≤ 5 nm. One method of achieving this high depth resolution is illustrated in Fig. 3.2. The primary ion beam is raster scanned in order to create a wide flat-bottomed crater. An electronic gating technique is applied such that the detector only accepts ions while the beam is scanning in the center region of the crater bottom. In this way, only the flat portion of the crater contributes to the signal received. Neutral primary ions, however, can not be raster scanned and are therefore able to sputter any portion of the specimen at any time. Electronic aperturing is of course unable to reduce the noise produced by neutral ions striking the crater sidewalls and the sample surface. This limits the resolution and sensitivity of SIMS, particularly for the case of an ion implanted profile where the sidewall impurity concentration may be several orders of magnitude higher than that on the crater floor.

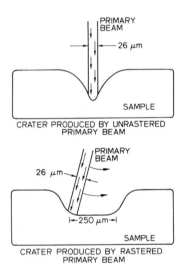

FIGURE 3.2. Illustration of craters produced by a rastered and unrastered beam (after Ref. 3.60).

For certain atoms such as sulfur, there is another limitation to the ultimate system detectability [3.21]. Diatomic oxygen has the same atomic mass number as monatomic sulfur and is therefore analyzed simultaneously. A high background of residual oxygen will therefore produce a false baseline on the SIMS output. Effects such as this due to molecular "feedthrough" can usually be minimized by improving the ultimate vacuum in the analysis chamber.

An example of a SIMS output is given in Fig. 3.3. for a Te implant in GaAs. The horizontal axis represents the number of Te-counts in the detector during a detecting time-interval, and the horizontal axis is the sputtering time. This data is converted to impurity concentration vs. depth through suitable calibration procedures.

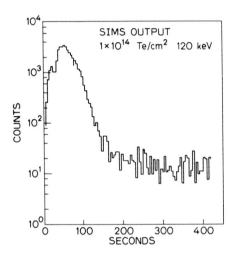

FIGURE 3.3.   SIMS output from a Te implanted GaAs sample (after Ref. 3.60).

The depth calibration is conveniently performed by measuring the depth of the sputtered crater by a step profiler. Since the total sputtering time is known, it is then easy to convert the sputtering time to depth. This procedure of course assumes that the sputtering rate is constant, which is usually the case for silicon.

When converting counts/interval to impurity concentration, the background count rate is first subtracted. That rate is obtained from the end of the SIMS output. For example, in Fig. 3.3, the background rate (for time > 200s) is about 0.5 counts/s. Final calibration is then performed by requiring that the integrated impurity dose be the same as the implanted dose.

Another way to calibrate concentrations is to use known calibration samples, which should be analyzed in the same run together with the unknown samples.

Thus the SIMS technique cannot provide <u>absolute</u> quantitative information about the number of impurities in the sample in question. If that information is needed (e.g., if the implanted dose is not accurately known), RBS analysis must be performed. This method is explained in the following section.

3.2.3 RBS and MeV Ion Channeling Analysis

As was indicated above, the Rutherford backscattering and channeling techniques can be used to determine absolute numbers of impurities in semiconductors, their lattice locations and crystalline quality of the host lattice. Unfortunately the study of impurities must, in practice, be limited to atoms which are heavier than the host atoms, as will be explained below. In the following we will give a short description of this technique in general, and thus show how it can be used for the three above-mentioned applications.

This section is aimed to help a reader unfamiliar with this method to better understand the results presented later in this chapter. Those wishing an in-depth description of the backscattering technique should consult Ref. 3.22.

3.2.3.1 General Description

A schematic diagram of a typical backscattering spectrometry system is given in Fig. 3.4. A Van de Graaff accelerator generates a monoenergetic MeV ion beam of He ions. Momentum analysis is made by a magnet to obtain a beam of ^{4}He nuclei at a well defined energy. This beam is then collimated by a set of slits, enters the target chamber and impinges upon the sample to be analyzed. The incident particles collide with the lattice and impurity atoms of the sample and some of them are backscattered from the near surface region through a single elastic collision. A solid state detector is used to collect and analyze the energy distribution of backscattered particles for a given direction. For a target surface perpendicular to the incident beam, any direction located at an angle greater than 90° from the incoming beam can be selected to collect the backscattered ^{4}He particles. A common choice is to place the detector at 170° where the back-scattered ^{4}He particles can be considered as having passed through the same depth of analyzed material on their way in and out of the sample. Detectors may also be placed at grazing angles when better depth resolution is needed [3.23].

The detector is typically a solid state surface barrier device which is not sensitive to the charge state of the collected particle. The electrical signal generated in the surface barrier detector is proportional to the incident particle energy. This

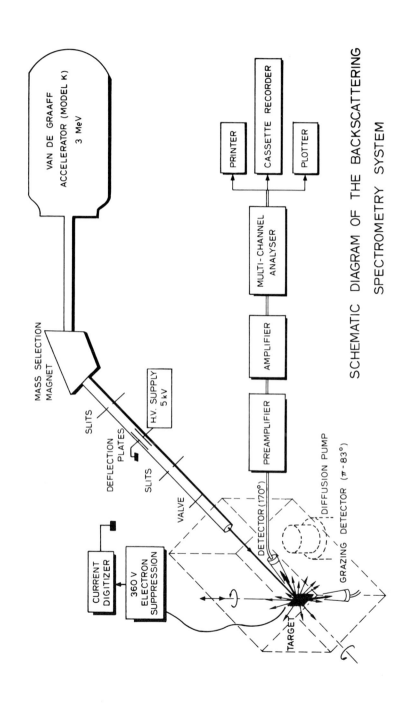

SCHEMATIC DIAGRAM OF THE BACKSCATTERING SPECTROMETRY SYSTEM

signal is amplified at several stages and then stored in a multi-channel analyzer. The final form of the data shows the number of backscattered events per channel as a function of energy. This spectrum can then be displayed on a plotter or printer.

The sample to be analyzed is held on a precision goniometer (typically 0.01° accuracy). Several constructions exist, with the typical system providing the following three degrees of freedom: a 360° rotation perpendicular to the incident beam, ± 90° sample tilt from its normal to the beam direction, and vertical translation.

3.2.3.2   Basic Physical Process: Shape of a Spectrum

As indicated above, the output of an RBS system is the number of backscattering events vs energy of the backscattered particles. The spectrum is digitized in the multi-channel analyzer to give the number of particles in each channel (typically 3-5 keV wide).

There are three factors governing the shape of an RBS spectrum for a given incident particle energy and flux. The kinematic factor, K_m, determines the fraction of energy the projectile retains in the elastic collision. The probability of collision yielding backscattering to the detector angle is proportional to the scattering cross section, σ. Finally, the projectile loses energy when it travels inside the target material; this energy loss is described by the stopping cross section, ε.

The kinematic factor depends on the backscattering angle and the mass ratio of the projectile and the target atom. For a constant projectile mass, the kinematic factor increases with increasing target mass. This is why trace impurities in Si can only be detected if their mass is larger than that of Si; in this case the signal from the impurities will be well separated from the silicon signal.

The scattering cross section for ^{4}He particles is proportional to the square of the target atom number. This is another reason why heavier impurities are more easily detected.

The stopping cross section, which determines how much energy the projectile loses while traveling in the target material, increases when the projectile energy decreases. This causes nonlinearity in depth vs energy calibration. (The concentration of implanted impurities in Si is always so low that the energy loss of the projectile is determined by Si material only.)

An example of the backscattering spectrum of 2.0 MeV ^{4}He$^+$ ions is given in Fig. 3.5 [3.24]. The sample analyzed was crystalline Si implanted with a nominal dose of 1.2×10^{15} As atoms/cm^2

at 250 keV. The plot gives, point by point, the number of counts in each channel (5.0 keV channel width). The silicon signal is a step with leading edge at 1.13 MeV. The particles scattered from inside the sample have a lower energy since energy is lost in going through the material. The As signal (plotted on amplified scale) appears at higher energies (1.5-1.6 MeV), since the kinematic factor is higher for As (mass = 75 amu) than for Si (mass = 28 amu). The As-signal is relatively higher than the Si-signal because the scattering cross-section is about a factor of 6 higher for As.

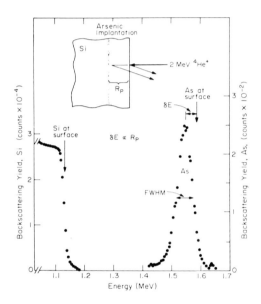

FIGURE 3.5. Energy spectrum of 2 MeV ^{4}He$^+$ ions backscattered from Si implanted to a ^{75}As dose of 1.2×10^{15} cm^{-2} at 250 keV. The vertical arrows indicate the energies of particles scattered from surface atoms of ^{28}Si and ^{75}As (from Ref. 3.24).

3.2.3.3 Total Number of Impurities

The total number of impurity atoms in the Si sample (i.e., the implantation dose) can be determined directly by RBS techniques. The impurity dose, N_i, is given by (Ref. 3.22, p.139):

$$N_i = \frac{A_i}{H_{Si}} \frac{\sigma_{Si}(E_o)}{\sigma_i(E_o)} \frac{\mathcal{E}}{[\varepsilon_o]_{Si}} \qquad (3.6)$$

where A_i is the total number of counts (area) of the impurity signal and H_{Si} is the height of the Si-signal at the edge; $\mathcal{E}$ is the width of a channel in the multi-channel analyzer; σ_{Si} and σ_i are the scattering cross sections for Si and the impurity atoms, respectively; and $[\varepsilon_o]_{Si}$ is the stopping cross section factor for the He particles in Si. H_{Si} is a machine parameter, and σ_{Si}, σ_i, and $[\varepsilon_o]_{Si}$ are tabulated in Ref. 3.22.

The RBS technique thus yields an <u>absolute</u> value for N_i. Unfortunately, this method is applicable only to impurities of higher mass than Si.

3.2.3.4   Channeling Technique: Thickness of Amorphous Layer, Crystalline Quality and Lattice Location of Impurity Atoms

The high resolution goniometers of a backscattering system can be used to align the principal crystalline orientation of the sample with the direction of the incident $^4He^+$ beam. If the sample has good crystalline quality, the yield will drop to 3-4% of the random value, since most of the host atoms lie in rows that terminate in a surface atom, and they are therefore hidden from the beam by it. This feature can be used to study the crystalline quality of the sample.

An example of a series of channeling spectra is given in Fig. 3.6. The Si sample was first amorphized with implantation and then subjected to low temperature thermal annealing for increasing times. The annealing caused the amorphous layer to regrow via solid phase epitaxy (SPE; see Section 3.3.2 below). The spectrum for the as-implanted sample has a high yield at energies from 1.13 MeV down to about 0.93 MeV, corresponding to particles scattered from the amorphous layer. Particles which go through the amorphous layer to the underlying crystalline material will partly channel, and the yield is markedly lowered. However, low angle scattering in the amorphous layer causes the beam to lose coherence, and the yield is higher than for a completely crystalline sample.

The energy scale of the particles scattered from silicon can be translated to a depth scale (Ref. [3.22], p. 59), as is shown in Fig. 3.6. Thus the thickness of the amorphous layer can be determined. Figure 3.6 shows the trailing edge of the amorphous Si signal moving to higher energies with increasing annealing time, corresponding to movement of the crystalline-amorphous interface closer to sample surface. Simultaneously, the yield from the underlying crystalline material is decreased because a thinner amorphous layer causes less incoherence in the helium beam.

Figure 3.6 shows that the crystalline structure of the sample was completely restored after 60 minutes of annealing. The quality of the crystal can now be judged by comparing the

yield of the crystalline sample (taken just behind the surface
peak) to the amorphous yield. This ratio is called the minimum
yield, λ_{min}. For a perfect Si sample of $\langle 100 \rangle$ orientation,
$\lambda_{min} = 3.4\%$. The sample illustrated in Fig. 3.6 has a minimum
yield of 6%; this indicates that some volume damage exists in
the sample. However, a part of the high yield is also due to
the high surface peak characteristic of a damaged surface layer
in regrown samples.

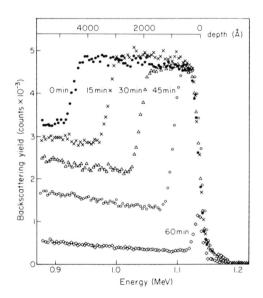

FIGURE 3.6.   Aligned spectra of 2 MeV ^{4}He ions for $\langle 100 \rangle$ Si.
The sample was self implanted to a total dose of 8×10^{15} cm^{-2} at
multiple energies ranging from 50 to 250 keV, and then annealed
at 500°C for different times (0–45 min) (from Ref. 3.25).

Finally, if the sample were implanted with impurities, such
as As, their lattice location can be determined by the channeling
technique. If the impurities are substitutional in the crystal,
they are "hidden" behind host atoms, and little or no signal is
then obtained from the impurities. It is customary to calculate
the ratio of the total impurity yield (signal area) to the implanted dose. If all impurities are substitutional, that ratio has
the value of λ_{min} (Ref. 3.22, p.269); anything above that indicates presence of interstitial impurities.

3.3 CW LASER INDUCED CRYSTALLIZATION OF AMORPHOUS SILICON

3.3.1 Introduction

It has been well established that when implantation amorphized silicon is furnace annealed at low temperatures (400–600°C), the annealing mechanism is solid phase epitaxy (SPE) which proceeds from the amorphous crystalline interface towards the sample surface [3.25]. However, simultaneously with the epitaxial regrowth, nucleation and growth of polycrystallites can take place in the amorphous layer [3.26]. When the polysilicon formation has proceeded far enough, the epitaxial regrowth will stop. Thus there is a limit on the thickness of the amorphous layer that can be regrown to single crystal silicon [3.27,3.28].

Since the early days of laser processing, the experiments have suggested that the mechanism of cw laser annealing of amorphous layers is SPE. However, due to the extreme rapidity of the phenomenon (annealing typically takes place in milliseconds or less), a clear proof for this has only recently been developed. Olson et al. [3.29] were able to observe in situ the cw laser induced regrowth of amorphous silicon. Their measurement technique utilizes interference effects within the amorphous layer, which has different optical properties than the underlying crystalline material. The regrowth can be observed by monitoring the reflectivity of the sample which is being annealed with a stationary beam.

In addition to SPE, the nucleation and growth of polycrystalline silicon has also been observed in laser processed amorphous layers. Gat [3.27] has shown that an attempt to anneal a deep (≈ 4000 Å thick) self implantation amorphized layer with a Kr+ cw laser leads always to formation of a polycrystalline layer (of unknown thickness) on top of the sample. Hess et al. [3.28] have found that while thin (≈ 1200 Å thick) UHV-evaporated amorphous layers could be epitaxially regrown (in solid phase), 5000 Å thick layers showed development of polycrystalline material. Little further characterization of the polycrystalline layer was performed in either of the above experiments.

Studies on the cw laser induced regrowth of amorphous silicon are the subject of this section. First we will give a short review of the kinetics of SPE as observed in low temperature furnace processing. Then we report on the observation of SPE and measurement of regrowth rates under typical laser annealing conditions. Finally, we give results on regrowth and nucleation of polycrystalline silicon in deep amorphous layers.

3.3.2 Low Temperature Regrowth Kinetics

The kinetics of SPE in the temperature range of 450–600°C have been extensively studied by Csepregi and coworkers [3.25, 3.30, 3.31]. The most immportant features are as follows.

The regrowth proceeds at a constant rate from the amorphous-crystalline interface towards the sample surface for all crystalline orientations except $\langle 111 \rangle$. The latter exhibits two distinct growth velocities, the first 1000 Å growing considerably more slowly than the rest of the layer [3.31]. The growth rate v_g is thermally activated, obeying the formula

$$v_g = v_{go} \exp{(-E_a/kT)} \qquad (3.7)$$

The activation energy, Ea, was measured to be, for all crystal orientations, [3.31]:

$$E_a = 2.35 \pm 0.1 \text{ eV} \qquad (3.8)$$

The pre-exponential terms for the most interesting crystal orientations are [3.31]:

$$v_{go} = 3.7 \times 10^{14} \text{ Å/s} \quad \langle 100 \rangle$$

$$= 1.4 \times 10^{14} \text{ Å/s} \quad \langle 110 \rangle$$

$$= 0.2 \times 10^{14} \text{ Å/s} \quad \langle 111 \rangle, \text{ slow}$$

$$= 0.35 \times 10^{14} \text{ Å/s} \quad \langle 111 \rangle, \text{ fast} \qquad (3.9)$$

A recent measurement on $\langle 100 \rangle$ Si gave the values of 2.85 $\pm$ 0.1 eV for E_a and 3.68×10^{17} Å/s for v_{go}.

The growth rate can be greatly enhanced by doping the amorphous layer with electrically active impurities, such as As, B, or P [3.30]. The n-type dopants do not change the activation energy, whereas B, while enhancing the regrowth still more than P and As, appears to decrease the activation energy. Table 3.1 gives the enhancement of the regrowth caused by P-doping at different concentrations.

3.3.3 Observation of cw Laser Induced SPE

Williams et al. [3.2] were the first to find evidence that the mechanism of cw beam induced crystallization of amorphous Si is SPE. These authors used glancing angle ion channeling measurements to measure amorphous layer thickness in [75]As-implanted and cw laser annealed silicon samples. They found that under suitable conditions, the amorphous—crystalline interface had moved a short distance towards the sample surface, while a higher laser power resulted in complete regrowth of the amorphous layer; this behavior is illustrated in Fig. 3.7.

TABLE 3.1. The Effect of Phosphorus Doping Concentration on the Rate of Solid Phase Epitaxial Regrowth of Silicon at 475°C (after Ref. 3.30).

Doping concentration $(10^{20}$ $cm^{-3})$	Regrowth rate (Å/min)	Enhancement factor
—	3.25	—
0.93	8.3	2.6
1.10	9.1	2.8
1.60	12.5	3.8
2.20	17.0	5.2
2.80	19.3	5.9

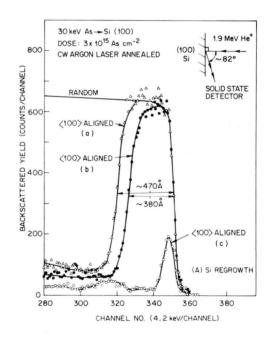

FIGURE 3.7. Glancing angle RBS random and aligned spectra for silicon which was implanted with As at 30 keV and annealed with a scanning Ar⁺ laser. The aligned spectra are for (a) as-implanted, (b) partially annealed, and (c) fully annealed samples (from Ref. 3.2).

While Williams et al. [3.2] found only one point of intermediate stage in the regrowth process, the technique introduced by
Olson et al. [3.4] allowed in situ observation of the growth of
the entire layer. Their technique is based on the interference
effects within the amorphous layer which has different optical
properties than the underlying crystalline material. That interference makes the sample reflectivity depend periodically on the
thickness of the amorphous layer, as illustrated in Fig. 3.8.
Thus, by monitoring the reflectivity of a sample being annealed
with a stationary laser beam, it is possible to observe the
regrowth. Figure 3.9 shows an example of a reflectivity vs time
plot measured by Olson et al. [3.4], which clearly demonstrates
that the regrowth is proceeding from the original interface
towards the sample surface.

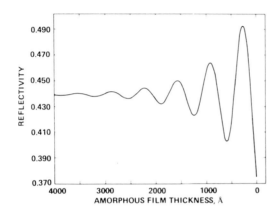

FIGURE 3.8. Calculated reflectivity (at 0.6328 μm) of amorphous Si film on crystalline Si substrate as a function of film
thickness. (Note: Thickness scale is reversed to simplify
comparison with experimental data, Fig. 3.9.) The refractive
indices used in calculations are 4.85-0.612i for amorphous Si
and 4.16-0.018i for crystalline Si (from Ref. 3.4).   ©1981 Elsevier
Science Publishing Co.

3.3.4 Rate of CW Laser Induced Regrowth

Lietoila et al. [3.32] have measured the epitaxial regrowth
rates under actual cw laser annealing conditions, using a scanning
beam and laser powers that yield a maximum surface temperature
close to 1000°C. This was accomplished by determining the scan
speed of a single laser scan required to regrow a known amorphous
layer at the center of the beam. The regrowth rates can be determined by using the formalism presented in Chapter 2 which gives
the effective dwell time for a scanning Gaussian beam. Special

attention was paid to the determination of the annealing temperature: one of the precautions was to leave the laser power unchanged after melting power calibration and regulate the annealing temperature by changing the sample holder temperature. This experiment is described below.

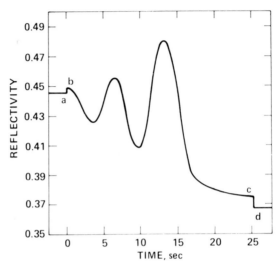

FIGURE 3.9.   The reflectivity of a HeNe probe vs time for a Si sample which is being annealed with a stationary Ar^+ laser (turned on at t = 0). The sample is <100> Si implanted to a ^{75}As dose of 5×10^{14} cm^{-2} at 158keV. The laser power is 7.5W and substrate temperature T_o = 460°C (from Ref. 3.4). ©1981 Elsevier Science Publishing Co.

3.3.4.1   Sample Preparation

The samples were of polished <100> Si.   The amorphous layer was formed by self implantation to avoid impurity effects.   The implantation energy and dose were 40 keV and $1 \times 10^{15} cm^{-2}$, respectively.   The samples were held at LN_2 temperature during implantation to prevent any self annealing.

The thickness of the as implanted amorphous layer was measured by using 2.2 $MeV^4 He^+$ channeling techniques (see Section 3.2), and was found to be 960 Å.

Prior to laser annealing,the samples were cleaned in methanol and D.I. water, the purpose of which is mainly to remove major dust particles.   Annealing was performed in the laboratory atmosphere.

3.3.4.2 Measurement of the Laser Beam Diameter

The cw laser scanning system described in Chapter I was used with a 136 mm focusing lens. The focussed beam radius was measured by determining the power required to reach surface melting of Si with the substrate temperature T_0 = 332°C. A scanning speed of 6 cm/s was used, which is slow enough to allow the steady state temperature calculations (Chapter II) to be used. The effect of scanning speed on temperature will be discussed later.

The incident power (discounting the losses in the beam shaping devices) required to cause surface melting of crystalline Si was found to be 5.0 W. The parameter p_a [see Eqs. (2.13,2.14)] has to be 1210 W/cm to reach the maximum temperature of 1412°C for T_0 = 332°C [Eqs. (2.7,2.13)]. The reflectivity of crystalline Si was measured to be 38% for the Ar^+ laser radiation; thus w = 26 μm is obtained.

We would like to stress that the beam radius determined above is used only to compute the dwell time of the spot. The calculation of the annealing temperature is based solely on the calibration against the melting power.

3.3.4.3 Procedure to Measure the Regrowth Rates

After the laser power required to reach melting was determined and the beam radius was calculated, the laser was kept at that power. The annealing temperature was then controlled by changing the substrate temperature, T_0. This was done because the beam radius is known to vary with laser power and a constant beam radius is essential for accurate determination of the annealing temperature.

For each substrate temperature T_0, two scanning speeds, v_1 and v_2, were determined so that the sample region underneath the center of the beam was just regrown at v_1, whereas v_2 was too fast to allow full regrowth. The results of the scan were judged optically: the regrown area appears dark against the lighter amorphous background [3.3]. The values of T_0 as well as v_1 and v_2 are given in the first two columns of Table 3.2.

3.3.4.4 Analysis of Results

The peak annealing temperatures for each substrate temperature, T_0, were obtained using the calculation presented in Chapter II. The scan speeds were quite low suggesting the use of Eqs. (2.7,2.13) derived for a stationary beam. However, first order corrections to the temperature due to scan speed were made by calculating the linear temperature Θ for the scanning beam [Eqs. (2.21, 2.12)] and applying the inverse Kirchhoff transform [Eq. (2.7)].

The Kirchhoff transform is not quite exactly valid if V (a parameter related to the scanning velocity: see Chapter II) deviates from zero. However, V gets quite small in this case: the change in the temperature due to scanning will be at most about 20°C, so this is not a critical point.

The peak annealing temperatures are now determined as follows. Since the melt temperature calibration was performed on crystalline silicon, while the actual samples were amorphous, the melting value, 1210 W/cm, for p has to be multiplied by the ratio of absorptivities (1-R) of amorphous and crystalline silicon. The reflectivity of the samples used here was measured to be 45%, so the effective p is 1073 W/cm for these samples. For each substrate temperature and scanning speed, the parameter V is calculated using Eq. (2.21). Equation (2.22) gives the coefficients η, which determine how much the linear maximum temperature Θ_{max} is reduced from the value for a stationary beam. Equation (2.13), with p = 1073 W/cm, is used to calculate the stationary $\Theta_{max,stat}$. $\Theta_{max,stat}$ is then divided by the value of η_{melt} for the melting calibration scan (.986, see Table 3.1), and multiplied by the corresponding value of η for each scan. The resulting value of Θ is finally converted to real temperature using Eq. (2.7). The results are given in Table 3.2.

Next the effective dwell times are determined using the formalism presented in Section 2.5. For any thermally activated process the effect of a single laser scan at the center of the scan line can be described using a constant temperature, equivalent to the <u>peak</u> temperature, T_{max}, and a reduced dwell time, t_{eff}:

$$t_{eff} = f \cdot \frac{2w}{v} \tag{3.10}$$

The dwell time reduction factor f was shown in Section 2.5 to depend mainly, but still not strongly, on the substrate temperature and the activation energy of the process. Values of f are given in Fig. 2.14. In this case, where the substrate temperatures are around 200°C, f = 0.26. We used the curve for E_a = 2.35 eV in Fig. 2.14; this choice is not critical, however, since f depends only weakly on E_a. The effective dwell-times calculated for each scan are given in Table 3.2.

The last step in the data reduction is the calculation of the regrowth rates using the known thickness of the amorphous layer and the effective dwell times. For each substrate temperature we obtain a lower and an upper limit for the regrowth rate,

TABLE 3.2. Results of the Study of the CW Laser Induced SPE Regrowth of <100> Si

T_0 (°C)	v_1a)(cm/s) v_2	$\kappa(T_0)$ (W/cm K)	V	η	Θ_{max} (K)	T_{max} (°C)	t_{eff} (ms)	$(10^6 \text{ Å}/s)$
174	1.5 2.3	0.86	0.003 0.005	0.998 0.996	357.0 356.3	797 795	0.88 0.57	1.1 1.7
184	1.5 3.0	0.84	0.003 0.007	0.998 0.995	367.3 366.2	825 822	0.88 0.44	1.1 2.2
193	3.0 4.5	0.81	0.007 0.010	0.995 0.993	375.3 374.6	847 845	0.44 0.29	2.2 3.3
202	3.0 6.0	0.80	0.007 0.014	0.995 0.990	384.6 382.6	872 866	0.44 0.22	2.2 4.4
211	6.0 9.0	0.78	0.014 0.021	0.990 0.985	391.8 389.6	891 886	0.22 0.15	4.4 6.4
332	6.0	0.59	0.019	0.986	578.1	1412	---	---

a) v_1 is the velocity at which complete regrowth was observed at the center of the beam; for v_2, the crystalline interface did not reach the sample surface.

corresponding to scan velocities v_1 and v_2, respectively. The lower and upper limits for the regrowth rate vs temperature are given in Table 3.2 and in Fig. 3.10. In the figure a bar is drawn to connect the limits obtained at each substrate temperature.

For comparison, Fig. 310 also shows the low temperature data given by Csepregi et al. [3.25, 3.30, 3.31] and in Ref. 3.59.

Figure 3.10 shows that the regrowth rates induced by laser annealing are one to two orders of magnitude higher than the rates extrapolated from furnace annealing data. This difference is too large to be explained by the uncertainties in the observed re-growth velocities (a factor of 1.5 - 2) or in the calculated temperatures ($\sim \pm$ 20°C). (The observed data are, however, too scattered to make any firm conclusions regarding the activation energy.) Thus, these measurements would indicate that the epi-taxial regrowth is enhanced by the laser radiation. It should be emphasized however, that the extrapolation of growth rates from the low temperature data (see Fig. 3.10) is subject to consider-able uncertainty, since the measurements performed in Refs. 3.25, 3.30, 3.31, and 3.59 were limited to a narrow temperature range.

3.3.5 Obstructed Regrowth of Thick Amorphous Layers

Hess et al. [3.28] have observed that under their experi-mental conditions, UHV deposited 0.5 μm thick amorphous Si layers could not be regrown with a scanning cw laser. This was shown to be a problem stemming from the layer thickness rather than layer preparation, since 0.1 μm thick, identically deposited layers grew fully with laser scanning.

Gat also observed [3.27] that implantation amorphized, 0.4 μm thick Si layers did not fully regrow but turned at least partly polycrystalline.

Hess et al. [3.28] postulated that spontaneous nucleation and growth of polycrystalline silicon would stop the epitaxial regrowth in thick layers. They suggested that doping of the amorphous material would make it possible to regrow thicker layers, since doping is known to enhance the epitaxial regrowth rate [3.30]. The latter hypothesis is very speculative: the growth of polycrystalline silicon may also be enhanced by doping. Also, it is not known whether the nucleation of polycrystallites originates at the sample surface or in the bulk of the amorphous material. In the former case, the ratio of the epitaxially regrown and polycrystalline layer thicknesses should be about the same if the thickness of the original amorphous layer is varied. But if the nucleation takes place in the bulk, the regrown thickness should stay constant, and the thickness of the

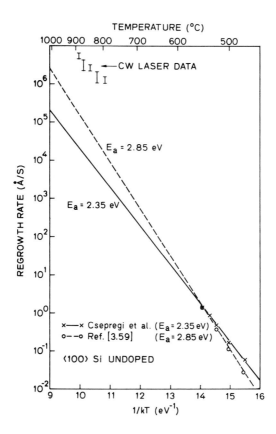

FIGURE 3.10.   The rate of cw laser induced solid phase epitaxial regrowth in ⟨100⟩ Si (after Ref. 3.32).   For comparison,
the low temperature growth data given in Refs. 3.25 and 3.59 are
also given.

remaining polycrystalline layer should vary with the original
amorphous layer thickness.

Lietoila [3.33] has conducted a study which sheds some
light on the above questions. Different amorphous layer thicknesses, doped and undoped, were used to determine where the
spontaneous nucleation originates and the effect of doping.
Results of this study are reported in the following.

3.3.5.1   Sample Preparation, Experimental Techniques

Silicon wafers of ⟨100⟩ orientation were amorphized by ion
implantation. Three different implantation schedules were carried

out to form two different thicknesses of undoped amorphous silicon and one P-doped layer. The implantation schedules are given in Table 3.3. The wafers were held at LN_2 temperature during implantation; and all implants were performed in the same implanter.

The laser annealing was performed using the system described in Chapter I with a 160 mm focusing lens. The substrate temperature was held constant, $T_o = 332°C$, and the processing temperature was controlled by changing the laser power. The beam radius for the different laser powers was determined by measuring the power required to melt the sample surface for elevated substrate temperatures (up to $T_o = 530°C$). This procedure is described in Section 3.3.4.2. and it was carried out to make the determination of the annealing temperature more accurate. The scan rate was kept constant, $v = 2.7$ cm/s, and the distance between adjacent scan lines was 10 μm in all cases.

The thicknesses of the amorphous layers were measured by 2.2 MeV $^4He^+$ channeling techniques as described in Section 3.2. For calculation of the layer thicknesses, the symmetrical mean energy approximation was used.

3.3.5.2 Undoped Amorphous Layers

The effect of laser power on the regrowth thickness was studied using samples subjected to the implantation schedule I (see Table 3.3). Those implants result in a continuous amorphous layer of 3740 Å as measured by channeling; the channeling spectrum is given in Fig. 3.11.

TABLE 3.3. Implantation Schedules for the Study of Laser Induced Epitaxial Regrowth. All implants were performed at LN_2 temperature (after Ref. 3.32).

Schedule	Ion	Energy (keV)	Dose (cm^2)
I	Si	25	3.0×10^{14}
		50	6.0×10^{14}
		90	1.1×10^{15}
		170	2.8×10^{15}
II	Si	100	1.0×10^{16}
III	P	25	3.0×10^{14}
		50	6.0×10^{14}
		90	1.1×10^{15}
		170	2.8×10^{15}

The laser powers used, the resulting maximum annealing temperatures, and results of regrowth are given in Table 3.6. The channeling spectra for a few cases are also given in Fig. 3.11. The fact that the channeling yield behind the random peak drops clearly as the amorphous layer gets thinner means that the regrown layer has good crystalline quality.

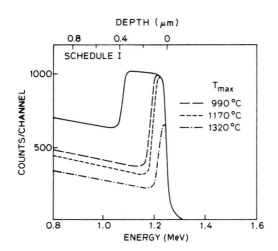

FIGURE 3.11. Channeling spectra for Si samples implanted according to Schedule I (see Table 3.3) and laser annealed at powers resulting in different maximum temperatures T_{max}. The solid line is the as-implanted spectrum (from Ref. 3.33).

In Table 3.4 are also included the calculated maximum thickness of regrowth, obtained using the extrapolated low temperature regrowth kinetics measured by Csepregi et al. [3.31] . The thickness was calculated at the beam center, including the contributions of the two adjacent scans; the same dwell time, 0.6 ms, was assumed for each. This is not quite correct, since the contributions of the adjacent scans come from slightly off the beam center. However, since the overlapping is tight ($\approx$ 83%), the error is not significant.

Table 3.4 shows that throughout the power range used, the thickness of regrowth is almost constant. The thickness of the remaining polysilicon layer decreases slightly with increasing laser power, but even for the highest power, more than 800 Å of Si has turned polycrystalline. That power resulted in a heavy network of sliplines on the sample surface, which is why no higher power was employed.

TABLE 3.4.   Results   of   Laser   Annealing   of the   Deep Si-implanted Amorphous Layer (Schedule I) (after Ref. 3.33).

P/P$_{melt}$	T$_{max}$	Amorphous/ PolySi Layer(Å)	Regrown Layer(Å)	Calculated Regrowth(Å) Cspregi et al. [3.25,3.31]	Ref. [3.39]
–	–	3740	–	–	–
0.65	990°C	1440	2300	180	1810
0.71	1070°C	1300	2440	680	9000
0.78	1170°C	1230	2510	2890	5.2×10^4
0.84	1240°C	1120	2620	6430	1.4×10^5
0.91	1320°C	830	2910	15700	4.1×10^5

The question as to whether the formation of polycrystalline silicon originates at the sample surface or in the bulk of the amorphous material was studied by laser scanning wafers implanted according to Schedule II (see Table 3.3).   This implant results in a continuous amorphous layer which is only 2570 Å thick or about the same as the regrowth for the above samples of Schedule I.   If the spontaneous crystallization originated at the sample surface, one would expect the ratio of the random and regrown layers to be the same as for the above thicker amorphous layers. But, if the nucleation started homogeneously in the bulk of the amorphized material, the regrowth of the thinner amorphous layers (Schedule II) should be complete.

Table 3.5, and Fig. 3.12, show that the regrowth of the samples, which originally had 2570 Å of amorphous material, is almost complete.   The remaining layer is too thin to be measured by channeling, since the FWHM of the surface peak is about the same as the system resolution (7 channels or 25 keV).   However, we can say that it is at most 200 Å, most likely less.

TABLE 3.5.   Channeling Results of As-implanted and Laser Annealed Wafers with the Shallower Si-implanted Amorphous Layer Schedule (II).   Results for a virgin sample are included for comparison (after Ref. 3.33).

P/P$_{melt}$	T$_{max}$	Amorphous/ Polysilicon Layer(Å)	Minimum Yield χ_{max}	Surface Peak χ_o
Virgin Sample		–	4.4%	8.8%
–	–	2570	–	–
0.78	1170°C	<200	7.6%	34%
0.84	1240°C	<200	5.8%	28%

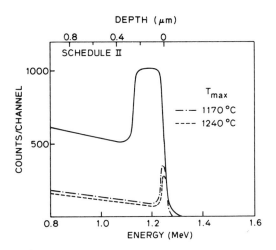

FIGURE 3.12. Channeling spectra for Si samples implanted according to Schedule II (Table 3.3). The solid line is the as-implanted spectrum (from Ref. 3.33).

The above results would thus suggest that the spontaneous nucleation of crystallites is a bulk, rather than surface phenomenon. This would mean that amorphous layers below a critical thickness can be completely regrown, whereas attempts to regrow more will result in only the critical thickness being crystalline: the rest will be polycrystalline material. The critical thickness will vary with laser power and other conditions, which is why the quantitative results given here should be used as guidelines only.

3.3.5.3 Phosphorous Doped Amorphous Layers

The possible effects of doping on the thickness of regrowth, were studied by annealing samples amorphized with a multiple-energy phosphorus implantation (Schedule III, Table 3.3) The result of that series of implants is a relatively flat P-concentration of about 1.7×10^{20} cm^{-3}, as is shown in Fig. 3.13 (calculated LSS-profile). This concentration would enhance the epitaxial growth rate by a factor of four at 475°C [3.30].

The laser scanning was performed under exactly the same conditions as in the previous section, using two different laser powers yielding the maximum annealing temperatures of 1170°C and 1270°C. The results are given in Table 3.6 and Fig. 3.14, and show that little improvement in the regrowth has taken place compared to undoped samples.

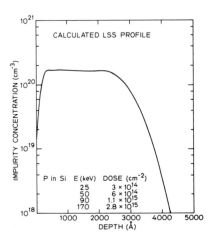

FIGURE 3.13. The calculated (LSS) ^{31}P-profile for the multiple-energy phosphorus implant (Schedule III, Table 3.3).

Thus the doping will not markedly increase the thickness which can be regrown with a scanning cw laser. This is not surprising, since the doping will enhance the growth of the poly-crystallites as much as the epitaxial regrowth, and the impurity atoms might also serve as additional nucleation sites.

3.3.6   Summary

The solid phase epitaxial regrowth of amorphous Si can proceed under typical laser annealing conditions at rates which are at least an order of magnitude higher than those extrapolated from low temperature furnace annealing data.

TABLE 3.6. Results of Laser Annealing for the P-Implantation Amorphized Samples.

P/P_{melt}	T_{max}	Amorphous/ Polysilicon Layer(Å)	Regrown Layer(Å)
–	–	3360	–
0.78	1170°C	710	2650
0.84	1240°C	470	2890

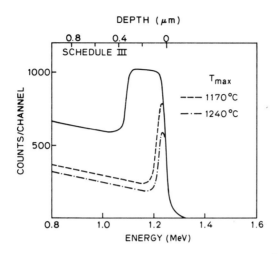

FIGURE 3.14. Channeling spectra for the ^{31}P-implanted samples (Schedule III, Table 3.3). The solid line is the as-implanted spectrum (from Ref. 3.33).

Attempts to laser anneal thick amorphous layers result in formation of polycrystalline silicon which stops the epitaxial regrowth. Comparison of regrowth data for different amorphous layer thicknesses would indicate that the spontaneous nucleation of polycrystallites takes place in the bulk of the amorphous material rather than at the sample surface. Doping does not markedly increase the maximum thickness of regrowth.

In the experiments reported in the previous sections, the maximum thickness of regrowth was between 2300 and 2900 Å, depending on laser power. This number is, however, likely to vary with sample preparation and other experimental conditions. Thus the results presented here give only the order of magnitude of the maximum regrowth.

3.4 ANNEALING OF ^{75}As IMPLANTED Si: $C < C_{ss}$

3.4.1 Introduction

Medium dose ^{75}As implanted Si samples are suitable for studies of the physical phenomena related to cw beam annealing. A ^{75}As dose of $\sim 1 \times 10^{15}$ cm^{-2} is sufficient to create an amorphous layer on the sample surface, and yet the as-implanted concentration is below the solid solubility, C_{ss}, of As in Si at typical processing temperatures. The benefit of using As is that RBS and channeling techniques can be applied. In this section, results are presented for the use of cw lasers and e-beams to anneal these medium dose [75]As implants.

3.4.2   Use of a CW Laser

A cw Ar ion laser has been applied to annealing of medium dose As implants by Gat et al. [3.1], Williams et al. [3.2] and Auston et al. [3.34]. The principal conclusion of the experiment by Williams et al. [3.2] was that the implantation amorphized layer regrows completely via SPE. This was established by channeling technique, as described in the previous section. In the following, we report on the other two experiments.

3.4.2.1   TEM, SIMS and Electrical Analysis of <100> Annealed Samples

TEM, SIMS and electrical studies were performed on <100> p-type Si samples which were [75]As implanted at 100 keV to a dose of $5x10^{14}$ cm^{-2} [3.1]. This implant creates an amorphous layer about 1000 µm thick. The samples were laser annealed using the system described in Chapter 1 with a 79 mm lens. The sample holder was at room temperature; the laser power providing optimum results was 7 W. Both single lines and overlapping areas were annealed. Control samples were annealed thermally at 1000°C for 30 min. in flowing N$_2$ gas.

Both laser and furnace annealed samples were analyzed by the TEM technique with the results given in Fig. 3.15. The thermally annealed sample [Fig. 3.15(a)] contains a relatively high concentration ($\sim 10^{10}$/cm^2) of residual damage in the form of dislocation loops (~ 300 Å average image diameter). In comparison, the concentration of defects in the laser annealed samples is considerably reduced. In the laser annealed samples, single crystal diffraction patterns [left inset Fig. 3.15(b)] were obtained. Thus the results indicate complete recrystallization of the amorphous material during laser treatment.

An interesting feature in Fig. 3.15(b) is the presence of a reasonably well defined boundary (dashed line) between the re-crystallized (C) and amorphous (A) zones at the edge of the laser scan line.

The [75]As distributions in the as-implanted as well as laser and thermally annealed samples were determined by SIMS. The results are given in Fig. 3.16 and show that, while thermal annealing has caused considerable diffusion of As, the impurity profile in the laser annealed sample is identical to the as-implanted distribution.

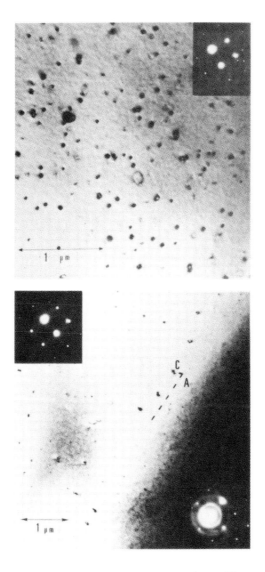

FIGURE 3.15. Electron micrographs of [75]As implanted Si subjected to thermal annealing at 1000°C for 30 min. (a) and laser annealing (b); inserts show diffraction patterns of the different regions (from Ref. 3.1).

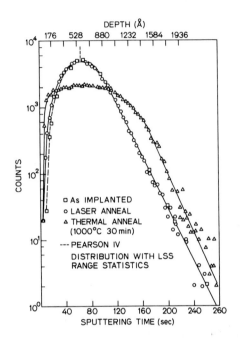

FIGURE 3.16. Relative concentration of ^{75}As as measured by SIMS in the implanted samples after thermal and laser annealing (from Ref. 3.1).

The sheet resistances in the laser and thermally annealed samples were found to be the same (180 $\Omega/\square$), indicating essentially 100% electrical activation in the laser annealed samples.

The results described above revealed the most striking features of cw laser annealing of ion implanted Si: complete recovery of the crystalline structure and 100% electrical activation with no redistribution of dopants.

3.4.2.2 Channeling Measurements on <100> and <111> Si

Auston et al. [3.34] used both <100> and <111> Si samples in their studies. The <111> Si samples were As-implanted at 50 keV to doses of 10^{14}, 10^{15} and 10^{16} cm^{-2}. The <100> samples were implanted to a dose of 10^{15} cm^{-2} at 30 keV.

The samples were laser annealed using an Ar ion laser (Spectra Physics Model 171) and a 100 mm focusing lens. The sample was held at room temperature on a moving table. The annealing beam was stationary. The laser powers used for different samples are given in Table 3.7.

TABLE 3.7.   CW Argon-ion Laser Annealing of As-implanted Si.

Orien-tation	As dose (cm^{-2})	Laser Power	Anneal strip width (μm)	χ_{min}(Si) (%)	χ_{min}(As) (%)	Resis-tivity ρΩ/
$\langle 111 \rangle$	10^{14}	9.0	37	3.5	10	930
$\langle 111 \rangle$	10^{15}	9.5	31	63	61	375
$\langle 111 \rangle$	10^{16}	9.5	30	13	40	60
$\langle 100 \rangle$	10^{15}	8.6	26	3.5	6.8	131
$\langle 100 \rangle$	10^{16}	8.0	30	3.5	10	46

Laser annealed samples of both crystalline orientations were analyzed by 1.9 MeV ^{4}He$^+$ channeling. Figure 3.17 shows representative spectra illustrating the orientation dependence of the annealing behavior; those spectra are from the samples receiving a dose of 10^{15} As/cm^2. For the $\langle 100 \rangle$ sample, the strong reduction in Si yield ($\chi_{min} \sim 3.5\%$) shows essentially complete recovery of the crystalline structure, which is in agreement with the TEM observations of Gat et al. [3.1] (see section 3.4.2.1 above). On the other hand, poor recovery of the $\langle 111 \rangle$ sample is indicated by the large residual damage peak near the surface ($\chi_{min} \sim 63\%$). However, the spectrum given in Fig. 3.17 clearly shows that some epitaxial growth has occurred even for the $\langle 111 \rangle$ sample, since the crystalline-amorphous interface has moved towards sample surface.

A summary of the channeling and resistivity measurements for all laser annealed samples is given in Table 3.7. Of the three $\langle 111 \rangle$ samples, only the one implanted with the lowest dose annealed satisfactorily, giving a minimum Si yield of 3.5% and 90% substitutionality for As. On the other hand, both $\langle 100 \rangle$ samples regrew completely, as expected.

The annealing behavior of $\langle 111 \rangle$ silicon can be well understood on the basis of the results given in this chapter above. It was shown in Section 3.3.2 that the low temperature epitaxial regrowth rate in the $\langle 111 \rangle$ direction is significantly slower than in the $\langle 100 \rangle$ direction. A similar difference is expected to exist also at laser annealing temperatures. The results given in Section 3.3.5 indicated that, simultaneously with the regrowth, nucleation and growth of polycrystallites takes place in the bulk of the amorphous material. The latter process eventually stops the epitaxial growth, when the density and size of the

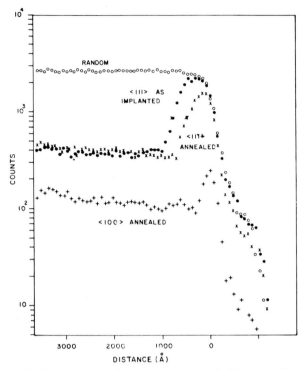

FIGURE 3.17. Channeling spectra of Si samples which were [75]As–implanted to a dose of 10^{15} cm^{-2} at 50 keV and laser annealed at powers indicated in Table 3.7 (from Ref. 3.34).

polycrystallites have grown large enough. The growth of poly-silicon should not depend on crystalline orientation, since the crystal structure is completely destroyed when the material is amorphized. Thus, the amount of maximum regrowth is expected to be significantly less for ⟨111⟩ Si. Indeed, Fig. 3.17 shows that only about 360 Å of ⟨111⟩ Si has regrown, whereas the maximum thickness of regrowth for ⟨100⟩ Si was shown in Section 3.3.5.2 to be almost 3000 Å.

The results presented above clearly indicate that, in practical cases, only ⟨100⟩ Si can be successfully annealed with a cw laser, if the implant dose is heavy enough to create an amorphous surface layer. But if the dose is so light that no amorphous layer is formed, both ⟨100⟩ and ⟨111⟩ Si can be cw laser annealed.

3.4.3 Use of Electron Beam Sources

Two types of cw e–beam sources have been used for anneal-ing purposes, namely an e–beam welder [3.35] and a modified SEM [3.36, 3.37]. The distinctive difference between these two systems is in the operating power: the welder is designed to yield currents high enough to melt sizable metal structures, whereas a SEM operates at very low currents. Thus, the beam diameter of the welder is typically about 0.1 mm or more, whereas the SEM beam has to be focused down to a few μm. The tight focusing in the SEM makes it possible to selectively anneal very small structures.

In what follows, annealing results obtained with the two types of e-beam sources are reported.

3.4.3.1 The E-beam Welder

Regolini et al., [3.35] have used an e-beam welder [Hamilton Model EBW (7.5)] to anneal ion implanted Si. The samples were <100> p-type Si, which were ^{75}As implanted at 100 keV to dose of 1.5×10^{15} cm^{-2}. The wafers were held at LN$_2$ temperature during implants.

Typical e-beam parameters used for annealing were 30 keV at 0.5 mA with the beam focused to a spot approximately 300 μm in diameter. The scanning speed was 2.5 cm/s, and the temperature of the sample holder remained below 50°C during the anneal. Control samples were thermally annealed at 575°C for 30 min.

Results of channeling studies for as-implanted and annealed samples are given in Fig. 3.18. The as-implanted spectrum was used to determine the implanted dose (1.5×10^{15} cm^{-2}) and the thickness of the amorphous layer.

The channeling spectrum for the e-beam annealed sample shows that a large number of As atoms (~25%) are on interstitial sites. The minimum Si yield for that sample is about 5%, indicating some residual damage in the lattice. The results for the thermally annealed sample are similar: 70% As substitutionality and 5% minimum Si yield. The modest results for the control samples can be explained by the low furnace annealing temperature, 575°C.

Figure 3.19 shows the results of differential Van der Pauw measurements for the e-beam annealed samples. Plotted in that figure is also the calculated Pearson impurity distribution, from which it is apparent that no diffusion of impurities has taken place during annealing. However, the electrical activation is incomplete at the peak of the concentration profile. The total active dose is 1.02×10^{15} cm^{-2}, which is ~ 67% of the implanted dose. This number agrees well with the result from channeling, 75% substitutionality.

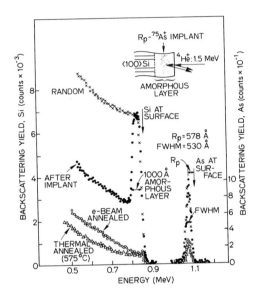

FIGURE 3.18. Aligned channeling spectra for 1.5 MeV ^{4}He ions incident on ⟨100⟩ Si implanted with ^{75}As to a dose of 1.5x10^{15} cm^{-2} at 100 keV and then subjected to different anneals (from Ref. 3.35).

Regolini et al., [3.35] also performed TEM studies on the e-beam annealed samples, with the results given in Fig. 3.20. Selected area diffraction pattern (left insert) indicates that complete recrystallization of the amorphous layer has occurred. However, the bright field micrograph indicates that residual defects in the form of small dislocation loops (40–100 Å image diameter) remain. The average loop concentration was estimated to be ~10^{14} cm^{-2}, which is significantly more than for the cw laser annealed samples studied in section 3.4.2.1. Removal of 280 Å of Si by anodic oxidation and etching reduced the loop density by about an order of magnitude, indicating that these loops are mostly associated with surface damage. After removal of the high concentration of loops, a distribution of rod-like structures ~500 Å in average length was observed. It is possible that those rods serve as precipitation sites for the As atoms.

3.4.3.2   The Modified SEM

The SEM was first applied by Ratnakumar et al., [3.37] to anneal ion implanted Si. They used a Cambridge Stereoscan S–4 scanning electron microscope which was modified for high current

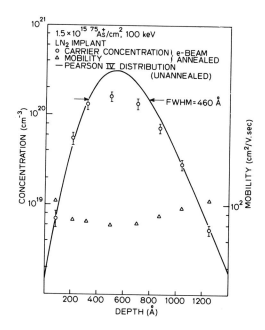

FIGURE 3.19.   Results of differential Van der Pauw measure-
ments for a sample which was implanted to a ^{75}As dose of 1.5×10^{15}
cm^{-2} at 100 keV and then e-beam annealed. The solid line shows
the calculated LSS profile for the ^{75}As implant (from Ref. 3.35).

operation. Even in the modified version, the current was limited
to a range of 15–30 μA, the voltage being 10–30 kV. The principal results of Ratnakumar et al., [3.37] were that complete
recrystallization and good electrical activity was obtained for
⟨100⟩ silicon implanted to an ^{75}As dose of 4×10^{15} cm^{-2} at
150 keV.

The SEM described above was also used by Regolini et al.,
[3.36] to anneal ⟨100⟩ silicon samples which were ^{75}As implanted
to a dose of 1.1×10^{15} cm^{-2} at 100 keV. The SEM annealing conditions in this case were 50–70 μA at 20 kV, with a scanning
speed of 10 cm/s. The estimated beam radius was 6 μm.

Results of differential Van der Pauw measurements for the
SEM annealed samples are shown in Fig. 3.21. The total active
dose obtained from stripping measurement is 9.6×10^{14} cm^{-2}, which
means that 14% of As is inactive. The partial inactivity of As
was also confirmed by channeling, (see Fig. 3.22): the fraction
of As atoms on non-interstitial sites was found to be 12%. Thus
the activation obtained with the SEM is better than that achieved
by the welder (67% activity, see Section 3.4.3.1).

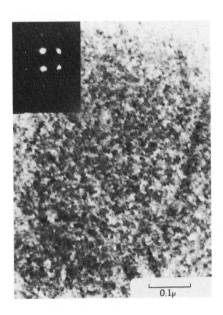

FIGURE 3.20.   TEM micrograph of an e-beam (welder) annealed, [75]As implanted sample (from Ref. 3.35).

The crystalline quality of the SEM annealed samples was studied by channeling and TEM. The minimum silicon yield (see Fig. 3.22) for the SEM annealed samples was found to be the same as for a virgin control sample, 4%. This indicates a perfect recovery of the crystalline structure. And indeed, the TEM analysis, with the results given in Fig. 3.23 shows that the annealed layer is completely crystalline with few remaining defects. The inset in Fig. 3.22(b) shows the electron-diffraction pattern for the dark-field image; the sharpness of the Kikuchi line confirms the good crystalline quality of the layer. The bright-field micrograph [Fig. 3.22(a)] reveals, however, the presence of some dislocation loops. The largest loops are ≤ 200 Å in diameter, and the estimated area density of the loops is 10^{10} cm^{-2}. This is four orders of magnitude less than the defect density produced by the e-beam welder (Section 3.4.3.1). However, the residual defect density in the cw laser annealed samples is still lower (Section 3.4.2.1).

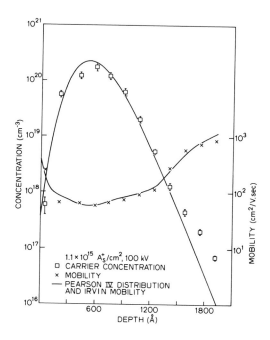

FIGURE 3.21.   Electron concentration and mobility vs. depth in a [75]As implanted, SEM-annealed sample. The solid lines show the calculated (LSS) [75]As distribution of the implant and the Irvin mobility calculated assuming complete activation of the as-implanted As distribution (from Ref. 3.36).

3.5   ANNEALING OF [75]As IMPLANTED Si: $C > C_{ss}$

3.5.1   Introduction

Numerous authors [3.39–3.44] have reported that pulsed laser annealing of ion implanted semiconductors can incorporate impurities on substitutional lattice sites in concentrations which greatly exceed the solid solubilities. Some authors have also observed that the supersaturated impurity concentrations are electrically active and thermally unstable [3.39,3.42–3.44]. The enhanced solubilities are commonly explained by "solute trapping" of impurities [3.45]. This explanation relies on the melting hypothesis for pulsed laser annealing: when the molten layer resolidifies via liquid phase epitaxy at high speed, impurities would be trapped on substitutional sites.

Since cw beam annealing of ion implanted Si takes place in solid phase, the formation of supersaturated impurity concentrations seems to be a great deal less likely. However, if the

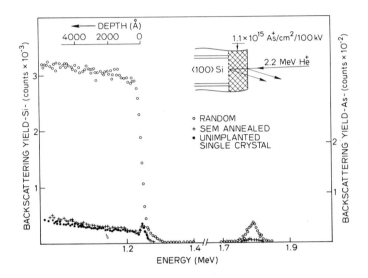

FIGURE 3.22.   Aligned spectra of 2.2 MeV ⁴He ions for ⟨100⟩
Si which was implanted with As and then e-beam (SEM) annealed
(from Ref. 3.36).

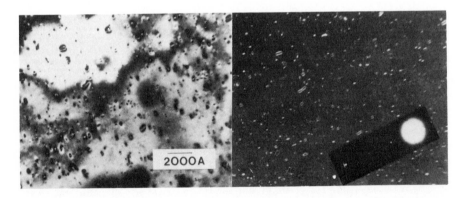

FIGURE 3.23.   TEM micrographs for ⁷⁵As implanted (100 keV,
1.1x10¹⁵ cm⁻²), SEM annealed ⟨100⟩ Si: (a) bright-field image
and (b) weak beam dark-field image (g, 3g, s⁺ conditions) and
electron diffraction pattern with spot sequence 000 (large spot),
220, 440 and 660 (from Ref. 3.36).

implant produces an amorphous layer, the circumstances are more favorable. The mechanism of SPE is broadly similar to liquid phase epitaxy : in both processes, there is an interface betweeen two phases which proceeds towards the sample surface.

In Section 3.3, it was shown that the rate of SPE under typical laser annealing conditions is quite high. The dwell time of the laser is typically short (<1 ms) and the cooling rate high, so one might expect that even cw beam processing is capable of forming supersaturated, metastable impurity concentrations. In this section we will show that cw laser annealing of As and P implants (which form an amorphous layer) indeed results in electrically active concentrations exceeding the solid solubilities. Furnace annealing causes those concentrations to relax to equilibrium values: this feature can be used to measure the solubilities of As and P in Si as electrically active dopants. We will also report on measurement of the electron mobilities in both metastable and equilibrium concentrations and on a study of the relaxation process of the metastable As concentrations. Finally, it is shown that cw e-beam annealing is also capable of forming metastable As concentrations.

3.5.2   Cw Laser Annealing of Metastable ^{75}As and ^{31}P Concentrations

Lietoila et al., [3.46-49] have used a cw Ar ion laser to form metastable As and P concentration in <100> Si. In what follows, a review of their experimental procedures and annealing results is given.

3.5.2.1 Sample Preparation, Experimental Methods.

The samples were of <100> 10-20 Ω cm p-type silicon. The implants were carried out at 0°C; the ^{75}As dose was 7.3×10^{15} cm^{-2} at 100 keV and the ^{31}P dose 9×10^{15} cm^{-2} at 45 keV. Both of these implants produce a maximum as-implanted concentration in excess of 10^{21} cm^{-3}.

The laser annealing was performed using the system described in Chapter I with a 250 mm focusing lens and a sample holder (substrate) temperature of T_O = 332°C. The beam radius was measured to be 60 μm by the technique described in Section 3.3.4.2 above. The laser scanning speed was 7 cm/s, and the distance between adjacent lines was 30 μm. The laser power was 86% of the power required for melting; this results in a calculated (see Chapter II) maximum temperature of about 1200°C.

3.5.2.2 Results of Laser Annealing

Laser annealing, as performed under conditions described in Section 3.5.2.1 above, results in practically complete electrical activation of both As and P dopants. This is shown by the sheet Van der Pauw measurements, as compared with the calculated (see Section 3.2.1.1) values of the sheet resistivity, R_s, average mobility, μ, and sheet carrier concentration, N_s, given in Table 3.8. The activation is further confirmed by differential Van der Pauw measurements, the results of which are given in Figs. 3.24 (As) and 3.25 (P). Figures 3.24 and 3.25 show that the electron concentrations follow closely the calculated LSS-profiles; and the electron mobilities agree well with the values obtained from the Irvin curve [3.50]. We would like to point out that the actual measured quantities are the resistivity and electron concentration versus depth; the mobility is calculated using those two quantities, and the result may not represent the carrier mobility in the conventional sense of the word at these high concentrations. (This point is further discussed in Section 3.5.4 of this chapter).

TABLE 3.8. Results from Sheet Van der Pauw Measurements for Laser Annealed Samples. The As implantation dose was 7.3×10^{15} cm^{-2} and the energy 100 keV; the P dose was 9×10^{15} cm^{-2} at 45 keV. For comparison we give the calculated values for R_s, μ and N_s for those implants, assuming complete activation and Irvin mobility. (after Ref. 3.46).

Ion	Measurements			Calculations		
	R_s ($\Omega/\square$)	μ (cm^2/V_s)	N_s (cm^{-2})	R_s ($\Omega/\square$)	μ (cm^2/V_s)	N_s (cm^{-2})
As	33	36	5.3×10^{15}	33	36	5.3×10^{15}
P	25	37	7.0×10^{15}	28	33	6.6×10^{15}

The effect of laser power on the electrical activation was studied for the As-implant. Sheet Van der Pauw measurements revealed that the activation was complete even at a power which was 75% of the melting power, corresponding to a maximum annealing temperature of 1050°C. Below this, unannealed areas were visible on the sample. Above the power normally used ($P/P_{melt} = 0.86$), slip lines started to form.

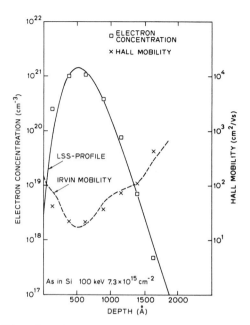

FIGURE 3.24. Results from differential Van der Pauw measurements for the laser annealed As-implanted sample (from Ref. 3.46).

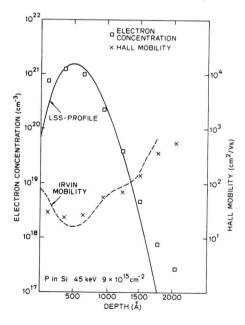

FIGURE 3.25. Results from differential Van der Pauw measurements for the laser annealed, P-implanted sample (see Ref. 3.46).

3.5.3   Thermal Treatment of Laser Annealed Concentrations

The active concentrations achieved by laser annealing are well below the values reported by Trumbore for both As and P [3.51]. However, the maximum active concentrations achieved by thermal diffusion are about 3×10^{20} cm^{-3} for As [3.52] and 4×10^{20} cm^{-3} for P [3.53]. Therefore, one could expect thermal annealing to cause deactivation of dopants in the laser annealed samples. Lietoila et al., [3.46-3.49] have shown that deactivation does occur. Long enough thermal annealing makes the active concentration relax to an equilibrium value, which allows the measurement of solubilities of As and P in Si as electrically active dopants. These observations [3.46-3.49] are reported below.

3.5.3.1   Electrical Deactivation

Thermal annealing of the laser annealed, As-implanted samples causes a significant decrease in the sheet carrier concentration, N_s, even at 400°C. This is illustrated in Fig. 3.26 where a plot of N_s vs. annealing temperature for isochronal, 5 minute anneals, is given.

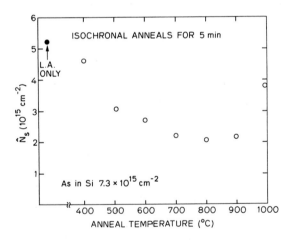

FIGURE 3.26.   The sheet carrier concentration, N_s, in [75]As-implanted, laser annealed samples as a function of temperature for isochronal (5 min.) anneals (from Ref. 3.46).

Figure 3.26 shows that when the temperature of isochronal annealing is increased, the sheet carrier concentration goes through a minimum, and then starts to increase. The increase is

due to impurity diffusion, which tends to lower the concentration
so that the non-active dopants are activated again. This phenomenon was confirmed by isothermal annealing at different temperatures for both the As- and P- implanted laser annealed samples.
The results are given in Figs. 3.27 (As) and 3.28 (P), and show
that after a rapid initial drop, the sheet carrier concentrations
start to slowly increase. (The exception was the lowest temperature for the P-samples, which gave a monotonic decrease of N_S;
this is probably due to out-diffusion of the impurities).

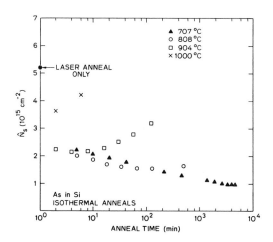

FIGURE 3.27. The sheet carrier concentration, N_S, in laser
annealed, [75]As-implanted samples as a function of annealing time
for different temperatures (from Ref. 3.46).

Since the initial decrease in the sheet concentration is so
much more rapid than the diffusion-assisted reactivation, the
behavior of N_S vs. time can be divided into two distinct regimes.
Lietoila [3.46] proposed the following model, illustrated in Fig.
3.29. While the dopants are deactivated [Fig. 3.29(a)], little
diffusion takes place. Eventually the equilibrium active concentration is reached, and only after that, significant diffusion
occurs [Fig. 3.29(b)]. During diffusion the maximum active
concentration remains unchanged, which results in an increase in
the sheet carrier concentration. Testing of this model is described in the following section 3.5.3.2.

Lietoila et al., [3.47,3.49] laser annealed a second time a
sample which was thermally annealed at 900°C for 2 min. The
maximum concentration showed partial recovery to 6×10^{20} cm^{-3},
suggesting that the solubility at the laser annealing temperature
(1200°C) is higher than at ~ 1000°C.

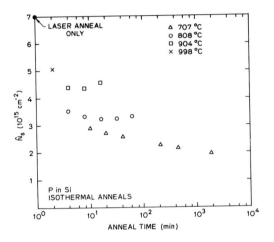

FIGURE 3.28. The sheet carrier concentration, N_s, in laser annealed, P-implanted samples as a function of annealing time for different temperatures (from Ref. 3.46).

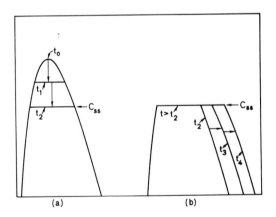

FIGURE 3.29. Schematic presentations of evolution of the active doping concentration. At t_0, the activation is complete in the laser annealed sample. During thermal annealing, the active concentration reaches thermal equilibrium by the time t_2. Diffusion, which is much slower, is significant only for times which are much greater than t_2. While diffusion takes place, nonactive dopants are reactivated, and the active concentration remains unchanged (from Ref. 3.46).

3.5.3.2 Measurement of Solid Solubilities

The results described above would suggest that the solubility of As and P as electrically active, non-complexed dopants in Si can be determined by letting the laser annealed concentrations relax to thermal equilibrium values. However, to assure that the measurement gives correct results, it has to be determined whether the amount of supersaturation affects the equilibrium concentration. Also, one has to test the assumption presented in the previous section that when diffusion broadens the doping profile the maximum active concentration does not change. Both of those features mean, of course, that the amount of non-active dopant does not affect the maximum active concentration.

Lietoila [3.46] studied the effect of the amount of super-saturation by laser and thermal annealing of different As doses. In addition to the dose of 7.3×10^{15} As/cm^2, 3×10^{15} and 1×10^{15} As/cm^2 were used (both at 100 keV), which give the maximum as-implanted concentrations of 6×10^{20} cm^{-3} and 2×10^{20} cm^{-3}, respectively. All samples were first laser annealed.

The two higher doses (7.3×10^{15} and 3×10^{15}) were thermally annealed for 15 minutes at 904°C. That thermal treatment is, according to Fig. 3.27, enough to pass the minimum in the sheet carrier concentration, which would indicate that the equilibrium concentration is reached. Results of differential Van der Pauw measurements are given in Fig. 3.30, and show that the maximum active concentrations for both doses are identical (2.3×10^{20} cm^{-3}). Thus, the amount of original supersaturation does not have an appreciable effect on the equilibrium concentration.

The low dose sample (1×10^{15} As/cm^2) was thermally annealed at 904°C for 5 and 10 minutes (the time was short to minimize diffusion effects), with the results of sheet Van der Pauw measurements given in Table 3.9. The maximum as-implanted concentration, 2×10^{20} cm^{-3}, is only slightly below the equilibrium value obtained above, and yet there is no decrease in the sheet concentration during thermal annealing. This further confirms that the maximum active concentration has a unique value which does not depend on the total As-concentration.

Lietoila [3.46] tested the assumption that once the minimum in sheet carrier concentration is passed, the equilibrium active concentration is reached and further diffusion does not change that concentration as long as there is inactive As left which can be reactivated. A laser annealed sample with the dose of 7.3×10^{15} As/cm^2 was annealed for 2 hours at 904°C. Figure 3.27 shows that this annealing results in a significant increase in sheet carrier concentration. However, differential Van der Pauw measurements (see Fig. 3.30) indicate that the maximum active

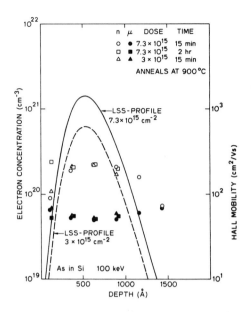

FIGURE 3.30.   Results from differential Van der Pauw measurements of As-implanted, laser annealed samples which were then
thermally annealed at 900°C (from Ref. 3.46).

TABLE 3.9   Results of Sheet Van der Pauw Measurements After
Laser and Thermal Annealing of a Sample Which Was As-implanted to
a Dose of 1×10^{15} cm^{-2} at 100 keV. This implant gives a maximum
concentration of 2×10^{20} cm^{-3} (after Ref. 3.46).

Annealing	R_S $\Omega/\square$	μ_n cm^2/V$_s$	N_S cm^{-2}
Laser only	100	76	8.2×10^{14}
Laser + 5 min. 900°C	90	86	8.2×10^{14}
Laser + 10 min. 900°C	89	86	8.2×10^{14}

concentration is unchanged compared with annealing for the shorter
time.

These results show that the solubilities of As and P as
electrically active, noncomplexed dopants in Si can be measured
by letting the metastable concentrations relax to equilibrium
values. Figures 3.27 and 3.28 are used to determine the times
required to reach the equilibrium. The results of differential
Van der Pauw measurements for samples which were thermally annealed to equilibrium are given in Figs. 3.31 (As) and 3.32 (P).

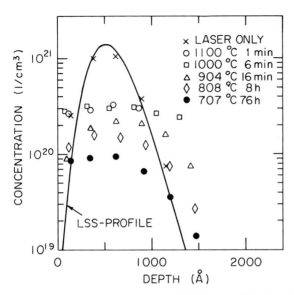

FIGURE 3.31. Results from differential Van der Pauw measurements for ^{75}As-implanted, laser annealed samples which were thermally annealed to equilibrium (from Ref. 3.48).

Figure 3.33 summarizes the results of the solubility measurements. The values for As at 1100°C and 1200°C are subject to some error, however. At 1100°C, the deactivation rate is so high that partial deactivation has probably taken place during the cooling down of the sample. The solubility value at 1200°C was obtained with the second laser scan (see Section 3.5.2.2 above), and it is not certain whether an equilibrium concentration was achieved during that scan.

We have included in Fig. 3.33 the solubility data originally reported by Trumbore in 1960 [3.51]. The difference in As solubilities is quite remarkable at about 1000°C. It should be noticed, however, that the actual measurement points in Ref. [3.51] are scattered and exist only at 1200°C and above. The curve quoted in many reference books is obtained by extrapolation of the few high temperature data points.

As for the phosphorus solubilities, the disagreement seems to increase with temperature. One possible reason is that phosphorus is deactivated fast enough to cause significant decrease in the active concentration during cooling from even 900°C. This does not seem very likely since Figs. 3.27 and 3.28 show that the rate of deactivation is about the same for As and P. Also, we would like to point out that the P-concentrations measured here are higher than those obtained by thermal diffusion [3.53].

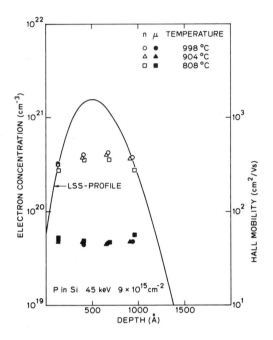

FIGURE 3.32.   Results from differential Van der Pauw measurements for P-implanted laser annealed samples which were thermally
annealed to equilibrium (from Ref. 3.46).

3.5.3.3 Discussion

The solubilities of As and P reported above were [3.46-3.49]
inferred from electrical measurements, and represent thus the
maximum electrically active concentrations. This information is
the most relevant for device processing applications.

As was mentioned in the previous section, the solubility
values determined by thermal annealing may be too low at the
highest temperatures because deactivation may have taken place
during cooling of the samples. However, the values given in
Fig. 3.33 are reliable upper limits for active concentrations
which in practice can be achieved by any thermal processes. In
fact, a production scale process would be expected to yield even
lower maximum concentrations than those given here, since cooling
is markedly slowed down by the thermal mass of the heavy furnace
boats and by the large number of wafers processed simultaneously.

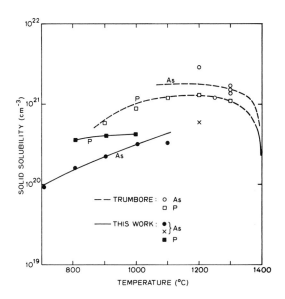

FIGURE 3.33. Measured solubilities for As (solid circles) and P (solid squares) in Si as electrically active dopants (Refs. 3.46, 3.48). The point at 1200°C (cross) was obtained by laser scanning a second time a thermally annealed sample (Ref. 3.51). Included also are the original data points given by Trumbore and the extrapolated lines presented in Ref. 3.51 (dashed lines).

3.5.4 Measurement of Hall Mobilities

The laser annealed samples provide a possibility to measure the electron mobilities at concentrations previously unattainable. Furthermore, the effect of nonactive dopants on the mobility can be studied in the concentrations which have relaxed to thermal equilibrium. Lietoila [3.46] has performed this study, with the results reported below.

The electron mobilities obtained by the Van der Pauw measurements performed for the laser and thermally annealed, As and P doped samples are given in Fig. 3.34. For comparison, the mobilities calculated from the Irvin resistivity curve [3.50] are included. We would like to point out that the Van der Pauw measurements give the mobility as a function of electron concentration, whereas the resistivity in the Irvin curve is given versus total impurity concentration. Therefore, it is most surprising that the agreement between the two measurements is so good. Moreover, the Van der Pauw measurements show that the mobility depends only on the electron concentration. The amount of nonactive dopant does not have any appreciable effect on the mobility, as can be seen in Fig. 3.34 by comparing the data obtained from laser annealed and laser and thermally annealed samples.

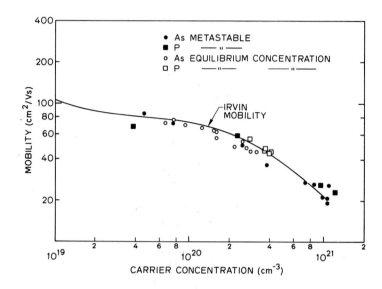

FIGURE 3.34.   Hall mobility of electrons in As and P implanted Si samples which were either laser annealed (metastable) or laser and thermally annealed (equilibrium) (from Ref. 3.46).

3.5.5   Characterization of Deactivation of As

The qualitative kinetics of As deactivation was reported in Section 3.5.3 above. In this section, results are given of some more detailed investigations of the deactivation process [3.46–3.49]. The samples for this study are the same as described in Sections 3.5.2–3.5.4 above.

3.5.5.1   TEM Studies

Results of TEM analysis [3.47] performed on laser and thermally annealed samples are given in Fig. 3.35. Laser annealing [Fig. 3.35(a)] results, as expected, in complete crystallization of the amorphous material. However, a relatively high concentration of crystal defects is left, mainly in the form of dislocation loops. Thermal annealing at 900°C results in a significant increase in the amount of crystalline damage. This is most likely due to the release of stress present in the supersaturated layer.

The amount of damage is markedly smaller in samples annealed at 1000°C after laser annealing [Fig. 3.35(d)]. This can be explained by defect annealing at 1000°C.

The annealing at 400°C [Fig. 3.35(b)] results in a formation of rod-shaped structures which are about 500 Å long and aligned along the <002> direction. These rods might be As precipitates, even though their volume is too small for direct As detection.

3.5.5.2 Channeling Measurements

Results from channeling measurements are given in Fig. 3.36 and Table 3.10. The sample which was only laser annealed shows a relatively high minimum yield of 9.7%: this is due to the remaining crystal damage also observed with TEM. Thermal annealing at 500°C and 900°C increase the minimum yield quite substantially. The result at 900°C is consistent with the TEM observations [see Fig. 3.35(c)]. However, at 400°C-500°C, TEM does not show such a high density of defects, indicating that the defect structures forming at those temperatures are mostly too small to be seen with TEM.

Comparison of the fractions of nonsubstitutional and electrically inactive As (see Table 3.10) reveals a surprising fact. At 500°C, the off axis As fraction is smaller than the inactive fraction. But at 900°C the amount of nonsubstitutional As is less than at 500°C, while the electrically inactive fraction has substantially increased. The gross difference in those fractions indicates that no precipitation in the usual sense of the word has occurred at 900°C. It is likely that deactivation has taken place via formation of complexes where As atoms are displaced off lattice sites only slightly, so that they cannot be detected with channeling without angular scan. This phenomenon has first been observed by Chu and Masters [3.54] in pulsed laser annealed samples. They used the angular scan to measure the displacement, which was found to be 0.15 Å after thermal annealing at 950°C. The angular scan method would not, however, yield any information about the structure of the complexes.

The fact that the off-axis As fraction is larger at 500°C than at 900°C, but still less than the electrical fraction, suggests that deactivation at 500°C may start with the formation of the complexes described above. The complexes would be, however, unstable at the lower temperatures, and would coalesce to form the rods seen by TEM.

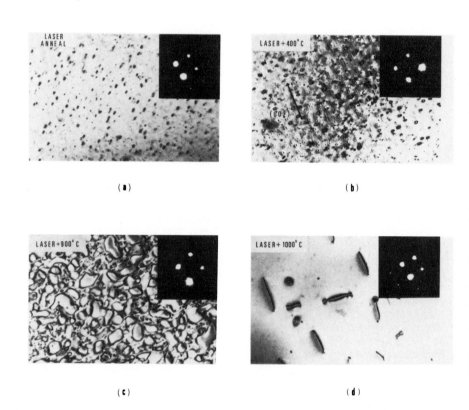

FIGURE 3.35.   TEM micrographs of the high dose $(7.3 \times 10^{15}$ cm$^{-2})$ As implanted samples subjected to laser anneal (a) and laser + thermal annealing (b)-(d) (from Refs. 3.46, 3.47).

TABLE 3.10.   Comparison ، Between Channeling and Electrical Measurements for As Implanted, Laser and Thermally Annealed Samples. Also given is the minimum yield, χ_{min}, for each sample (after Ref. 3.49).

Annealing	χ_{min}	Off-axis As	Electrically inactive As
Laser only	9.7%	7.8%	5.5%
Laser + 5 mins. 500°C	23.5%	30.3%	50.0%
Laser + 16 mins. 900°C	20.2%	18.9%	71.0%

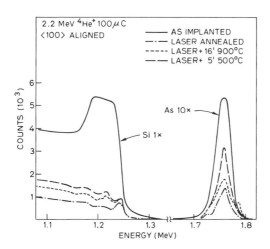

FIGURE 3.36. Channeling spectra of the high dose (7.3×10^{15} cm^{-2}) ^{75}As implanted samples which were laser and thermally annealed (from Ref. 3.48).

3.5.5.3 Activation Energy of the As Deactivation

The activation energy for the As deactivation process was determined by Lietoila et al. [3.48-3.49] by measuring the time required to reduce the sheet concentration of a laser annealed sample by 7% at various temperatures. For practical reasons, the measurement had to be restricted to a temperature range of 350-410°C; above this range the process is too quick to allow accurate measurements.

Results of the activation energy measurements are given in Fig. 3.37, and yield E_a = 2.0 eV. This value is close to the energy of vacancy formation in Si [3.55] which would support the assumption that As deactivation is a reaction which involves vacancies.

3.5.6 E-beam Annealing of Metastable ^{75}As Concentrations

3.5.6.1 Use of an E-Beam Welder

The e-beam welder described in Section 3.4.3.1 has been used by Regolini et al [3.56] to anneal metastable As concentrations. The samples were of <100> Si implanted to an As dose of 10^{16}/cm^{-2} at 100 keV. The e-beam annealing was performed at 31 kV using a current of 0.45 ma; the estimated beam diameter was 300 μm.

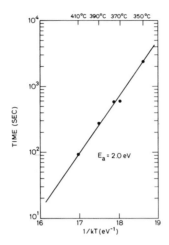

FIGURE 3.37.   The Arrhenius plot of the time required to decrease the sheet concentration by 7% in the As implanted laser annealed samples (from Ref. 3.48).

Control samples were annealed thermally at 500°C for 30 min. and also using a cw Ar ion laser.

Results of the differential Van der Pauw measurements performed on annealed samples [3.56] are given in Fig. 3.38.   Both beam annealing methods have resulted in maximum carrier concentrations of about 1.1×10^{21} cm^{-3}.   However, annealing is not complete at the peak of the as–implanted concentration (the calculated LSS profile is given with a solid line in Fig. 3.38).   As expected on the basis of the results given in the previous sections, thermal annealing has resulted in poor activation, the maximum electron concentration being only 3×10^{20} cm^{-3}.   None of the three annealing methods had caused any appreciable impurity diffusion.

3.5.6.2   Use of a Modified SEM

As mentioned in Section 3.4.3.2, a modified SEM was used by Ratnakumar et al. [3.37] to anneal a heavy As implant of 4×10^{15} cm^{-2} at 150 keV in Si.   Control samples were annealed in a furnace at 900°C and 1000°C for 30 min.

The electron concentrations in the e–beam and thermally annealed samples [3.37] were determined by spreading resistance measurements [3.38], with the results given in Fig. 3.39.   E–beam annealing (curve labeled "SEBA") has produced a maximum electron concentration at 7.5×10^{20} cm^{-3}, which clearly exceeds the concentrations achievable by thermal annealing [3.52].   On the other

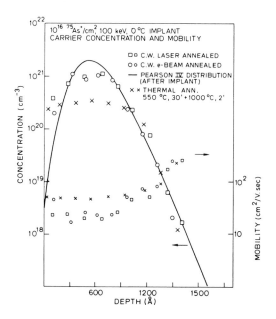

FIGURE 3.38. Results of differential Van der Pauw measurements for Si samples which wer ^{75}As implanted to a dose of 10^{16} cm^{-2} and then e-beam (welder), cw Ar^{+} laser or thermally annealed (from Ref. 3.56).

hand, both thermal anneals have caused significant dopant redistribution which is why the electrons concentrations are below 10^{20} cm^{-3}.

3.6   HIGH RESOLUTION SELECTIVE E-BEAM ANNEALING OF As IMPLANTED SILICON

The modified SEM described in Section 3.4.3.2 was used by Ratnakumar et al. [3.37] to demonstrate the potential for a high resolution localized annealing. The samples were <100> Si implanted to a dose of 5×10^{14} cm^{-2} at 100 keV. Results of selective e-beam annealing [3.37] are shown in Fig. 3.40, which is an optical micrograph of an annealed area. The area was scanned twice, with the line and frame scans interchanged for the second scan. The separation between successive scan lines is ~2 μm. Due to localized annealing by the finely focussed beam, the two scans produce a grid of annealed areas. In the picture, the light areas correspond to the amorphous material and the dark areas to recrystallized silicon. Annealed islands less than 1 μm^2 are shown in the figure. Sheet resistance measurements also indicate

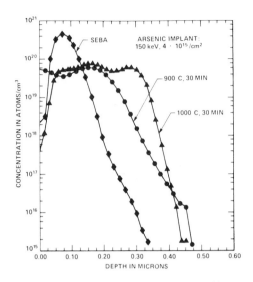

FIGURE 3.39. Doping profiles from spreading resistance measurements for e-beam (SEM) and thermally annealed, [75]As implanted
Si (from Ref. 3.37).

the very narrow widths of the annealed stripes. Linewidths in
the range of 40-100 mm were observed [3.37]. Hence scanning
e-beam annealing using a regular SEM introduces the intriguing
possibility of selective annealing of submicron areas.

3.7 ANNEALING OF B IMPLANTED SI

In contrast to the thorough studies performed on cw beam
annealed As-implanted Si, very little work on boron implanted Si
has been reported in the literature. One of the earliest papers
on cw laser annealing was that of Gat et al. [3.57] reporting
the use of a Kr ion laser to anneal boron implanted Si, but it
remained for a long time the only one on that subject. Much later,
Yep et al. [3.58] used a scanning e-beam for the same task. The
principal results of those two experiments are reported in the
following sections 3.7.1 and 3.7.2.

3.7.1 Use of a CW Kr Ion Laser

Gat et al. [3.57] implanted <100> n-type Si with boron to a
dose of 2×10^{15} cm^{-2} at 35 keV. This dose is not sufficient to
drive the silicon amorphous but damages the surface layer to a
depth of ~ 0.25 µm. Laser annealing was performed using a Kr

ion laser in a multimode operation (the main output is at λ = 0.647 μm). The laser power was 6 W and the estimated beam radius was 11 μm (79 mm focussing lens). The scan rate was 9.8 cm/sec and the sample holder was held at 178°C. Control samples were thermally annealed at 1000°C for 30 min. The annealed samples were then analyzed by electrical measurements, SIMS and TEM.

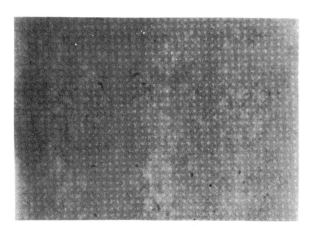

FIGURE 3.40.   Micrograph of submicron single crystal islands in amorphous Si formed by annealing with a finely focused electron beam (SEM) scanning in a grid pattern of 2x2 μm spacing (from Ref. 3.37).

3.7.1.1 Electrical Measurements

The electrical sheet resistance measurements [3.57] were performed using a 2 point probe station. The station was originally designed for spreading resistance measurements [3.38], which allows well controlled stepping of the probes across the sample surface. Results of this analysis are given in Fig. 3.41 for one thermally annealed sample (SIB 155) and two laser annealed samples (7 S 36 and 7 S 36). The thermally annealed sample shows a probe to probe resistance of 70 Ω, and very good lateral uniformity. In the laser annealed sample (7 S 37), the probes were stepped from a non-annealed area to the annealed regions in a direction <u>perpendicular</u> to the scan lines. The average probe to probe resistance in the annealed area is 100 Ω , with some periodic fluctuation. The fluctuation is due to nonuniformity in the temperature produced by the raster scanned beam. The areas under the center of the beam are exposed to higher temperature than the areas between two scans. Stepping of the probes in sample 7 S 36 was performed <u>along</u> the scan lines, and the resistivity is now constant, ~ 120 Ω.

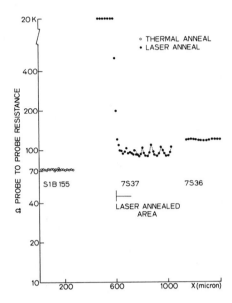

FIGURE 3.41.   Surface spreading resistance measurements performed on ^{11}B implanted [2×10^{15} cm^{-2}, 35 keV] Si subjected to cw Ar$^+$ laser and thermal annealing (from Ref. 3.57).

The fact that the probe to probe resistance in the laser annealed samples is up to 50% higher than in the thermally annealed control sample indicates that the electrical activation of impurities is not quite complete, but about 70–80%. This is further confirmed by the non-uniformity of the resistivity in the laser annealed sample 7 S 37.

3.7.1.2   SIMS Measurements

The boron distributions in the laser and thermally annealed samples were determined by SIMS, and as a calibration, an as-implanted sample was also analyzed in the same run. The results are given in Fig. 3.42 and show that, while thermal annealing has caused very significant motion of boron atoms, laser annealing has completely preserved the as-implanted profile.

3.7.1.3   TEM Analysis

The crystalline quality of laser and thermally annealed samples was studied by TEM, with the results given in Figs. 3.43 and 3.44. It should be emphasized that the samples were <u>crystalline</u>, but damaged after implantation. The as-implanted damage is not, however, resolvable by TEM. The thermally annealed samples (Fig. 3.43) contains large defects such as loops and dipoles, which are formed during annealing by coalescence of the small defects created by the implant.

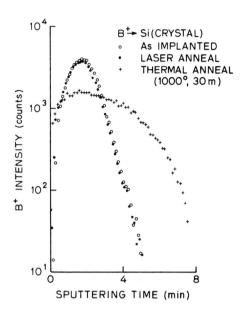

FIGURE 3.42. Relative boron profiles determined by SIMS in as-implanted, laser annealed and thermally annealed, ^{11}B implanted [2×10^{15}, 35 keV] Si (from Ref. 3.57).

The laser annealed sample, Fig. 3.44 shows two distinct regions. The center of the scan (labelled "Laser scanned region") is completely free of defects. There is no trace of rod-shaped structures or dislocation loops, typically found in boron implanted samples thermally annealed at temperatures > 700°C. However, at the edge of a scan line (labelled " unscanned region" in Fig. 3.44), we see small dislocation loops whose concentration increases with increasing distance from the center of the scan. This means that the laser induced temperature in the edge region was such as to allow some growth of the defects during the scan, but was too low to produce a defect-free crystalline structure.

3.7.2 Use of a Scanning Low Resolution E-Beam

Yep et al. [3.58] have used a broadly rastered, low resolution electron beam to anneal ^{11}B-implanted diode structures. The

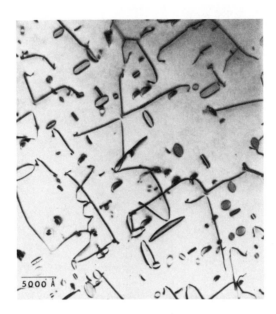

FIGURE 3.43.   Transmission electron micrograph of a B-
implanted sample which was thermally annealed at 1000°C for 30
min. (from Ref. 3.57).

device properties are beyond the scope of this chapter, but since
the experimental apparatus of Yep et al. incorporates some interesting features, we report on their experiments.

Diodes were fabricated on <100> n-type Si wafers (2" diame-
ter) by implanting B to a dose of 1×10^{14} cm^{-2} at 50 keV. The
diode areas were defined to be 16 mil (0.41 mm) diameter dots by
a 0.63 μm thick SiO_2 layer.

The electron beam annealing was performed using an accelerating voltage of 5.2 keV and a current of 52 mA. The annealing
system is illustrated in Fig. 3.45. The electron beam was deflected electrostatically ±1.5" in the x-direction and 0.75" in
the y-direction. In addition, the sample stage was mechanically
scanned at 0.13 inch/s in the y-direction. This arrangement was
effectively a 1.5"x 3" bar of electrons annealing the wafer. The
annealing temperature was measured to be 850 ±20°C by an optical
pyrometer, and the exposure time for a given point was 3-4 s.
Annealing was completed by a single pass across the whole wafer,
which only took about 20 s. Control wafers were annealed at 850°C
and 1000°C for 30 min in flowing nitrogen.

The B doping profiles in the annealed samples were determined
by SIMS, with the results given in Fig. 3.46. While both furnace

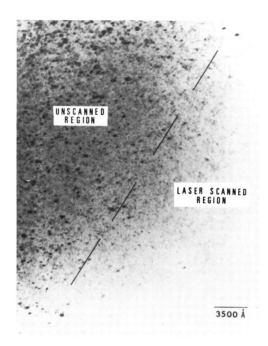

FIGURE 3.44. Transmission electron micrograph of B-implanted, laser annealed Si. The area labelled "laser scanned" is at the center of the scan, and the "unscanned region" refers to the edge of a single scan (from Ref. 3.57).

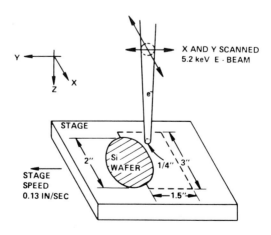

FIGURE 3.45. Experimental setup for the low resolution e-beam annealing system (from Ref. 3.58).

anneals have broadened the B profile by 1200–1600 Å, the profile has shifted only about 450 Å during e-beam annealing. Figure 3.46 also shows the electrically active B profile, as determined by spreading resistance measurements. The measured area under this curve shows that the activation of boron by e-beam annealing is about 75%.

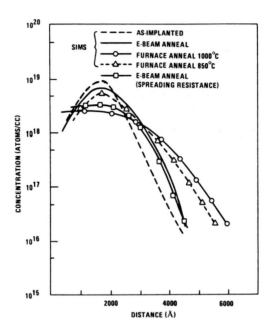

FIGURE 3.46.   SIMS [11]B profile for the sample as-implanted, after e-beam anneal and after furance anneal at 1000°C for 30 min. For comparison, the electrically active [11]B profile obtained by spreading resistance after e-beam anneal is shown (Ref. 3.58).

In conclusion, the e-beam annealing method by Yep et al. [3.58] differs significantly from those presented in Sections 3.4.3, 3.5.6 and 3.6. The scan frequency was 550 Hz in the x-direction and 330 Hz in the y-direction. While the latter systems feature a narrow, low power beam, leading to short dwell times but long processing time for a wafer, Yep et al. utilize a broad beam with a long exposure time (~ 4s) and fast wafer processing. The long dwell time permits the temperature to be relatively low, about 850°C. Even though the results of Yep et al. [3.58] are obviously not yet optimal (the electrical activation was only ~ 75%), their experiment demonstrates the potential of a wide beam to combine diffusion free annealing and a high throughput capacity.

REFERENCES

3.1 Gat, A., Gibbons, J. F., Magee, T. J., Peng, J., Deline, V. R., Williams, P., and Evans, C. A., Jr., Appl. Phys. Lett. 32, 276 (1978).

3.2 Williams, J.S., Brown, W. L., Leamy, H. J., Poate, J. M., Rodgers, J. W., Rousseau, D., Rozgonyi, G. A., Shelnutt, J. A., and Sheng, T. T., Appl. Phys. Lett. 33, 542 (1978).

3.3 Gat, A., Lietoila, A., and Gibbons, J. F., J. Appl. Phys. 50, 2926 (1979).

3.4 Olson, G. L., Kokorowski, S. A., Roth, J. A., and Hess, L. D., in "Laser and Electron-Beam Solid Interactions and Materials Processing" (J. F. Gibbons, L. D. Hess and T. W. Sigmon, eds.), p. 125. North-Holland (1981).

3.5 Gat, A., Gibbons, J. F., Magee, T. J., Peng, J., Williams, P., Deline, V., and Evans, C. A., Jr., Appl. Phys. Lett. 33, 389 (1978).

3.6 Lietoila, A., Gibbons, J. F., Magee, T. J., Peng, J., and Hong, J. D., Appl. Phys. Lett. 35, 532 (1979).

3.7 Lietoila, A., Gibbons, J. F., and Sigmon, T. W., Appl. Phys. Lett. 36, 765 (1980).

3.8 Regolini, J. L., Sigmon, T. W., and Gibbons, J. F., Appl. Phys. Lett. 35, 114 (1979).

3.9 See, e.g., White, C. W., Christie, W. H., Appleton, B. R., Wilson, S. R., and Pronko, P. P., Appl. Phys. Lett. 33, 662 (1978).

3.10 Thomas, G., "Transmission Electron Microscopy of Materials," Wiley (1971).

3.11 Van der Pauw, L. J., Philips Technical Review 20, 220 (1958/1959).

3.12 Johansson, N. G. E., Mayer, J. W., and Marsh, O. J., Solid State Electron. 13, 317 (1970).

3.13 Bube, R.H., "Electronic Properties of Crystalline Solids," p. 379, Academic Press (1974).

3.14 Barber, H. D., Lo, H. B., and Jones, J.E., J. Electrochem. Soc. 123, 1404 (1976).

3.15 Regolini, J. L., unpublished.

3.16. Dearhaley, G., Freeman, J. H., Nelson, R. S., and Stephen, J., "Ion Implantation," p. 539, North-Holland (1973).

3.17 Hofker, W.K., Werner, H.W., Oosthoek, D.P., and de Grefte, H. A. M., Appl. Phys. 2, 265 (1973).

3.18 Williams, P., Lewis, R. K., Evans, C. A., Jr., and Harley, P. R., Analytical Chem. 49, 1399 (1977).

3.19 Williams, P., and Evans, C. A., Jr., Appl. Phys. Lett. 30, 559 (1977).

3.20 Dobrott, R. D., NBS Special Publication 400-23, ARPA/NBS Workshop IV, Surface Analysis for Silicon Devices, held at NBS, Gaithersburg, Md., 23-24 April 1975, p. 31.

3.21 Ibid, p. 45

3.22 Chu, W.-K., Mayer, J.W., and Nicolet, M.-A., "Backscattering Spectrometry," Academic Press (1978).

3.23   Williams, J. S., Nuclear Instruments and Methods 126, 205 (1975).

3.24   Sigmon, T. W., Chu, W.-K., Muller, H., and Mayer, J. W., J. Appl. Phys. 5, 347 (1975).

3.25   Csepregi, L., Mayer, J. W., and Sigmon, T. W., Phys. Lett. 54A, 157 (1975).

3.26   Zellama, K., German, P., Squelard, S., Bourgoin, J. C., and Thomas, P. A., J. Appl. Phys. 50, 6995 (1979).

3.27   Gat, A., Ph.D. Thesis, Stanford University, Department of Electrical Engineering (1979).

3.28   Hess, L. D., Roth, J. A., Anderson, C. L., and Dunlap, H. L., "Laser–Solid Interactions and Laser Processing – 1978," (S. D. Ferris, H. J. Leamy, and J. H. Poate, eds.) AIP Conf. Proc. 50, 496, American Institute of Physics (1979).

3.29   Olson, G. L., Kokorowski, S. A., McFarlane, R. A., and Hess, L. D., Appl. Phys. Lett. 37, 1019 (1980).

3.30   Csepregi, L., Kennedy, E. F., Gallagher, T. J., Mayer, J. W., and Sigmon, T. W., J. Appl. Phys. 48, 4234 (1977).

3.31   Csepregi, L., Kennedy, E. F., Mayer, J. W., and Sigmon, T. W., J. Appl. Phys. 49, 3906 (1978).

3.32   Lietoila, A., Gold, R. B., and Gibbons, J. F., Appl. Phys. Lett. 53, 1169 (1982).

3.33   Lietoila, A., Ph.D. Thesis, Stanford University, Department of Applied Physics (1981), Chapter IV.

3.34   Auston, D. H., Golovchenko, J. A., Smith, P. R., Surko, C. M., and Venkatesen, T. N. C., Appl. Phys. Lett. 33, 539 (1978).

3.35   Regolini, J. L., Gibbons, J. F., Sigmon, T. W., Pease, R. F. W., Magee, T. J., and Peng, J., Appl. Phys. Lett. 34, 410 (1979).

3.36   Regolini, J. L., Johnson, N. M., Sinclair, R., Sigmon, T. W., and Gibbons, J. F., in "Laser and Electron–Beam Processing of Materials," (C. W. White and P. S. Peercy, eds.), p. 297, Academic Press (1980).

3.37   Ratnakumar, K. N., Pease, R. F. W., Bartelink, D. J., Johnson, N. M., and Meindl, J.D., Appl. Phys. Lett. 35, 463 (1979).

3.38   D'Avanzo, D. C., Rung, R. D., and Dutton, R. W., Stanford Electronics Laboratories Technical Report No. 5013-2 (1977).

3.39   White, C. W., Narayan, J., and Young, R. T., AIP Conference Proceedings 50, 275 (1979).

3.40   Fogarassy, E., Stuck, R., Grob, J. J., Grob, A., Siffert, P., in "Laser and Electron–Beam Processing of Materials," (C. W. White and P. S. Peercy, eds.), p. 117, Academic Press (1980).

3.41   Rimini, E., in "Laser and Electron Beam Processing of Electronic Materials," ECS Proc. 80, 270, Electrochemical Society (1980).

3.42   Tamura, M., Natsuaki, N., and Tokuyama, T., in "Laser and Electron-Beam Processing of Materials," (C. W. White and P. S. Peercy, eds.), p. 247, Academic Press (1980).

3.43   Chu, W. K., Mader, S. R., and Rimini, E., in "Laser and Electron-Beam Processing of Materials," (C. W. White and P. S. Peercy, eds.) p. 253, Academic Press (1980).

3.44   Pianetta, P., Amano, J., Woolhouse, G., and Stolte, C. A., "Laser and Electron-Beam Solid Interactions and Materials Processing," (J. F. Gibbons, L. D. Hess and T. W. Sigmon, eds.), p. 239, North Holland (1981).

3.45   Jackson, K. A., Gilmer, G. H., and Leamy, H. J., in "Laser and Electron-Beam Processing of Materials," (C. W. White and P. S. Peercy, eds.), p. 104, Academic Press (1980).

3.46   Lietoila, A., Ph.D. Thesis, Stanford University, Department of Applied Physics (1981). Chapter V.

3.47   Lietoila, A., Gibbons, J. F., Magee, T. J., Peng, J., and Hong, J. D., Appl. Phys. Lett. 35, 532 (1979).

3.48   Lietoila, A., Gibbons, J. F., and Sigmon, T. W., Appl. Phys. Lett. 36, 765 (1980).

3.49   Lietoila, A., Gibbons, J. F., Sigmon, T. W., Magee, T. J., Peng, and Hong, J. D., [Ref. 3.41], p. 350.

3.50   Sze, S. M., "Physics of Semiconductor Devices," p. 43, Wiley (1969).

3.51   Trumbore, F. A., Bell Syst. Tech. J. 39, 205 (1960).

3.52   Tsai, M.Y., Morehead, F.F., Baglin, J.E.E., and Michael, A. E., J. Appl. Phys. 51, 3230 (1980).

3.53   Fair, R. B., J. Electrochem. Soc. 125, 323 (1978).

3.54   Chu, W. K., and Masters, B. J., "Laser-Solid Interactions and Laser Processing - 1978," (S. D. Ferris, H. J. Leamy, and J. H. Poate, eds.) AIP Conf. Proc. 50, 305, American Institute of Physics (1979).

3.55   See, e.g., Bennerman, K. H., Phys. Rev. 137A, 1497 (1965); and Van Vechten, J. A., and Turmond, C.D., Phys. Rev. B14, 3551 (1976).

3.56   Regolini, J. L., Sigmon, T. W., and Gibbons, J. F., Appl. Phys. Lett. 35, 114 (1979).

3.57   Gat, A., Gibbons, J. F., Magee, T. J., Peng, J., Williams, P., Deline, V., and Evans, C. A., Jr., Appl. Phys. Lett. 33, 389 (1978).

3.58   Yep, T. O., Fulks, R. T., and Powell, R. A., in "Laser and Electron-Beam Solid Interactions and Materials Processing" (J. F. Gibbons, L. D. Hess and T. W. Sigmon, eds.) p. 345, North-Holland (1981).

3.59   Lietoila, A., Wakita, A., Sigmon, T. W., and Gibbons, J.F. J. Appl. Phys. 53, 4399 (1982).

3.60   Lidow, A., Ph.D. Thesis, Stanford University, Dept. of Applied Physics (1977).

3.61   Nissim, Y. I., Ph.D. Thesis, Stanford University, Dept. of Electrical Engineering (1981).

CHAPTER 4

Electronic Defects in CW Transient Thermal Processed Silicon

N. M. Johnson

XEROX CORPORATION
PALO ALTO RESEARCH CENTER
PALO ALTO, CALIFORNIA

4.1 INTRODUCTION

Directed energy sources such as lasers and electron beams and incoherent light sources have been proposed as alternatives to conventional furnace annealing for recrystallizing the amorphous layer created by high-dose ion implantation in single-crystal silicon. These energy sources have also been used to crystallize silicon thin films on insulating amorphous substrates in order to obtain semiconducting material for electronic device fabrication. In both applications the purpose of transient thermal processing is to produce single-crystal material of high crystalline perfection. With cw energy sources, recrystallization of implanted amorphous layers occurs by solid-phase epitaxy, with lower densities of extended defects (e.g., stacking faults) than can be achieved by conventional furnace annealing. However, materials studies with techniques such as capacitance transient spectroscopy and luminescence reveal high densities of residual defects in and near the recrystallized layers, which give rise to deep levels in the silicon forbidden-energy band. This chapter reviews the current state of knowledge in the ongoing study of residual defects in recrystallized bulk silicon and introduces the subject of electronic defect evaluation in crystallized-silicon thin films.

The high degree of perfection and control which has been achieved in silicon integrated-circuit fabrication must be retained while realizing the potential advantages of beam recrys-

tallization. Only then will the unique features of spatial and
temporal localization of energy be appreciated. The stringency
of this requirement becomes evident when it is noted that one of
the major achievements underlying the success of present-day
silicon integrated-circuit processing is the minimization and
control of electrically active defects in single-crystal silicon.
Impurities such as oxygen and carbon can be tolerated at concentrations of the order of 10^{18} cm^{-3}, which is typical for
Czochralski-grown silicon, because they can be rendered electrically inactive. On the other hand, a metallic impurity such as
gold can significantly alter the electrical properties of a
device at concentrations of less than 10^{11} cm^{-3} through the
introduction of deep traps and recombination centers. For transient thermal processing of silicon, the issue centers on the
removal of electronic defects associated with lattice damage,
such as that created by ion implantation.

This chapter first examines the problem of residual defects
which result from the use of cw transient thermal sources to
recrystallize the amorphous layer created by high-dose ion-
implantation in silicon. This topic is of central importance in
silicon integrated-circuit processing due to the extensive use
of ion implantation to introduce dopant impurities (e.g., As, P,
and B) for spatially controlling the free-carrier concentration.
The displacement damage created by ion implantation is conventionally removed by furnace annealing, which also electrically
activates the implanted impurities through their placement on
substitutional lattice sites. As discussed in Chapter 3, directed
energy sources such as laser and electron beams can be used for
this purpose. The potential advantages of energy beams are
derived from their spatial and temporal features. For example,
regions of silicon can be selectively annealed thereby preventing
possible electrical degradation of other regions not requiring
thermal cycling. It has been shown that implanted layers can be
recrystallized with low densities of extended defects and with
complete electrical activation of the dopant. Cw transient
thermal sources, including incoherent light sources, offer the
additional feature that the implanted amorphous layer is recrystallized without dopant redistribution. This is a consequence
of recrystallizing the amorphous layer by solid-phase epitaxial
regrowth on a time scale of seconds or less, which precludes significant dopant diffusion. In addition, it is generally found that
free-carrier mobilities in recrystallized layers equal their
values in conventionally processed bulk single-crystal silicon.
The above features of cw transient thermal processing are particularly applicable in silicon submicron device fabrication and
have motivated numerous studies of residual electronic defects
and their dependence on transient annealing conditions.

Beam-crystallized silicon thin films are of technological
interest as a materials/processing system for fabricating high-
performance thin-film transistors (TFT) and circuits. As discus-

sed in Chapter 5, such devices have been fabricated on insulating layers over single-crystal silicon wafers and could lead to vertical integration in silicon integrated-circuit technology. High-performance TFTs have also been fabricated on bulk glass substrates and may be applicable for switching and logic circuitry in large-area displays. For a TFT technology, crystalline silicon thin films offer the significant feature of utilizing conventional silicon microelectronic processing techniques in the fabrication of thin-film devices and circuits. The potential impact of these rapidly emerging device technologies underscores the need for comprehensive evaluation of the electronic properties of crystallized-silicon thin films, and in particular, characterization of electronic defects and their correlation with materials processing and device operation.

The chapter is organized as follows: general experimental considerations of device fabrication and measurements of electronic defects are discussed in Section 4.2. Electronic defects in cw beam annealed bulk silicon are reviewed in Section 4.3. This section is concerned both with residual defects arising from the incomplete removal of ion-implantation damage by beam annealing and with the generation of defects under such annealing conditions. Laterally distributed defects are examined in Section 4.4, which includes consideration of the effect of patterned dielectric overlayers in scanned beam annealing. Section 4.5 presents results from deep level studies in isothermal transient annealed silicon. Section 4.6 reviews post-recrystallization processing techniques for further removal of residual defects. The emerging topic of beam-crystallized silicon thin films is discussed in Section 4.7 with results for electronic defects in silicon layers on both thin film and bulk amorphous substrates. A summary with conclusions is presented in Section 4.8.

4.2   DEVICE PROCESSING AND MEASUREMENT TECHNIQUES

In bulk single-crystal silicon, several forms of energy beams have been used to recrystallize the amorphous layer created by high-dose ion implantation. The features of cw beam processing include spatial selection of annealed regions, negligible dopant redistribution, and retention of majority-carrier transport properties. As discussed in Chapter 1, these features have been demonstrated with scanned cw lasers, scanned electron beams, incoherent light sources, and line-source (shaped) electron beams. However, from electrical measurements that are specifically sensitive to minority-carrier transport and from direct measurements of deep levels it has been found that all of the above forms of beam annealing leave residual electronic defects, with densities in excess of those obtained by conventional furnace annealing. This section reviews general considerations in the design and fabrication of test devices for the characterization of electronic defects in transient thermal annealed silicon.

4.2.1   CW Transient Thermal Processing

The principal diagnostic technique that has been used for electronic defect evaluation is capacitance transient spectroscopy performed on current-rectifying devices. Both Schottky-barrier and p-n junction diodes have been used for this purpose. Schematic cross-sectional diagrams of these structures are shown in Fig. 4.1, with the amorphized layer resulting from high-dose ion implantation represented by the cross-hatch patterns. The Schottky diode (Fig. 4.1a) is the simplest device to fabricate, requiring only the deposition of a metallic rectifying contact onto the silicon surface. However, the Schottky contact cannot be used to directly evaluate implanted dopants because the high doses required to amorphize silicon ($>10^{14}$ cm^{-2}) produce a degenerately doped surface layer which destroys the current rectifying feature of the Schottky barrier. To utilize this structure, self implantation may be used to create lattice displacement damage, of the form accompanying implantation of the commonly used dopants in silicon, without altering the shallow dopant concentration. Alternatively, the Schottky diode is well suited for studying induced defects in the absence of implantation. Dopant implants are conveniently studied with p-n junction test devices in which the heavily doped surface layer permits the fabrication of structures which closely approximate one-sided step junctions. The p-n junction structure can be realized in two configurations: the mesa diode (Fig. 4.1b) and the oxide-passivated planar diode (Fig. 4.1c). As test devices, these two configurations differ solely in the form of the junction perimeter. And while the planar p-n junction is the more technologically relevant configuration, the mesa structure is often the more convenient to fabricate for defect studies.

For capacitance transient spectroscopy it is essential to have back electrical contacts which remain ohmic at low temperatures, since the measurement is performed over a range of temperatures typically from 78 K to above room temperature. This is readily achieved through the use of semiconducting epilayers on degenerately doped substrates. On bulk single-crystal silicon the back surface may be implanted with the substrate dopant, which is then electrically activated with a furnace anneal (e.g., 800°C, 30 min, N$_2$), in order to produce a degenerately doped layer for ohmic contact. To further facilitate electrical connection to the back contact, a metallic thin film (e.g., aluminum) can be deposited over the back surface of the wafer and then sintered (at, e.g., 450°C, 30 min, forming gas). In studies of scanned-beam annealing, preparation of back contacts can be completed prior to front surface processing.

Transient thermal processing is used to recrystallize the amorphous layer created by ion implantation and to electrically

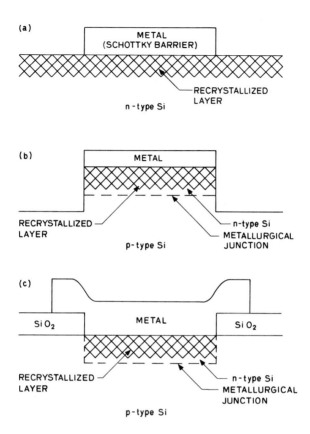

FIGURE 4.1.   Schematic diagrams of two-terminal current rectifying devices for electronic defect evaluation in recrystallized bulk silicon: (a) Schottky-barrier diode, (b) p-n junction (mesa) diode, and (c) oxide-passivated planar p-n junction diode. The cross-hatch areas represent material driven amorphous by ion implantation and subsequently recrystallized.

activate implanted dopants. Cw beam processing is performed with a focused laser or electron beam by scanning the energy beam over the silicon surface in a raster pattern. Energy beams shaped to form a scanned line source permit recrystallization over large areas in a single pass. In scanned beam processing, the silicon wafer is locally heated with dwell times of the deposited energy typically of the order of milliseconds. With isothermal transient annealing (Section 4.5), the entire wafer is annealed at an essentially constant temperature for times on the order of seconds.

Test devices are completed by vacuum evaporating metallic thin films over the front surface of the wafer and using conventional photolithography to define the electrodes. For p–n junction diodes, plasma etching or wet chemical etching is used to form the mesa structure, and on planar diodes the junction is defined by chemically etched windows in the silicon-dioxide passivation layer. For control devices, conventional furnace annealing replaces transient thermal processing in the fabrication of Schottky or p–n junction test devices.

4.2.2 Measurement Techniques

Several electrical and physical measurement techniques have been used to evaluate residual electronic defects, and their possible correlation with structural imperfections, in transient thermal recrystallized silicon. Both capacitance transient spectroscopy and luminescence have been used to directly probe deep levels in the silicon bandgap. However, its high sensitivity and direct electrical detection of residual deep levels have rendered capacitance-transient spectroscopy the most widely used diagnostic technique for electronic-defect evaluation. Indirect observations of residual defects, by detection of minority-carrier recombination, have been made with current/photocurrent-voltage measurements and with electron-beam-induced currents. Structural information for correlation with electrical measurements has been obtained from Nomarski-interference microscopy, transmission-electron microscopy, and x-ray topography.

Most of the results reviewed in this chapter were obtained by capacitance transient spectroscopy. The basic method is termed deep-level transient spectroscopy (DLTS) [4.1] and yields the energy levels and densities of electronic defects. Many of the results presented here were obtained from DLTS measurements performed in the constant-capacitance mode, which is particularly applicable when measuring trap densities comparable to the shallow dopant concentration. A detailed analysis of constant-capacitance DLTS (CC-DLTS) measurements of bulk-semiconductor defects has been presented elsewhere [4.2]. In addition, various DLTS techniques have been developed to obtain spatial depth profiles of defect levels [4.1-4.3].

The essential features of the DLTS technique are illustrated in Fig. 4.2 with an energy-band diagram for a Schottky-barrier structure with a single trap level at energy E_t in the bandgap of an n-type semiconductor. The DLTS measurement involves first the application of a reverse bias V_R to establish a depletion layer of width W_D. The trap level intersects the quasi-Fermi energy E_{FS} at X_D. Traps located below E_{FS} are filled with electrons and those above are empty. Voltage pulses periodically

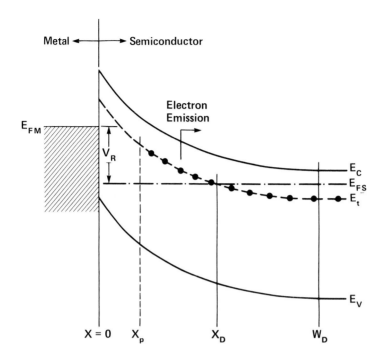

FIGURE 4.2. Schematic energy-band diagram for a Schottky-barrier structure with a single discrete defect level at an energy E_t in the silicon bandgap. The distance W_D is the depletion width for a reverse bias V_R, and X_D is the depth below the silicon surface at which the defect level intersects the quasi-Fermi level. The distance X_p is the depth at which the trap level intersects the quasi-Fermi energy during a pulse bias V_p.

reduce the depletion width in order to populate additional traps. After returning to the quiescent bias, these trapped electrons are thermally emitted to the conduction band and swept out of the depletion layer. This gives rise to a capacitance transient in the transient-capacitance mode or a voltage transient in the constant-capacitance mode of measurement.

Analysis of the DLTS signal provides a characterization of deep levels in the semiconductor. The DLTS signal is obtained by forming the difference of the transient response (either capacitance or voltage transient) measured at two delay times t_1 and t_2 after a charging pulse. For an exponential transient, the delay times define an emission rate constant e_0 which is given by the expression [4.1]

$$e_o = (t_2-t_1)^{-1} \ln(t_2/t_1). \tag{4.1}$$

A DLTS spectrum is obtained by recording the DLTS signal over a range of temperatures. An emission peak appears at that temperature for which the emission rate of a given trap, e_n, equals e_o. The emission rate constant can be varied to obtain the emission rate over a range of temperatures. Then the activation energy for thermal emission is obtained from an Arrhenius analysis of the emission rate, which from considerations of detailed balance may be expressed as

$$e_n = \sigma_n v_n N_c \exp[-(E_c - E_t)/(kT)], \tag{4.2}$$

where σ_n is the electron capture cross section, v_n is the mean thermal velocity for electrons, N_c is the effective density of states in the semiconductor conduction band, E_c is the conduction-band minimum, k is Boltzmann's constant, and T is the absolute temperature. The DLTS analysis thus yields the activation energy for thermal emission of charge carriers and the cross-section for capturing free carriers. In addition, the amplitude of the DLTS signal is proportional to the trap concentration and can be used to measure the spatial depth profile of deep levels.

4.3 CW BEAM PROCESSED BULK SILICON

4.3.1 Recrystallized Silicon

Electronic defects remaining after beam recrystallization of an implanted amorphous layer may be considered to arise from either of two origins. Ion implantation introduces displacement damage to depths beyond the amorphized surface layer. A beam annealing schedule can be sufficient to recrystallize the amorphous layer by solid phase epitaxy while only partially removing this more deeply penetrating lattice damage. On the other hand, beam irradiation can generate both electronic and extended structural defects as a consequence of the high thermal gradients and rapid quenching which are consequences of the spatial and temporal features of beam recrystallization. A clear distinction between these two origins can be obscured by defect interactions; that is, the form, distribution, and density of induced defects may be altered by the presence of displacement damage and impurities.

Residual electronic defects in cw beam-recrystallized silicon were first evaluated in Schottky-barrier diodes which were processed with a scanned cw Ar-ion laser [4.4]. Epitaxial silicon wafers with n-type conductivity were implanted at room temperature with SiH^+ to create displacement damage, of the form accompanying the implantation of dopants, without altering the shallow dopant concentration. The ions were implanted at 80 keV to a dose of $2x10^{15}$ cm^{-2}, which is sufficient to drive the silicon amorphous to a depth of ~120 nm. The implanted species was chosen to avoid nitrogen contamination (N_2^+) which can occur during $^{28}Si^+$ implantation. CC-DLTS emission spectra are presented in Fig. 4.3 for an as-laser-annealed diode. The two emission spectra were recorded with different combinations of reverse and pulse biases, which define different spatial observation windows for trap detection. The bounds of the spatial intervals quoted in Fig. 4.3 (i.e., ΔW_D) are the steady-state depletion depths under reverse and pulse biases, which were computed from the measured device

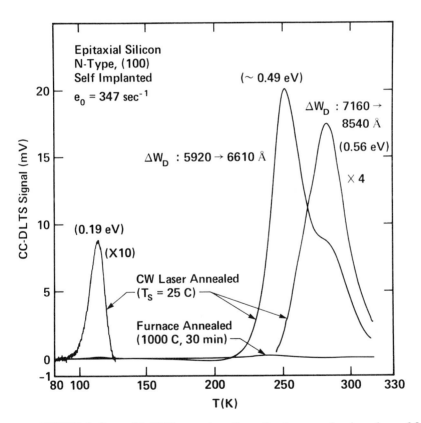

FIGURE 4.3. CC-DLTS spectra for electron emission in self-implanted cw laser-annealed silicon and from a furnace-annealed control device.

capacitance. The intervals indicate the relative depths of the
observation windows for the two spectra. The electron emission
spectra are dominated by two defect levels which are located
near the middle of the silicon band gap. The emission spectra
recorded with different observation windows indicate that the
ratio of the densities of the two defect levels varies rapidly
with depth. The activation energies were obtained from measurements over spatial intervals in which a single emission peak
dominated the spectrum. The activation energy for the high-
temperature peak is 0.56 eV. The spatial separation of the two
traps is primarily due to a rapid decrease with distance in the
density of the shallower of the two midgap levels. The spatial
separation is less complete for this shallower level, giving
rise to an uncertainty in the activation energy which is found
to be ~0.49 eV. A broad emission signal of reverse polarity
at intermediate temperatures is ascribed to a high trap concentration near the silicon surface which contributes to a high
degree of charge compensation. At low temperatures a third
emission peak of low intensity is detectable, with an activation
energy of 0.19 eV; this level is removed by a forming-gas anneal
as discussed below. (See Section 4.3.2 for further discussion
of this level.) Also shown in Fig. 4.3 is the featureless emission spectrum for a control diode which was furnace annealed at
1000°C (30 min, N_2).

CC–DLTS emission spectra are shown in Fig. 4.4 for a diode
which received a forming-gas anneal (15% H_2, 85% N_2, 450 C, 30
min.). These spectra are dominated by the same two midgap defect
levels as found above, with the level at ~0.49 eV decaying more
rapidly than the 0.56 eV level with depth into the silicon substrate. However, there is an additional emission peak with an
activation energy of 0.28 eV which is comparable in magnitude to
the midgap peaks. This level is well resolved in the emission
spectrum and can be unambiguously profiled by the double-correlation CC–DLTS technique. The defect-density distribution is shown
in Fig. 4.5. The horizontal bars mark the spatial intervals over
which average defect densities were measured, and the vertical
bars denote the uncertainties in the densities due to uncertainty
in the measurement of the net emission signal. The defect density
decreases monotonically with depth below the laser-irradiated surface, over the investigated interval. Also shown is the projected
range for Si^+ implanted at 77 keV [4.5]. This approximates the
projected range of silicon for the actually implanted species,
SiH^+, if it is assumed that the ionized molecule dissociates
into silicon and hydrogen atoms at the silicon surface, with
equal particle velocities at the instant of separation.

High densities of deep levels remain after cw laser annealing. The 450-C anneal effectively removed the level at 0.19 eV
and introduced another level at 0.28 eV. The densities of all

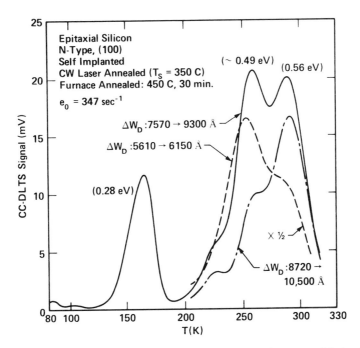

FIGURE 4.4.   CC–DLTS emission spectra for a self–implanted laser–annealed diode which also received a 450°C furnace anneal.

levels decrease with depth into the silicon substrate from $\geqslant 10^{15}$ cm^{-3} in the near–surface region, as shown for the level at 0.28 eV in Fig. 4.5.   The projected range for Si$^+$ reveals that the measured defects reside in material that was not driven amorphous by ion implantation.   Diodes fabricated with cw laser–processed Czochralski–grown silicon produce the same results as shown above for epitaxial material [4.6].

Electronic defect levels in scanned electron–beam annealed (SEBA) silicon are shown in Fig. 4.6.   The test devices were p–n junction diodes fabricated by implanting As$^+$ (100 keV, 5x10^{14} cm^{-2} into Czochralski–grown p–type (B,1-3 Ω-cm) silicon.   The ion energy and dose were sufficient to drive the silicon surface amorphous to a depth of ~100 nm.   Scanned electron–beam annealing was performed with both a commercial electron–beam welder [4.7] and a scanning electron microscopy [4.8], with the substrates at room temperature.   The p–n junction diodes were completed by depositing aluminum over the recrystallized suface, using photolithography and plasma etching to define the mesa structure, and sintering (450°C, 30 min. forming gas).   CC–DLTS spectra in Fig. 4.6 are for hole emission in the depletion layer of the p–type substrate.   The hole trap at 0.28 eV dominates the emission

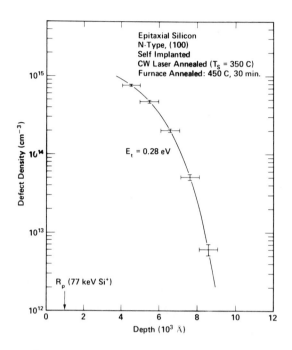

FIGURE 4.5.   Spatial depth profile of the defect level at $E_c-0.28$ eV in self-implanted laser-annealed silicon.

spectra, with an additional peak at 0.36 eV in the SEM-annealed device. These are activation energies for thermal emission of holes to the silicon valence band. Neither level is detectable in the furnace-annealed control diode. The defect densities were found to be in the 10^{13} cm^{-3} range to greater depths than a micron below the metallurgical junction [4.9].

Results for cw Ar-ion laser-fabricated p–n junctions are shown in Fig. 4.7 [4.10]. Boron-doped silicon wafers were implanted with As$^+$ (100 keV, 4×10^{15} cm^{-2}), cw laser annealed with the substrate at 350 C, and fabricated into mesa diodes; the diodes were not sintered. In Fig. 4.7 four hole traps are detectable with activation energies as follows: 0.10 eV, 0.20 eV, 0.28 eV, and 0.45 eV; these are denoted as H(0.10), H(0.20), H(0.28), and H(0.45), respectively. The H(0.28) level is the same level shown in Fig. 4.6 for sintered diodes and was not present when the substrate temperature was set to 250 C during laser annealing. The dominant hole trap at 0.45 eV was found to be unstable at room temperature. The defect density decays with time, with the level transmuting to the shallower level at 0.10 eV. By comparing these defects in laser-annealed silicon with

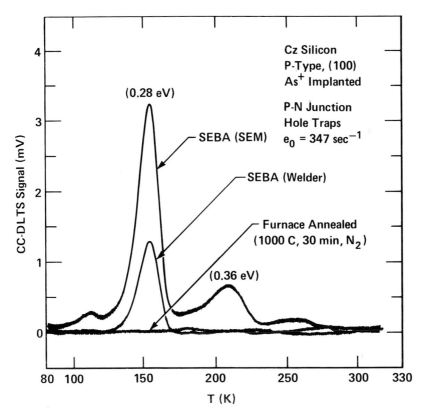

FIGURE 4.6. CC-DLTS spectra for hole emission in As^+-implanted scanned electron-beam annealed silicon. The spectrum for a furnace-annealed p-n junction diode is also shown.

defects produced by furnace annealing followed by rapid cooling, and with other published results, the laser induced defects were identified as interstitial Fe and Fe-B pairs.

A comparison of electronic defect levels in laser-annealed and scanned-electron-beam-annealed silicon is shown in Fig. 4.8. [4.11]. The test devices were As^+-implanted (100 keV, 4×10^{15} cm^{-2}) p-n junction diodes, fabricated with the mesa structure, which were annealed with either a cw Ar-ion laser at a substrate temperature of 250° C or with a scanning electron microscope. Figure 4.8 shows a DLTS spectrum immediately after diode preparation. For cw laser annealing, the hole emission spectrum is dominated by a level at 0.45 eV, as seen previously. As shown with spectra (a) and (b), the density of this level increases rapidly with laser power P, where P_M is the power required to melt the silicon.

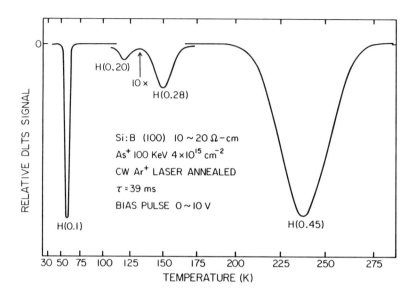

FIGURE 4.7. Typical capacitance transient spectrum of p-type Si implanted with As$^+$ and annealed by cw Ar-ion laser. (N. H. Sheng and J. L. Merz, Ref. 4.10.)

These laser powers are below that required to generate slip dislocations as discussed in Section 4.3.3. When the laser power is sufficiently high to produce slip dislocation, the dominant defect level is found to depend on substrate temperature and appears at 0.43 eV for a temperature of 250° C. The SEBA results for slip-free annealing reveal only a weak signal at E_v+0.40 eV, which does not depend strongly on electron-beam power. As discussed above, the 0.45 eV level is not stable at room temperature and is associated with interstitial iron.

4.3.2 Beam-Induced Defects

Recrystallization of implanted amorphous layers on single-crystal silicon involves the epitaxial regrowth of the amorphous layers, with electrical activation of the implanted dopant, and removal of displacement damage in the near-surface region which remains crystalline during ion implantation. It has also been observed that beam irradiation of single-crystal silicon can induce or quench in electronic defects [4.9,4.12,4.13]. This is illustrated in Fig. 4.9 with the electron-emission spectrum for an electron-beam annealed Schottky-barrier diode. The diode was not implanted and received the same beam-annealing schedule used to recrystallize the As$^+$-implanted layer in Fig. 4.6. The electron-emission spectrum for a SEBA-annealed diode is compared

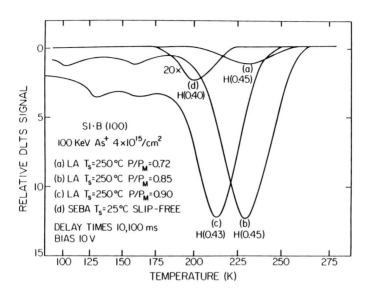

FIGURE 4.8. Capacitance transient spectrum of p-type Si implanted with As$^+$ and annealed by either a cw Ar-ion laser (LA), using three different values of laser power, or a scanning electron beam (SEBA). T_S is the substrate temperature during beam annealing. (N. H. Sheng et al., Ref. 4.11.)

to that for unannealed material. The spectrum for beam-annealed silicon is dominated by two emission peaks with activation energies of 0.19 and 0.44 eV. Shoulders on both peaks indicate additional unresolved emission centers. In the unannealed control, no emission peaks are detectable on the indicated scale of sensitivity. Similar quenched-in deep levels have been observed in cw laser irradiated silicon, where it was further determined that the low-temperature emission peak has an electric-field dependent activation energy with a zero-field value of 0.22 eV [4.14]. The 0.44 eV level has been ascribed to the phosphorus-vacancy complex [4.9,4.13-4.15]. The spatial distributions of the quenched-in defects differ in electron-beam irradiated [4.9] and cw laser-irradiated silicon [4.13], but in both cases the defects are located in the near-surface region of the silicon.

The essential observation is that even without displacement damage from ion implantation the localized heating and rapid quenching, which are inherent features of beam recrystallization, can provide the stimulus for defect formation during laser or electron-beam annealing. However, it has been further observed, with cw laser annealing, that beam-induced damage can be strongly influenced by the crystalline quality of the substrate before laser annealing [4.12]. In addition, near the point of melting,

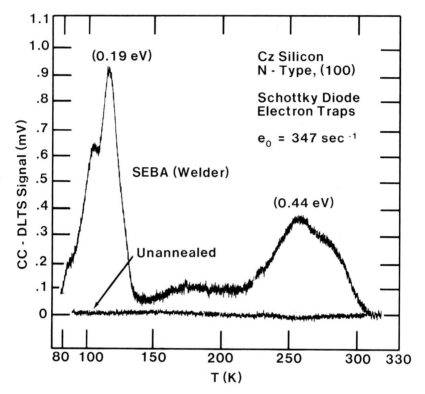

FIGURE 4.9. CC–DLTS spectrum for electron emission in unimplanted scanned electron-beam annealed silicon. Also shown is the spectrum for the unannealed material.

beam-induced slip dislocations are generated as discussed in the next section.

4.3.3 Minimization of Beam Recrystallization Defects

Critical optimization and control of annealing conditions are required for successful beam processing. To illustrate, with a scanned electron beam too low a beam current (for a given accelerating voltage) yields incomplete recrystallization of the implanted amorphous layer, while beam currents near that required for melting generate slip dislocations, which can be readily viewed in an optical microscope as a cross grid of slip lines over the annealed area. These extremes have also been documented in scanned laser annealed silicon. The photoluminescence efficiency of the recrystallized material varies dramatically depending upon the creation or suppression of dislocations. For example, beyond a critical exposure time, at a constant power near that required

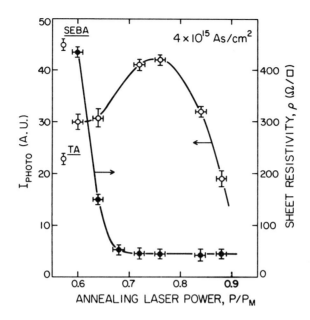

FIGURE 4.10. Dependence of short-circuit photocurrent I_{photo} in arbitrary units and sheet resistivity ρ on laser power P for a diode implanted with 4×10^{15} As$^+$/cm^2 and annealed with a 100 μm laser spot and 6 μm step. Laser power is normalized to the power required for melting, P_0. (M. Muzita et al., Ref. 4.17.)

for melting, slip appears and the luminescence is dominated by dislocation-related defect levels [4.16]. Such studies have further revealed that the range of laser power over which good quality annealed material can be achieved is limited [4.11,4.17], as further discussed below. Finally, electronic deep levels associated with dislocations in deformed silicon have been detected by capacitance transient spectroscopy [4.18].

Optimum annealing conditions for a cw Ar-ion laser are shown in Fig. 4.10 [4.17,4.19]. Single-crystal silicon with p-type conductivity was implanted with As$^+$ at 100 keV to a dose of 4×10^{15} cm^{-2}. The laser annealing conditions were selected for laterally uniform recrystallization, with laser power a variable. After annealing, the sheet resistance ρ and short-circuit photocurrent I_{photo} of mesa p-n junction diodes were measured over a range of laser powers P up to that required for melting, P_0. The sheet resistance depends on majority-carrier properties, while the photocurrent is sensitive to those of the minority carriers. As shown in Fig. 4.10, for $P \geqslant 0.68P_0$ the sheet resistivity achieves a minimum value of 45 Ω/square, corresponding to 100% electrical activation of the implanted arsenic.

On the other hand, the relative magnitude of the photocurrent increases with laser power to a peak value at $0.76P_0$ and then decreases at higher laser powers.

The effect of laser power on electronic-defect density has been directly evaluated with DLTS [4.11]. In Fig. 4.11 are plotted defect concentrations versus beam power for scanned-laser and scanned-electron beam annealing. The defect concentration of scanned-electron-beam-annealed diodes is lower than that of laser-annealed diodes and is comparatively insensitive to electron-beam power. However, for laser-annealed diodes, the defect concentration increases more than one order of magnitude with increasing laser power. For high-power laser-annealed devices, the defect concentration is comparable to the substrate dopant concentration. Results from EBIC are also shown in Fig. 4.11. The EBIC charge-collection efficiency decreases as the defect concentration increases. It was found in this study [4.11] that laser-induced defects can extend several microns below the implanted junction.

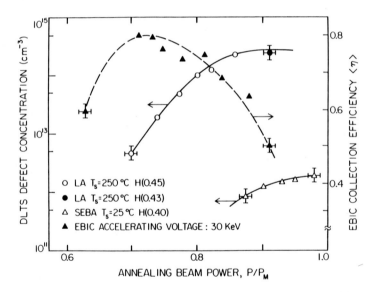

FIGURE 4.11. Defect concentration measured by DLTS, as a function of annealing beam power. The broken curve displays the EBIC charge-collection efficiency as a function of laser power. [N. H. Sheng et al., Ref. 4.11.]

4.4 LATERAL NONUNIFORMITIES IN SCANNED-BEAM ANNEALING

This section discusses some practical limits in optimizing cw beam annealing conditions for minimum densities of residual defects in the recrystallized silicon layer. First, the specific issue of lateral variations in defect density due to scan line separation is discussed with results from luminescence and electron-beam induced current measurements. Then the topic of scanned-beam annealing of silicon wafers with patterned dielectric overlayers is addressed with results from transmission electron microscopy.

4.4.1 Scan Line Separation

Defect luminescence has been used to evaluate cw laser-annealed silicon [4.16,4.20,4.21]. In particular, it has been found that when the lateral separation between adjacent laser-scan lines is reduced, the luminescence intensity decreases [4.20]. This is demonstrated in Fig. 4.12 with luminescence spectra for n-type Czochralski-grown silicon which was implanted with SiH^+ at 80 keV to a fluence of $2x10^{15}$ cm^{-2} and recrystallized at a substrate temperature of 350° C with the scanned beam of a cw Ar-ion laser. Sample A was annealed at 5.2 W with a 40 µm diameter spot scanned at 12 cm/sec and with a 15 µm separation between lines. The two lower spectra correspond to samples laser annealed at 11.5 W, with an 80 µm diameter spot and a scan speed of 5 cm/sec. For sample B the separation of the scan lines was 30 µm and for sample C it was 15 µm. Although the spectra are similar, the defect luminescence intensity varies by a factor of 30, with sample C giving the weakest signal. Inspection of the samples in an optical microscope showed that A and B have discernible lateral nonuniformity consisting of stripes between adjacent scan lines, which is not present in sample C. These samples were also evaluated with electron spin resonance. In each sample the characteristic dangling bond resonance of amorphous silicon [4.21] was observed with volume densities (calculated with an estimated thickness of 180 nm for the initially amorphized layer) of $4.1x10^{17}$, $1.1x10^{17}$, and $0.4x10^1x10^{17}$ cm^{-3} for samples A, B, and C, respectively. Thus, as the lateral separation between adjacent scan lines was reduced, the luminescence intensity and spin density decreased and the nonuniformity of the surface was also reduced. These results indicate that the defects are associated with an imperfectly crystallized interface between adjacent laser annealed zones. It was suggested that the luminescence is possible due to extended defects such as dislocations or grain boundaries.

Lateral nonuniformities in cw laser annealed silicon have been directly demonstrated with electron beam-induced currents

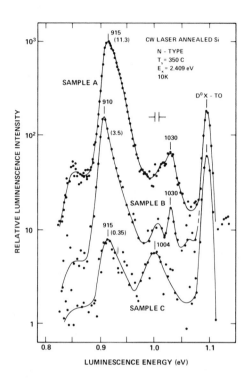

FIGURE 4.12. Luminescence spectra of n-type silicon, showing the effect of changing the laser annealing conditions.
Details of the preparation of samples A, B, and C are given in the
text.

[4.17]. A typical EBIC image is shown in Fig. 4.13. The sample
was implanted with As[+] at 100 keV to a fluence of 6×10^{14} cm^{-2} and
annealed with an Ar-ion laser beam with a 40 μm diameter spot
and 30 μm step. After annealing, mesa diodes were fabricated
by standard photolithographic techniques, and ohmic metal contacts
were formed on both sides of the diode. The metallurgical p-n
junction was location 0.2-0.3 μm below the implanted surface.
The EBIC measurement was performed with a 5 keV accelerating
voltage which resulted in an electron range of ~0.3 μm. The EBIC
image arises from the diffusion of beam-generated minority carriers to the junction where they drift in the electric field of
the depletion layer, yielding an induced current in the external
circuit. This current is monitored as the electron beam is
rastered over the specimen to produce an image which is displayed
on a cathode ray tube. The upper and lower portions of Fig. 4.13
display the standard secondary electron image of a diode. In
this mode the laser-annealed surface appears to be smooth and
defect free. However, when the EBIC mode is switched on (center

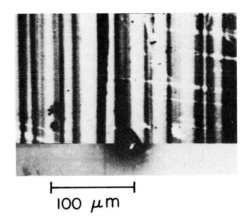

├────────────┤
100 μm

FIGURE 4.13. EBIC micrograph (center) and secondary emission image (top and bottom) excited by 5 keV beam. The sample was implanted with 6×10^{14} As$^+$/cm^2 and annealed with 40 μm spot and 30 μm step. In the EBIC mode, dark regions correspond to low charge collection. [M. Mizuta et al., Ref. 4.17.]

of Fig. 4.13), high-contrast dark lines appear which are parallel to the direction of the laser scan. The EBIC dark-line spacing is of the order of the laser-scan step; the deviations from constant spacing are considered to be associated with complex overlap effects [4.17]. Further, it was found that improved results were obtained for a large-diameter beam (100 μm), small scan step (6 μm), and slow scan speed (6 cm/sec), which is consistent with the photoluminescence results described above.

The dark-stripe features have also been observed in x-ray topographs of cw Ar-ion laser-annealed, slip-free silicon, with a one-to-one correspondence between the dark stripes in the EBIC image and in the x-ray topograph [4.22]. With TEM it has been determined that the dark-stripe regions contain a high density of dislocation loops due to interstitial atoms [4.23]. It has been further determined that CO_2 laser annealing (at 10.6 μm wavelength) does not differ significantly from cw Ar-ion laser annealing (0.514 μm wavelength) in that the use of a scanned hot spot to anneal implantation damage seems always to be accompanied by the production of residual interstitial dislocation loops, which act to decrease minority-carrier lifetime [4.24]; this issue is further discussed in Section 4.4.2. The above identified limitations of raster-scanned point sources can be overcome by using scanned-line sources, where the length of the line source is greater than the maximum lateral dimension of the surface area to be recrystallized.

4.4.2 Patterned Dielectric Overlayers

Most studies of electronic defects in cw beam-recrystallized silicon have been conducted on material which was uniformly amorphized by ion implantation and recrystallized over lateral dimensions large compared to the examined area. Isolating the recrystallization process in this way serves to focus attention on basic materials issues. However, the actual application to integrated-circuit fabrication introduces additional constraints which must be addressed if beam recrystallization is to replace conventional furnace annealing. For example, in integrated-circuit processing only small well-defined areas of a silicon wafer are ion implanted to high doses, such as the source and drain contacts in a metal-oxide-semiconductor field-effect transistor (MOSFET). The implanted area is often defined by a window etched in a dielectric thin-film overlayer. An example is shown in Fig. 4.1(c). The planar p-n junction diode is bounded by a silicon-dioxide layer, which electrically passivates the silicon surface. This is a more complex structure for beam recrystallization since the oxide step at the edge of the implanted amorphous layer alters both the spatial distribution of energy deposited by the scanned beam and the temperature profile (or heat dissipation) during recrystallization.

Oxide-passivated p-n junction diodes have been examined after scanned electron-beam annealing with a modified SEM [4.25]. In Fig. 4.14 is shown a Nomarski interference optical micrograph for an SEM-annealed region of an implanted amorphous layer bounded by a silicon-dioxide overlayer. The starting material was Czochralski grown <100>-oriented silicon with p-type conductivity (B, 1÷3 Ω-cm). A silicon-dioxide layer was thermally grown to a thickness of 260 nm and photolithographically patterned and chemically etched to expose areas of silicon for ion implantation. The entire wafer was then implanted with As^+ at 100 keV to a dose of 1×10^{15} cm^{-2}. The SEM annealing conditions were as follows: beam voltage of 20 kV, line-scan rate of 10 msec/line, frame-scan rate of 40 sec/frame, rastered area of 1.10x1.15 mm, beam diameter of 12÷14 μm, substrate temperature of 25°C, and a beam current of 78 μA. The beam current was ~90% of that required to melt the exposed silicon. In Fig. 4.14 the SEM-recrystallized region of the implanted silicon is distinguished from the amorphized area by the difference in reflectivity. In Fig. 4.15 is shown a bright-field TEM micrograph of the boundary between a region of implanted and recrystallized silicon and an oxidized region; the rastered electron beam overlapped the oxide layer, and the oxide was removed for TEM. The dark band running diagonally across the micrograph locates the boundary between the recrystallized and oxide-passivated regions and was identified as a zone of amorphous material.

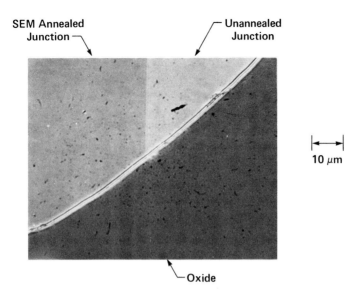

FIGURE 4.14. Nomarski interference optical micrograph of an oxide-passivated SEM-annealed p-n junction in silicon.

In samples annealed at higher beam currents, the boundary contained a high density of dislocations. In a MOSFET such a defective zone would be particularly detrimental to device operation since it would be situated at the critical interface between the source or drain and the channel. Diodes fabricated in the above material were also evaluated by DLTS, a spectrum from which is shown in Fig. 4.16. The dominant hole trap has an activation energy of 0.28 eV, as discussed in Section 4.3, with a density of $\sim10^{14}$ cm^{-3} at a depth of 1 μm below the metallurgical junction. No hole emission peak is detectable on the same scale of sensitivity in the furnace-annealed diode.

The results presented above illustrate the added difficulties which can be encountered in the practical application of beam recrystallization in silicon integrated-circuit processing. It is not anticipated that nonuniform recrystallization arising from patterned dielectric overlayers will be alleviated with scanned-line sources. However, this is an area of cw beam processing requiring further investigation.

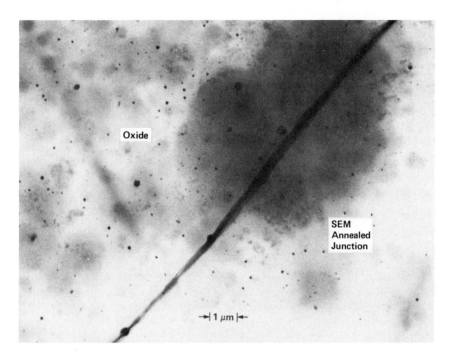

FIGURE 4.15. Bright-field TEM micrograph of the boundary in single-crystal silicon between a region of exposed silicon which was implanted with As^+ (100 keV, 1×10^{15} cm^{-2}) and a region which was covered with a thermally grown layer of silicon dioxide (260 nm thick) during implantation and SEM recrystallization. The silicon-dioxide layer was removed for TEM.

4.5   ISOTHERMAL TRANSIENT ANNEALING

As reviewed in Chapter 1, transient isothermal annealing techniques have recently been introduced to activate ion implanted dopants through solid-phase regrowth. The implanted silicon wafer is irradiated by an incoherent light source (e.g., tungsten halogen lamp, arc lamp, or graphite heater), and the damage is annealed on a time scale of seconds. The combination of large-area illumination and long dwell times (as compared to fast pulse or scanned laser annealing) insure that the wafer is heated essentially uniformly. The technique results in minimum redistribution of implanted dopants, complete electrical activation, and uniform annealing across the wafer. Residual electronic defects after transient isothermal annealing have also been investigated [4.26, 4.27].

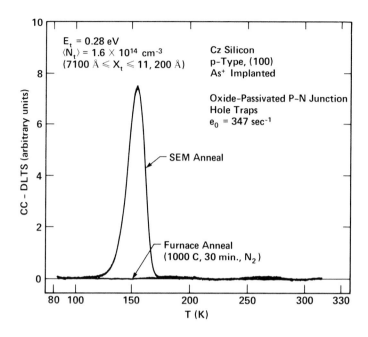

FIGURE 4.16. CC–DLTS spectrum for hole emission in an oxide-passivated SEM-annealed p–n junction diode. Also shown is the spectrum for a furnace-annealed control diode.

A comprehensive study of electronic defects in silicon after transient isothermal annealing has recently been reported which includes a comparison of different annealing systems and clearly demonstrates the importance of quench rate on the density of residual defects [4.27]. A typical trap-level spectrum for a boron-doped epitaxial silicon wafer after transient thermal annealing is shown in Fig. 4.17; the wafer was not ion implanted. After an anneal at 900°C for 20 seconds, a large DLTS peak appears at a temperature of ~187 K. From the DLTS analysis, the trap concentration $N_T = 10^{15}$ cm^{-3}, the trap ionization energy $E_T = E_{v2} + 300$ meV, and and the capture cross-section $\sigma_p = 2 \times 10^{-15}$ cm^2; this appears to be the same as the $E_V + 0.28$ eV level discussed in Section 4.3. The defect level is not detectable in the starting (reference) material, but is present in low concentration after a conventional furnace anneal. In unimplanted Czochralski-grown and float zone grown silicon, the level was also not detectable after the transient thermal annealing (i.e., $N_T < 1 \times 10^{11}$ cm^{-3}), while in arsenic-implanted (5×10^{14} As$^+$ cm^{-2}, 100 keV) p–n junction diodes fabricated in Czochralski-grown silicon, the level is present after transient annealing at a concentration of 5×10^{12} cm^{-3}. Transient thermal annealing of n-type (phosphorus-doped)

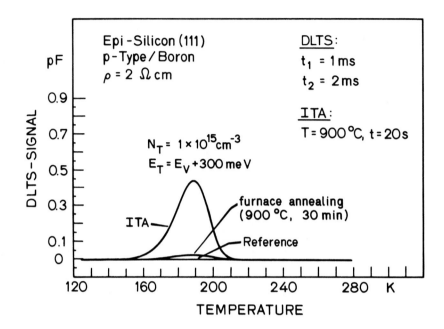

FIGURE 4.17. DLTS spectrum for a boron–doped epitaxial silicon sample after transient thermal annealing (ITA). Also shown are spectra for a furnace annealed control sample and for the unannealed (reference) material.

silicon wafers was also investigated. No significant DLTS signal was detected between 50 K and 300 K, from which it was concluded that no electrically active trap levels with concentrations greater than 10^{11} cm^{-3} are induced in the upper half of the silicon bandgap under isothermal transient annealing conditions.

The dependence of trap concentration on the principal annealing parameters was also investigated [4.27]. In Fig. 4.18 is shown the effect of annealing temperature on the trap concentration for three different annealing systems. The samples in the SEL and AG 210 annealing systems were cooled at a rate of ~100°C/sec; the cooling rate in the Varian IA 200 system was not controllable. All three curves peak at a temperature of ~1000°C. The relative displacements of the three curves is considered to be primarily due to differing cooling rates for samples in the three systems, even for nominally the same speci-fied rate. This is further discussed below.

The trap concentrations induced after isothermal transient annealing in a single system with different ramp down rates is

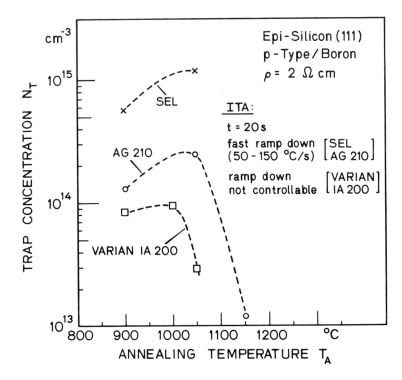

FIGURE 4.18. Trap concentration of the 300 meV level versus annealing temperature. The three curves are measured with different systems as indicated.

shown in Fig. 4.19. The rates of sample cooling are not well defined because of problems in the measurement of sample temperature under transient conditions. The dashed curve (solid circles) and solid curve (open circles) were taken on separate samples from the same wafer. The open circles represent measurement points taken at a slow ramp down of 5°C/sec, and the solid circles are points taken at a ten times faster rate. Additionally, furnace anneals were performed for comparison. The results reveal that the production of traps is strongly reduced by a slow ramp down of the temperature, independent of the anneal method.

Further observations on the behavior of the 300 meV level are as follows [4.27]: The defect concentration decreases with increasing distance from the surface, which may be caused by the temperature gradient during cool down or may indicate that the surface is a possible source of the defect. The level can be removed by a furnace anneal at temperatures above 400°C. The trap level is induced by isothermal transient annealing after

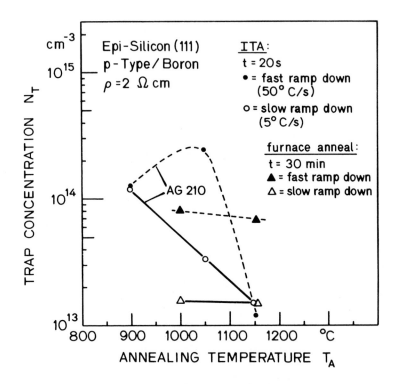

FIGURE 4.19. Comparison of residual trap concentrations of the 300 meV level after slow and fast ramp down rates. The circles and triangles represent points of transient thermal annealing and furnace annealing, respectively.

As^+, B^+, and Ar^+ implantation into boron-doped silicon and also after B^+ implantation in Al-doped silicon and transient annealing. It is not present after transient annealing of $^{30}Si^+$ implanted samples. It was proposed that the defect giving rise to the 300 meV level is a complex involving boron and the lattice vacancy.

4.6 POST-RECRYSTALLIZATION PROCESSING

Can residual electronic defects from transient thermal processing be removed with other techniques which are compatible with integrated-circuit processing and which retain the inherent features of beam recrystallization (as enumerated in Section 4.1)? This possibility may be anticipated since the defects arise from incomplete removal of displacement damage from ion implantation and the rapid quenching which is characteristic of transient thermal annealing. As a consequence, the annealing kinetics of the residual deep levels should be similar to those found for centers introduced by ionizing radiation where it has been shown that most of the defects can be removed at low temperatures (e.g., ≤ 600°C) [4.15,4.28].

4.6.1 Furnace Anneals

A low-temperature furnace anneal is commonly used for contact alloying in integrated-circuit processing. In Section 4.3 it was noted that an anneal at 450°C was effective in removing the electron trap at 0.19 eV in self-implanted Schottky diodes. The effect has also been demonstrated with hole traps in p-n junction diodes [4.29]. Mesa diodes [Fig. 4.1(b)] were fabricated by implanting As$^+$ into B-doped (1-3 Ω-cm) <100>-oriented Czochralski-grown silicon and recrystallizing with a scanning cw Ar-ion laser; the substrate was maintained at room temperature (T_s = 25°C) during beam annealing. Charge-emission spectra are shown in Fig. 4.20. The spectrum in Fig. 4.20(a) is for a diode which received no further thermal processing after laser recrystallization. The spectrum is dominated by the hole emission center at 0.28 eV, which was discussed previously. Also evident is a broad emission peak of reverse polarity near room temperature. In a p-n junction, the reverse polarity signifies minority-carrier emission [4.1], which in this case arises from electron emission from traps in the upper half of the silicon bandgap. Similar broad emission peaks of reverse polarity were observed in furnace-annealed mesa diodes, while such a feature is absent in oxide-passivated diodes with the planar structure (see Fig. 4.16). These observations suggest that the electron emission is largely due to edge effects in the mesa structure. Figure 4.20(b) shows the spectrum for a diode which was processed simultaneously with that of Fig. 4.20(a) except for the addition of a 450°C furnace anneal which was performed immediately after laser annealing. Both diodes were then aluminum metallized and plasma etched to define the mesa structure. In Fig. 4.20(b) the hole trap at 0.28 eV is absent and the broad emission signal near room temperature has been significantly reduced in intensity. In addition, a new electron emission peak appears corresponding to an electron trap at E_c - 0.28 eV; that is, the narrow peak of reverse polarity identifies a deep level with an activation energy of 0.28 eV for thermal emission of electrons to the silicon conduction band. The electron emission intensity is slightly enhanced by using a forward-bias voltage pulse for trap filling which injects electrons (minority carriers) into the p-region of the contact from the n$^+$ layer [see Fig. 4.1(b)]. This suggests that the electron traps are located near the metallurgical junction. This new electron emission peak appears at nearly the same temperature as the hole emission peak and is apparently masked by the more intense hole emission in Fig. 4.20(a). The

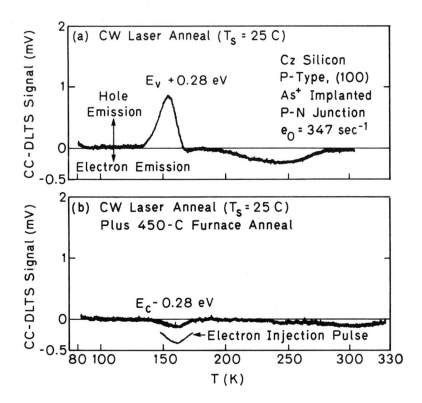

FIGURE 4.20. CC-DLTS emission spectra for cw laser-annealed silicon diodes. In (a) the diode received only a laser appeal, and in (b) the diode also received a 450°C furnace anneal.

removal of the hole trap uncovers the residual electron trap, which requires higher anneal temperatures for removal, as demonstrated below. It was found that the observed reduction in defect density after the low-temperature furnace anneal is accompanied by a decrease in the reverse-bias leakage current in the above diodes. In addition, the observed reduction in defect density realized by a 450°C furnace anneal after laser recrystallization suggests that a similar reduction should be achievable by laser annealing with a heated substrate. However, diodes which were cw laser annealed with a substrate temperature of 350°C display the same dominant hole trap at 0.28 eV, with defect densities comparable to those of diodes with substrates at room temperature during laser recrystallization.

The effect of high-temperature furnace annealing after cw laser recrystallization is illustrated in Fig. 4.21 [4.6]. The results are for Schottky-barrier diodes on epitaxial silicon

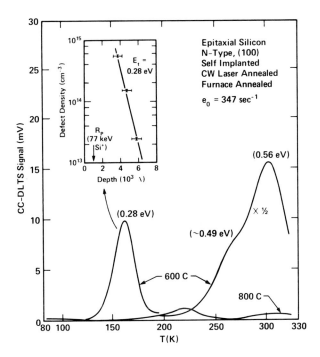

FIGURE 4.21. Electron emission spectra for epitaxial silicon which was furnace annealed after self implantation and cw laser annealing. The inset shows the defect distribution for the level $E_C - 0.28$ eV in the specimen annealed at 600°C.

which were first self-implanted and laser annealed and then furnace annealed at 600°C and 800°C before metallization. In specimens annealed at 600°C, the emission spectrum is dominated by the same three levels shown previously in Fig. 4.4. Except for the level at ~0.49 eV there is little change in the trap concentrations after a 600°C anneal. This level has significantly decreased in density relative to the others, which suggests a low dissociation energy for the corresponding (unidentified) defect complex. After an 800°C anneal all defect densities are significantly reduced. The inset in Fig. 4.21 shows the spatial distribution of the electron trap at 0.28 eV after the 600°C anneal. The profile was measured by the double-correlation CC-DLTS technique [4.2], and the horizontal bars mark the spatial intervals over which the local average defect concentrations were measured. Also shown in the inset is the projected range of $^{28}Si^+$ [4.5], which approximates the projected range of silicon for the actually implanted species, SiH^+. The defect density decreases rapidly with depth, again illustrating that the residual defects are localized near the recrystallized surface but extend into material not driven amorphous by ion implantation.

Pulsed-laser annealing studies offer further examples of high-temperature furnace annealing of residual defects. A capacitance-transient spectrum typical of the defect states introduced by melting silicon with either a 1.06 μm or 0.53 μm Nd:YAG laser is shown in Fig. 4.22 [4.30]. The two major peaks labeled A and B represent states in the silicon bandgap at 0.19 eV and 0.33 eV, respectively, below the conduction-band minimum. The relative concentrations of these defects were found to vary in different samples, but the same defect states were consistently observed within a surface layer of thickness equal to the depth of melting. A density maximum was observed for the 0.33 eV level at high laser powers which suggested that it is quenched-in during the pulsed laser processing. It was also suggested that the 0.19 eV level may be due to sodium in-diffusion from the surface [4.31]. The stability of these prominant defect states was examined with a series of 20 min isochronal anneals. Defect state concentration (averaged over a depth of 0.7 μm) as a function of annealing temperature in a hydrogen atmosphere is plotted in Fig. 4.23. Annealing begins at 400°C and is complete by 700°C. The results illustrate that conventional furnace anneals of 500-700°C can remove the dominant residual defects after transient thermal processing.

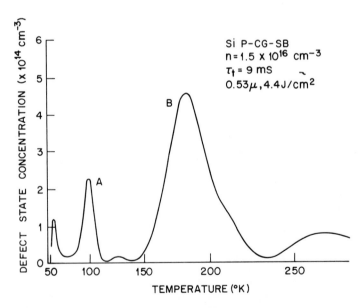

FIGURE 4.22. Capacitance transient spectrum typical of n-type silicon melted with a Nd:Yag laser at wavelengths of 1.06 μm or 0.53 μm. Defect state A is identified as E(0.19 eV), and B is E(0.33 eV). [J. L. Benton et al., Ref. 4.30.]

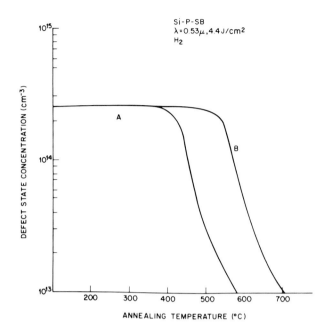

FIGURE 4.23. Schematic representation of 20 min, isochronal annealing in hydrogen ambient of the A and B defect states. [J. L. Benton et al., Ref. 4.30.]

4.6.2 Hydrogenation

Hydrogen passivation of pulsed-laser induced defects is illustrated in Fig. 4.24 [4.32]. Spectrum (a) shows post-recrystallization defect concentrations of $\sim 10^{14}$ cm^{-3}. Superimposed on the DLTS spectrum is a thermally stimulated capacitance (TSCAP) scan. The steps in capacitance indicate a high concentration of defects and appear when the thermal energy is sufficient to empty the traps. The increase in capacitance over the original substrate value of 50 pF indicates an increase in carrier concentration corresponding to the introduction of donors by the laser processing. There is no significant change in these measurements after a 4 hr anneal at 300°C in molecular hydrogen [Fig. 4.24(b)]. In contrast, monatomic hydrogen neutralizes the electrically active defects at low temperatures. The DLTS spectrum in Fig. 4.24(c) is featureless and the TSCAP scan is constant and equal to the bulk-silicon value. Both measurements indicate that the hydrogen plasma passivates the quenched-in recombination centers. Equivalent results were obtained with plasma anneals at 100°C and 300°C and times of 10 min to 4 hrs. If the defect reaction

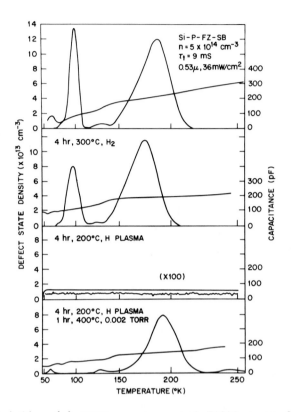

FIGURE 4.24. (a) DLTS spectrum and TSCAP scan for laser-melted phosphorus-doped crystalline silicon. (b) DLTS spectrum and TSCAP scan of laser-melted silicon taken after a subsequent 4 hr, 300°C anneal in H_2 showing no change in defect-state density. (c) DLTS and TSCAP spectra showing passivation of electrically active defects after treatment in a 0.38 Torr hydrogen plasma at 200°C for 4 hr. (d) DLTS and TSCAP spectra produced following hydrogen plasma treatment and a subsequent hydrogen evolution, .002 Torr vacuum anneal at 400°C for 1 hr. [J. L. Benton et al., Ref. 4.32.]

is limited by hydrogen in-diffusion, it should be effectively instantaneous at these temperatures for a depth of 1 μm. A final experiment to evolve the hydrogen after plasma treatment is shown in Fig. 4.24(d). The reappearance of electrically active defects after a vacuum anneal at 400°C for 1 hr at .002 Torr is evident. The data again reveal defect densities of $\sim 10^{14}$ cm^{-3}. The changes in the spectral features relative to the pre-annealed state are not yet understood. However, the results demonstrate that hydrogen passivation can be used to neutralize residual electronic defects at temperatures of 100–300°C.

4.7 BEAM-CRYSTALLIZED SILICON THIN FILMS

As reviewed in Chapter 5, crystallization techniques such as laser melting are used to increase the grain size of deposited silicon films to dimensions that are comparable to or greater than those of the active device, thereby minimizing the detrimental effects of grain boundaries on device operation. Residual grain and subgrain boundaries intersecting metallurgical p-n junctions in beam-crystallized silicon films have been shown to provide efficient paths for dopant diffusion during silicon processing [4.33,4.34], and in thin-film transistors grain boundaries have been shown to reduce the channel mobility [4.35-4.38]. However, crystallization techniques currently under development should permit large-scale device fabrication in single-crystal silicon thin films (see, for example, Refs. 4.38-4.44).

In this section results are presented from investigations of residual electronic defects in beam-crystallized silicon films on amorphous substrates. To date three studies have been reported which specifically focus on the characterization of electronic defects in crystallized silicon thin films [4.45-4.47]. All three used the metal-oxide-silicon structure and concluded that the Si-SiO$_2$ interface is the dominant source of residual electronic defects in this material. Deep levels at this interface are generally found to be continuously distributed in energy. Since this feature has not been encountered in previous sections, the measurement of continuous deep-level distributions is introduced in this section and the arguments underlying the above conclusion are presented.

4.7.1 Cw Ar-Ion Laser Crystallized Films

Crystallized-silicon thin films are processed on insulating substrates and generally contain residual grain boundaries. These features introduce unique requirements for electronic defect evaluation. The test device must necessarily be a thin-film structure, and the semiconductor-insulator interface will generally be an integral part of the device. The conventional current-rectifying devices (i.e., Schottky-barrier and p-n junction diodes) used in previous sections are not as readily implemented in thin semiconducting films because lateral placement of electrical contacts generally results in a non-negligible series resistance [4.48]. This effect can be minimized through proper selection of device geometry and semiconductor conductivity as discussed in Section 4.7.3. A unique approach compatible with beam crystallization is shown in Fig. 4.25. Since the semiconducting layer can be processed on an insulating thin film, a metal-oxide-semiconductor (MOS) structure can be fabricated which possesses negligible

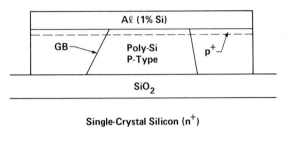

FIGURE 4.25. Schematic cross-section of an inverted MOS capacitor consisting of a beam-crystallized silicon thin film on thermally grown silicon dioxide over single-crystal silicon.

series resistance. Such a configuration has been used for capacitance-voltage (C-V) measurements of cw laser-crystallized silicon thin films [4.49].

The MOS structure in Fig. 4.25 has been used for DLTS evaluation of defects in cw Ar-ion laser-crystallized silicon films on oxidized bulk silicon [4.45]. The device is an inverted MOS capacitor consisting of a silicon thin film deposited and crystallized on a thermally grown layer of silicon dioxide. The single-crystal silicon substrate was degenerately doped for n^+ conductivity (resistivity $\leqslant$ .003 Ω cm) and provided the gate electrode in this inverted structure. For device fabrication, 0.65 μm of polycrystalline silicon was deposited by low-pressure chemical vapor deposition onto a 200 nm-thick thermal oxide. Next, the silicon film was implanted with boron at 100 keV to a dose of 1×10^{12} cm^{-2}. The continuous film was then encapsulated with 30 nm of silicon nitride, also by low-pressure chemical vapor deposition, in preparation for laser processing. Crystallization was performed with a scanned cw Ar-ion laser with the substrate heated to 500°C; typical beam conditions were as follows: laser beam power 5 W, beam diameter 80 μm, scan speed 50 cm/s, and scan line separation 20 μm. After crystallization, the silicon-nitride layer was removed (with phosphoric acid at 180°C). Next, the silicon layer was implanted with boron (10 keV, 5×10^{14} cm^{-2}) and furnace annealed (900°C, 30 min) to form a shallow p^+ layer for ohmic contact to the "backside" of the semiconducting film. The MOS capacitors were completed by vacuum depositing an Al-Si alloy (1% Si) over the crystallized silicon film, photolithographically patterning the metal layer to define back electrodes (500 μm dia.), plasma etching the exposed silicon film to form the mesa structure depicted in Fig. 4.25, and finally sintering the devices at 450°C in forming gas (15% H$_2$ and 85% N$_2$) for 20 min. The high-dose boron implant was found to be essential in order to maintain an ohmic back contact over the wide temperature range required for DLTS, and the Al-Si alloy was used to minimize contact alloying during sintering.

A crystallized-silicon layer as shown in Fig. 4.25 offers three distinct sources of electronic defect: (1) point and extended defects within grains, (2) defects at grain boundaries, and (3) defects at the semiconductor-insulator interface. Only the first of these is generally encountered in recrystallized bulk silicon. The latter two differ from the first in that the deep levels are spatially localized and generally distributed continuously in energy in the silicon bandgap. Continuous deep-level distributions have been demonstrated for grain boundaries in bulk polycrystalline silicon [4.50-4.51] and are generally found at the Si-SiO$_2$ interface on bulk single-crystal silicon [4.52]. In general, the above three souces of electronic defect will be detected in a DLTS measurement. As illustrated in previous sections, point defects in single-crystal silicon contribute emission peaks in a DLTS spectrum. On the other hand, a continuous deep-level distribution will produce a nonzero DLTS signal over a wide range of temperatures, corresponding to carrier emission from the distributed deep levels, which results in a smoothly varying, often featureless, DLTS spectrum. In general, the above three sources of emitted carriers will be superimposed in a given DLTS spectrum.

In Fig. 4.26 is shown a TEM micrograph of a cw laser-crystallized silicon layer in an inverted MOS capacitor. As is typical for cw laser crystallization of continuous silicon films, the grains are elongated in the direction of the laser scan with widths in the range of 1 to 5μ, [4.53,4.54].

In Fig. 4.27 is shown a CC-DLTS spectrum for the p-type laser-crystallized silicon film. The spectrum was obtained by biasing the capacitor in depletion and periodically pulsing the gate voltage into strong accumulation in order to populate deep levels throughout the film with majority carriers (in this case, holes). The emission signal over the entire investigated temperature range was found to be essentially saturated with respect to voltage-pulse amplitude. Failure to achieve saturation could result from partial band bending in the vicinity of grain boundaries during pulse biasing due to high defect densities. In this regard the test structure shown in Fig. 4.25 is nonideal for characterizing grain boundaries since they extend through the depletion layer of the silicon thin film and therefore do not experience a uniform trap filling condition. Saturation of the emission signal suggests that the grain boundaries are not the dominant contribution of emitted holes. In Fig. 4.27 the emission signal is nonzero from the lowest investigated temperature to above room temperature; the upper end of the temperature range generally corresponds to emission of holes from defect

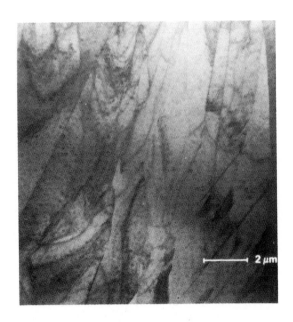

FIGURE 4.26.   TEM micrograph of a silicon thin film which
was crystallized with a scanned cw Ar-ion laser.

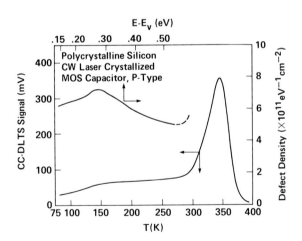

FIGURE 4.27.   CC-DLTS spectrum and defect distribution
obtained from an inverted MOS capacitor fabricated on a cw laser-
crystallized silicon thin film. The spectrometer delay times for
hole-emission detection were t_1 = 2 ms and t_2 = 4 ms. The defect
distribution was obtained with an assumed capture cross-section
for holes of 1×10^{-14} cm^2.

levels near midgap. Such spectra are representative of carrier emission from deep levels which are continuously distributed in energy. Such defect distributions exist at insulator-semiconductor interfaces and have been extensively investigated in MOS devices [4.55]. These studies have shown that in an MOS capacitor a high-temperature emission peak, as displayed in Fig. 4.27 and corresponding to emission from states near midgap, need not arise from a discrete energy level but rather can result from the onset of minority-carrier emission (in the present case, electrons) through a slowly varying continuous distribution of states at the semiconductor midgap. Thus, over the entire temperature interval there is no definite evidence of carrier emission from discrete levels. It was therefore proposed from the above considerations that defect levels at the $Si-SiO_2$ interface are the dominant source of emission centers in the crystallized silicon thin film, with possible secondary contributions from grain boundaries and defects within grains. This issue is discussed in more detail in Section 4.7.3.

For the purpose of numerically estimating the defect density, if it is assumed that interface states are the sole contribution to the emission signal, the spectrum in Fig. 4.27 can be analyzed directly to obtain the defect distribution, which is shown in the same figure. The high-temperature peak in the emission spectrum is not included in the defect distribution because the analysis assumes that majority-carrier emission dominates the DLTS transient response [4.55]. The energy scale was calculated with an assumed value of $1x10^{-14}$ cm^{-2} for the hole capture cross section, which is based on similar measurements of interface states in MOS capacitors on p-type bulk single-crystal silicon [4.56]. The defect density varies slowly in the mid-10^{11} eV^{-1} cm^{-2} range, with a slight peak in the distribution at approximately 0.3 eV above the silicon valence-band maximum. Against such a high background signal from interface defects, the density of a spatially uniform discrete level would have to be $\geqslant 1x10^{14}$ cm^{-3} to be detectable.

4.7.2   Strip-Heater Crystallized Silicon Films

Residual electronic defects have also been investigated in graphite strip-heater crystallized silicon thin films [4.46]. For electrical characterization inverted MOS capacitors were fabricated as described above. High-frequency capacitance-voltage characteristics yielded effective fixed charge densities in the oxide of $\leqslant 2x10^{11}$ cm^{-2}. Trap-emission spectra, recorded with CC-DLTS on both p-type and n-type capacitors, indicated a continuous distribution of deep levels throughout the silicon bandgap. Since the films were essentially single crystal with a low density of subgrain boundaries [4.40,4.41], the $Si-SiO_2$ interface was

considered to be the principal source of this deep-level continuum. The effective interface-state density was found to
be $< 2.5 \times 10^{10}$ eV^{-1} cm^{-2} near midgap which closely approaches
the state-of-the-art for thermally oxidized bulk single-crystal
silicon. An electron emission peak with an activation energy
of ~ 0.35 eV was also detected above the background continuum
and may be the signature of a bulk defect with an equivalent
spatially uniform density of $\sim 1 \times 10^{13}$ cm^{-3}, although the defect
was not chemically or microscopically identified.

A contaminant which has been detected in graphite strip-
heater crystallized silicon is carbon [4.57]. It has been
reported [4.28] that this impurity when incorporated in the
silicon lattice as an interstitial contributes a hole-emission
deep level at 0.27 eV and as a carbon interstitial-carbon substitutional complex yields a hole-emission level at 0.36 eV. The
absence of any clearly defined peaks in DLTS hole-emission spectra
for graphite strip-heater crystallized silicon [4.46] as summarized above, argues against the presence of high concentrations
of electrically active carbon in such films.

4.7.3   Laser-Crystallized Silicon Films on Fused Silica

Residual electronic deep levels in fully processed cw laser-
crystallized silicon thin films on fused silica have been measured by transient capacitance spectroscopy, supplemented with
capacitance-voltage techniques [4.47]. The test devices were
MOS thin-film capacitors with p-type conductivity. A schematic
cross-section of the capacitors on fused silica is shown in Fig.
4.28. A scanning CO_2 laser was used to crystallize silicon films
(0.5 μm thick) on bulk fused silica substrates with techniques
previously shown to produce single-crystal material containing
low densities of linear defects in areas (termed "islands") that
were pre-selected for subsequent device fabrication [4.38, 4.42,
4.43, 4.58]. The dopant for the p^+ regions beneath the substrate
contacts was introduced by ion implantation (B^+, 1×10^{15} cm^{-2}, 50
keV) into the crystallized silicon islands prior to oxidation.
The islands were then oxidized in dry O_2 at 1000°C to grow a 130-
nm thick gate oxide; this also served to drive in and activate the
contact dopant. Dual ion implantation of boron (i.e., 2.8×10^{11}
cm^{-2} at 60 keV and 4×10^{12} cm^{-2} at 170 keV) followed by a 900°C
anneal (30 min in N_2) was used to dope the silicon film p-type
in the gate region with a nearly uniform boron profile in the
space charge layer immediately beneath the thermal oxide and a
high dopant density near the back surface of the film. This
dopant profile was essential to minimize the series resistance
of the silicon substrate in these laterally contacted devices.
After opening contact windows in the oxide, an Al-Si alloy (1 %
Si) was vacuum evaporated and photolithographically patterned

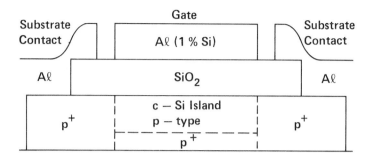

Fused Silica Substrate

FIGURE 4.28.   Schematic diagram of an MOS capacitor fabricated in a laser-crystallized silicon thin film on bulk fused silica.

to define gate and substrate contacts. The gate area was determined by an island width of 40 μm and a gate width of 50 μm. The laser-crystallization conditions were not optimized for this island width, which is larger than that used in the previous studies referenced above, and consequently the silicon films generally contained some residual subgrain boundaries and microtwins. The MOS capacitors were completed with a sinter in forming gas at 450°C for 30 min. The separate electrical contacts to the substrate on opposite sides of the gate electrode in Fig. 4.28 were used to verify the low resistance of the laterally contacted substrates and were commonly connected for capacitance measurements. From C-V measurements the dopant concentration was estimated to be 2×10^{16} cm^{-3}, which is in good agreement with the average near-surface boron concentration as determined from SUPREM III process simulations [4.59].

The test capacitors were evaluated with transient capacitance spectroscopy. Hole-emission DLTS spectra are shown in Fig. 4.29. The transient capacitance measurements were performed at fixed temperatures over the range from 80 to 300 K. At each temperature, the C-V characteristic was recorded and used to select the gate bias required to maintain the same equilibrium device capacitance in depletion, C_D, over the entire temperature range, and transient waveform averaging was required to attain a useful trap emission sensitivity. The gate-voltage adjustment was required to maintain a constant C_D since the C-V characteristic of an MOS device shifts with temperature due to the temperature dependence of the Fermi energy and the presence of interface states [4.52]. This shift is essentially parallel to the voltage axis and results in the high-frequency capacitance changing with

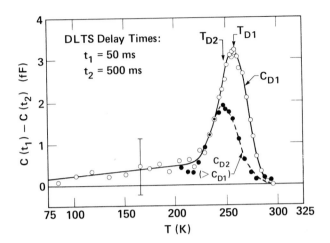

FIGURE 4.29.   DLTS spectra for a laser-crystallized thin-film silicon MOS capacitor. The spectra were recorded with two different equilibrium device capacitances in depletion, C_D. The error bar represents twice the standard deviation of the DLTS signal, which is formed as the difference between the average capacitances measured at delay times t_1 and t_2 after a trap-filling voltage pulse.

temperature if the depletion gate bias is held constant; this is avoided by holding C_D constant with temperature, which simplifies the interface-state analysis [4.55].

The two spectra in Fig. 4.29, recorded with different values of C_D, have a further significance. The emission peak at the high-temperature end of the DLTS spectra in Fig. 4.29 could be the signature of a discrete energy level situated near midgap, or it could arise from the continuous distribution of deep levels which is revealed by the smoothly varying, non-zero signal over the rest of the spectrum. As discussed previously, point defects in single-crystal silicon contribute emission peaks in a DLTS spectrum, while a continuum will produce a non-zero DLTS signal over a wide range of temperatures corresponding to carrier emission from the distributed deep levels. The dominant source of a continuous deep-level distribution in the present material is the Si–SiO$_2$ interface. In order to decide between bulk versus interface defects as the origin of the emission peak, the second spectrum was recorded with its equilibrium device capacitance in depletion, C_{D2}, chosen such that $C_{D2} > C_{D1}$. As seen in Fig. 4.29, the results are (1) a lower peak height and (2) the peak shifts to a <u>lower</u> temperature (i.e., $T_{D2} < T_{D1}$). Increasing C_D reduces the magnitude of the silicon surface potential with corresponding decreases in the width of the silicon space-charge layer and in the magnitude of the electric-field intensity at every point within this depletion layer [4.60]. If the emission peak arose from a discrete level spatially distributed in the bulk of the silicon, then the reduced depletion width would mean fewer defect centers contributing to the emission signal and consequently a lower peak height, as is observed. However, this would not cause the temperature of the emission peak to shift.

For hole emission from a bulk discrete level, the change in the electric field would have to mediate the shift of the peak temperature when C_D is changed. This follows from the fact that the trap filling conditions were identical for the two spectra and from the experimental determination that the DLTS signals which comprise the spectra were saturated with respect to both gate-voltage pulse amplitude and pulse width. Thus, it is the emission cycle, rather than the capture cycle, that is being affected by changing C_D. For hole emission from a discrete level, the peak appears at a temperature T_p which is related to the DLTS delay times and trap parameters as follows:

$$e_p(T_p) \equiv \gamma_p \exp[-E_A/(kT_p)] = (t_2-t_1)^{-1} \ln(t_2/t_1), \qquad (4.3)$$

where e_p^{-1} is the hole emission time, γ_p is the attempt-to-escape frequency for trapped holes, E_A is the activation energy for thermal emission, and k is the Boltzmann constant. Since the right-hand side of Eq. (4.3), which is commonly termed the DLTS emission rate window (as discussed in Section 4.2.2), is a constant, a change in T_p must be compensated by a change in E_A and/or γ_p. Furthermore, since T_p depends logarithmically on γ_p, a strong functional dependence of γ_p on field would be required, which is not generally observed for discrete deep levels. However, electric-field modulation of the activation energy is a well-documented phenomenon. Specifically, Poole-Frenkel emission from a neutral center, which for hole emission would be an acceptor-like defect, yields the following field-dependent activation energy (for a one-dimensional coulombic defect potential) [4.61]:

$$E_A = E_0 - q \sqrt{[qF/(\pi \varepsilon_s)]}, \qquad (4.4)$$

where F is the magnitude of the electric field, E_0 is the zero-field activation energy, q is electronic charge, and ε_s is the silicon permittivity. (A defect center that is neutral when empty and charged when occupied will not experience this effect, as discussed in Section 4.3.2.) Equation (4.4) predicts that the activation energy increases with decreasing electric field. From Eq. (4.3), this requires that the peak temperature increase, rather than decrease, with an increase in C_D. Another mechanism which would yield the same qualitative conclusion is high-field tunnel emission which has been documented for several point defects in III-V semiconductors (see, for example, Ref. 4.62). Thus, the experimental results are inconsistent with field-modulated hole emission from a discrete level.

The dependence on C_D of the high-temperature emission peak in Fig. 4.29 is consistent with the effect of surface generation on the DLTS measurement of interface states near the silicon mid-gap [4.55]. In this case, at temperatures corresponding to the DLTS detection of majority-carrier emission (i.e., holes) from interface states near midgap, minority-carrier emission from those states becomes a competitive process and can introduce a peak in a DLTS spectrum due to surface generation of charge through a slowly varying continuous deep-level distribution. The dependence on C_D is due to the change of the surface potential with gate voltage. The surface potential determines the intersection of the Fermi level with the interface which specifies the equilibrium occupation of the interface states; states above the Fermi level do not change occupancy during the capacitance transient (in p-type capacitors) and therefore cannot contribute to the DLTS signal. Reducing the magnitude of the surface potential, by increasing the equilibrium depletion capacitance, pinches off the surface generation process which both attenuates the emission peak and shifts it to lower temperatures, as displayed in Fig. 4.29.

Based on the arguments presented above, the entire DLTS spectrum in Fig. 4.29 may be ascribed to a continuous distribution of deep levels at the $Si-SiO_2$ interface. Then the interface-state density can be estimated with a standard analysis from the DLTS signal at temperatures below the emission peak where hole emission dominates [4.55]. For example, the DLTS signal just below the peak (i.e., near 210 K) yields a value of approximately 6×10^{10} eV^{-1} cm^{-2} for the density of interface states at energies near the silicon midgap. Against this background signal from interface states, the density of a spatially uniform discrete level would have to be $> 1 \times 10^{14}$ cm^{-3} to be detectable in the capacitors examined in this study. To conclude, the results of this study provide a defect-specific verification of the device-grade quality of CO_2 laser-crystallized silicon thin films on fused silica.

4.8   SUMMARY AND CONCLUSIONS

From the experimental studies reviewed in Sections 4.3-4.6, the following general observations may be drawn regarding electronic defects in cw transient thermal annealed bulk single-crystal silicon:

1.   Residual electronic defects remain after transient thermal annealing with all investigated forms of energy sources. The defect densities are in excess of those obtained by conventional furnace annealing.

2. These defects are located in the near-surface region. In recrystallized silicon the defects extend with decreasing density into material not driven amorphous by ion implantation.

3. Even in material which has not been ion implanted, electronic defects are quenched in during transient thermal annealing.

4. Beam-annealing conditions (e.g., beam power and scan line separation) can be optimized for recrystallizing implanted amorphous layers with a minimum density of residual defects. However, the use of a raster-scanned point source to anneal implantation damage seems always to be accompanied by the production of laterally-distributed residual defects.

5. Laterally nonuniform recrystallization can also result from wafer topography (e.g., patterned dielectric overlayers), which introduces lateral variations in energy deposition and heat dissipation during beam recrystallization.

6. Post-recrystallization processing can be used to reduce the density of residual defects. Furnace anneals in the range of 400-700°C and hydrogen passivation at temperatures of 100-300°C have been shown to be effective in removing deep levels.

From the standpoint of electronic defects, of the several cw transient thermal processing techniques that have now been investigated, isothermal transient annealing is the most promising as a replacement for conventional furnace annealing in silicon integrated-circuit technology. Of course, this technique most closely approaches a furnace anneal in its thermal treatment of a silicon wafer. This is further reflected in the similarity of the deep levels which result from these two annealing procedures and in the convergence of their residual defect densities as the quench rate in isothermal transient annealing is reduced. On the other hand, directed energy sources, such as scanned laser and electron beams, may offer unique advantages for materials modifications other than recrystallization as well as provide a controlled means of introducing defects in silicon for fundamental studies.

For beam-crystallized silicon thin films on amorphous substrates, the results presented in Section 4.7 indicate that the $Si-SiO_2$ interface is the dominant source of residual electronic defects in this materials system. Since the measurements were performed on MOS devices in which the interface was a source of

detectable defects, interface states established a high background
signal for detecting defects in the bulk of the silicon thin
films. This background required that the density of a spatially
uniform bulk defect be greater than typically 1×10^{14} cm^{-3} to be
detectable. This effective sensitivity to bulk defects is rather
low for studies directed toward the correlation of residual
defects with materials processing and device operation. Novel
deep-level spectroscopic techniques can be combined with multiple
terminal MOS devices (e.g., transistors) to achieve high sensitivity for bulk defect detection, in part by excluding any contribution from the interface [4.63]. It may be anticipated that
such techniques and devices will be extended to electronic-defect
characterization in beam-crystallized silicon thin films.

REFERENCES

4.1   Lang, D. V., J. Appl. Phys. 45, 3014 and 3023 (1974).
4.2   Johnson, N. M., Bartelink, D. J., Gold, R. B., and J. F.
       Gibbons, J. Appl. Phys. 50, (1979).
4.3   Lefevre, H. and Schulz, H., Appl. Phys. 12, 45 (1977).
4.4   Johnson, N. M., Gold, R. B., and J. F. Gibbons, Appl. Phys.
       Lett. 34, 704 (1979).
4.5   Gibbons, J. F., Johnson, W. S., and Mylroie, W. W., "Pro-
       jected Range Statistics, Semiconductors and Related Mat-
       erials," 2nd ed. Halstead, New York (1975).
4.6   Johnson, N. M., Gold, R. B., Lietoila, A., and Gibbons,
       J. F., "Laser-Solid Interactions and Laser Processing -
       1978," (S. D. Ferris, H. J. Leamy, and J. M. Poate, eds.)
       AIP Conference Proceedings, pp. 550-555. American Insti-
       tute of Physics, New York (1979).
4.7   Regolini, J. F., Gibbons, J. F., Sigmon, T. W., Pease, R. F.
       W., Magee, T. J., and Peng, J., Appl. Phys. Lett. 34, 410
       (1979).
4.8   Ratnakumar, K. N., Pease, R. F. W., Bartelink, D. J.,
       Johnson, N. M., and Meindl, J. D., Appl. Phys. Lett. 35,
       463 (1979).
4.9   Johnson, N. M., Regolini, J. L., Bartelink, D. J., Gibbons,
       J. F., and Ratnakumar, K. N., Appl. Phys. Lett. 36, 425
       (1980).
4.10   Sheng, N. M. and Merz, J. L., "Laser and Electron Beam In-
       teractions with Solids," (B. R. Appleton and G. K. Celler,
       eds.) pp. 313-318. Elsevier, New York (1982).
4.11   Sheng, N. H., Mizuta, M., and Merz, J. L., Appl. Phys.
       Lett. 40, 68 (1982).
4.12   Sheng, N. H., Mizuta, M., and Merz, J. L., "Laser and
       Electron-Beam Solid Interactions and Materials Process-
       ing," (J. F. Gibbons, L. D. Hess, and T. W. Sigmon, eds.)
       pp. 155-162. Elsevier, New York (1981).
4.13   Chantre, A., Kechouane, M., and Bois, D., "Laser and Elec-
       tron-Beam Interactions with Solids," (B. R. Appleton and G.
       K. Celler, eds.) pp. 325-330. Elsevier, New York (1982).
4.14   Chantre, A., Kechouane, and Boise, D., "Defects in Semi-
       conductors II," (S. Mahajan and J. W. Corbett, eds.), pp.
       547-551. Elsevier, New York (1983).
4.15   Kimerling, L. C., IEEE Trans. Nucl. Sci. NS-23, 1497
       (1976).
4.16   Uebbing, R. H., Wagner, P., Baumgart, H., and Queisser, H.
       J., Appl. Phys. Lett. 37, 1078 (1980).
4.17   Mizuta, M., Sheng, N. H., Merz, J. L., Lietoila, A., Gold,
       R. B., and Gibbons, J. F., Appl. Phys. Lett. 37, 154 (1980).
4.18   Kimerling, L. C. and Patel, Appl. Phys. Lett. 34, 73 (1979).
4.19   Mizuta, M., Sheng, N. H., and Merz, J. L., Appl. Phys.
       Lett. 38, 453 (1981).

4.20   Street, R. A., Johnson, N. M., and Lietoiola, A., "Laser and Electron-Beam Processing of Materials," (C. W. White and P. S. Peercy, eds.) p. 435. Academic Press, New York (1980).

4.21   Street, R. A., Johnson, N. M., and Gibbons, J. F., J. Appl. Phys. 50, 8201 (1979).

4.22   Rozgonyi, G. A., Baumgart, H., and Phillipp, F., "Laser and Electron-Beam Solid Interactions and Materials Processing," (J. F. Gibbons, L. D. Hess, and T. W. Sigmon, eds.), pp. 193-199. Elsevier, New York (1981).

4.23   Baumgart, H., Hildebrand, O., Phillipp, F., and Rozgonyi, G. A., "Proceedings of 2nd Oxford Conference on Microscopy of Semiconductor Materials." Institute of Physics, London (1981).

4.24   Baumgart, H., Phillipp, F., and Leamy, H. J., "Laser and Electron-Beam Interactions with Solids," (B. R. Appleton and G. K. Celler) pp. 355-360. Elsevier, New York (1982).

4.25   Johnson, N. M., Sinclair, R., and Moyer, M. C., IEEE Trans. Elec. Dev. ED-27, 2199 (1980).

4.26   Benton, J. L., Celler, G. K., Jacobson, A. C., Kimerling, L. C., Lischner, D. J., Miller, G. L., and Robinson, McC., "Laser and Electron-Beam Interactions with Solids," (B. R. Appleton and G. K. Celler, eds.) pp 765-770. Elsevier, New York (1982).

4.27   Pensl, G., Schulz, M., Johnson, N. M., Gibbons, J. F., and Hoyt, J., "Energy Beam-Solid Interactions and Transient Thermal Processing," (J. C. C. Fan and N. M. Johnson, eds.) in press. Elsevier, New York (1984).

4.28   Kimerling, L. C., "Radiation Effects in Semiconductors 1979." Inst. Phys. Conf. Ser. 31, Bristol (1977).

4.29   Johnson, N. M., Bartellink, D. J., Moyer, M. D., Gibbons, J. F., Lietoila, A., Ratnakumar, K. N., and Regolini, J. L., "Laser and Electron-Beam Processing of Materials," (C. W. White and P. S. Peercy, eds.) pp. 423-429. Academic Press, New York (1980).

4.30   Benton, J. L., Doherty, C. J., Ferris, S. D., Kimerling, L. C., Leamy, H. J., and Celler, G. K., "Laser and Electron-Beam Processing of Materials," (C. W. White and P. S. Peercy, eds.) pp. 430-434. Academic Press, New York (1980).

4.31   Kimerling, L. C. and Benton, J. L., "Laser and Electron-Beam Processing of Materials," (C. W. White and P. S. Peercy, eds.) pp. 385-396. Academic Press, New York (1980).

4.32   Benton, J. L., Doherty, C. J., Ferris, S. D., Flamm, D. L., Kimerling, L. C., and Leamy, H. J., Appl. Phys. Lett. 36, 670 (1980).

4.33   Johnson, N. M., Biegelsen, D. K., and Moyer, M. D., Appl. Phys. Lett. 38, 900 (1981).

4.34   Maby, E. W., Atwater, H. A., Keigler, A. L., and Johnson, N. M., Appl. Phys. Lett. 43, 482 (1983).

4.35   Kamins, T. I. and Von Herzen, B. P., IEEE Elec. Dev. Lett.
        EDL-2, 313 (1981).

4.36   Ng, K. K., Celler, G. K., Povilonis, E. I., Frye, R. C.,
        Leamy, H. J., and Sze, S. M., IEEE Elec. Dev. Lett. EDL-2,
        316, (1981).

4.37   Frye, R. C. and Ng, K. K., "Grain Boundaries in Semicon-
        ductors," (C. H. Seager, G. E. Pike, and H. J. Leamy, eds.)
        pp. 275-286. Elsevier, New York (1982).

4.38   Biegelsen, D. K., Johnson, N. M., Nemanich, R. J., Moyer,
        M. D., and Fennel, L. E., "Laser and Electron-Beam Inter-
        actions with Solids," (B. R. Appleton and G. K. Celler,
        eds.) pp. 331-336. Elsevier, New York (1982).

4.39   Fan, J. C. C., Geis, M. W., and Tsaur, B.-Y., Appl. Phys.
        Lett. 38, 365 (1981).

4.40   Maby, E. W., Geis, M. W., LeCoz, L. Y., Silversmith, D. J.
        Mountain, R. W., and Antoniadis, D. A., IEEE Elec. Dev.
        Lett. EDL-2, 241 (1981).

4.41   Geis, M. W., Smith, H. I., Tsaur, B.-Y., Fan, J. C. C.,
        Maby, E. W., and Antoniadis, D. A., Appl. Phys. Lett. 40,
        158 (1981).

4.42   Hawkins, W. G., Black, J. G., and Griffiths, C. H., Appl.
        Phys. Lett. 40, 319 (1982).

4.43   Fennell, L. E., Moyer, M. D., Biegelsen, D. K., Chiang, A.,
        and Johnson, N. M., "Energy Beam-Solid Interactions and
        Transient Thermal Processing," (J. C. C. Fan and N. M.
        Johnson, eds.). Elsevier, New York (1984) in press.

4.44   Atwater, H. A., Thompson, C. V., Smith, H. I., and Geis,
        M. W., Appl. Phys. Lett. 43, 112 (1983).

4.45   Johnson, N. M., Moyer, M. D., and Fennell, L. E., Appl.
        Phys. Lett. 41, 562 (1982).

4.46   Johnson, N. M., Moyer, M. D., Fennell, L. E., Maby, E. W.,
        and Atwater, H., "Laser-Solid Interactions and Transient
        Thermal Processing of Materials," (J. Narayan, W. L. Brown,
        and R. A. Lemons, eds.) pp. 491-497. Elsevier, New York
        (1983).

4.47   Johnson, N. M. and Moyer, M. D., "Comparison of Thin Film
        Transistors and SOI Technologies," (H. W. Lam and M. J.
        Thompson, eds.) Elsevier, New York (1984) in press.

4.48   Wiley, J. D. and Miller, G. L., IEEE Trans. Elec. Dev.
        EDL-22, 265 (1975).

4.49   Kamins, T. I., Lee, K. F., and Gibbons, J. F., IEEE Elec.
        Dev. Lett. EDL-1, 5 (1980).

4.50   Seager, C. H., Pike, G. E., and Ginley, D. S., Phys. Rev.
        Lett. 43, 532 (1979).

4.51   Werner, J., Jantsch, W., Froehner, K. H., and Quiesser, H.
        J., "Grain Boundaries in Semiconductors," (C. H. Seager,
        G. E. Pike, and H. J. Leamy, eds.). Elsevier, New York
        (1982).

4.52   Nicollian, E. H. and Brews, J. R., "MOS Physics and Tech-
        nology." Wiley, New York (1982).

4.53   Gat, A., Gerzberg, L., Gibbons, J. F., Magee, T. J., Peng, and Hong, J. D., Appl. Phys. Lett. 33, 775 (1978).

4.54   Johnson, N. M., Biegelsen, D. K., and Moyer, M. D., "Laser and Electron-Beam Solid Interactions and Materials Processing," (J. F. Gibbons, L. D. Hess, and T. W. Sigmon, eds.) pp. 463–470. Elsevier, New York (1981).

4.55   Johnson, N. M., J. Vac. Sci. Techn. 21, 303 (1982).

4.56   Johnson, N. M., Biegelsen, D. K., and Moyer, M. D., J. Vac. Sci. Techn. 19, 390 (1981).

4.57   Pinizzotto, R. F., Lam, H. W., and Vaandrager, B. L., Appl. Phys. Lett. 40, 388 (1982).

4.58   Johnson, N. M., Biegelsen, D. K., Tuan, H. D., Moyer, M. D., and Fennell, L. E., IEEE Elec. Dev. Lett. EDL-3, 369 (1982).

4.59   Chiang, A., Zarzycki, M. H., Meuli, W. P., and Johnson, N. M., "Energy Beam-Solid Interactions and Transient Thermal Processing," (J. C. C. Fan and N. M. Johnson, eds.). Elsevier, New York (1984).

4.60   Sze, S. M., "Physics of Semiconductor Devices," 2nd ed. Chap. 7. Wiley, New York (1981).

4.61   Milnes, A. G., "Deep Impurities in Semiconductors," Chap. 5. Wiley, New York (1973).

4.62   Makram-Ebeid, S., "Defects in Semiconductors," (J. Narayan, and T. Y. Tan, eds.) pp. 495–501. Elsevier, New York (1981).

4.63   Collet, M. G., Solid-State Elec. 18, 1077 (1975).

CHAPTER 5

Beam Recrystallized Polycrystalline Silicon: Properties, Applications, and Techniques

K. F. Lee*, T. J. Stultz†, and James F. Gibbons

STANFORD ELECTRONICS LABORATORIES
STANFORD UNIVERSITY
STANFORD, CALIFORNIA

Heavily doped polycrystalline silicon (polysilicon) is a material that is widely used in present day silicon integrated circuit technology for gates and interconnection lines in MOS integrated circuits. Initial interest in the beam annealing of polysilicon arose because of its potential as a process that could be used to obtain resistivity reduction in as–deposited films. Experiments undertaken to explore that potential are described in Sec. 5.2. These experiments showed that beam-recrystallized polysilicon films have electronic properties that closely approximate those of single crystal material. This result ultimately led to successful attempts to fabricate MOS transistors and integrated circuits directly in beam recrystal-lized polysilicon [5.1-5.4]. A substantial interest now exists in the potential of beam recrystallized silicon-on-insulators (SOI) as a substrate for integrated circuit fabrication.

Our development of this subject will begin with a discussion in Sec. 5.1 of the mechanisms by which thin films of polycrystal-line Si are recrystallized by a scanning circular cw laser beam. In Sec. 5.2 we will consider the basic majority carrier electrical properties of the recrystallized material and establish the limits

---

*Present address: AT&T Bell Laboratories, Holmdel, New Jersey 07733.
†Present address: TS Associates, San Jose, California 95128.

on resistivity reduction that are possible with this process. In Sec. 5.3 we will describe the basic electronic properties of recrystallized films intended for use as MOSFET substrates, leading to a discussion of the electrical characteristics of devices fabricated directly in these films in Sec. 5.4. We conclude the chapter with a discussion of a variety of improvements in the recrystallization process that have been developed to eliminate various defects in device and film properties that are obtained when a circular beam is used for the recrystallization process.

5.1   RECRYSTALLIZATION OF THIN POLYCRYSTALLINE FILMS WITH A SCANNING CW LASER

In this section we describe first the annealing of thin polysilicon films irradiated by a cw argon laser, with emphasis on the relation between laser power and the surface morphology, crystal structure, dopant distribution and majority carrier electrical properties of the annealed films.

Three laser power levels (low, medium, and high) were identified in early studies directed toward understanding the recrystallization mechanisms in polysilicon films. The general effects are described in Fig. 5.1. It can be seen from this figure that complete dopant activation occurs at low power levels, similar to those leading to complete dopant activation in ion-implanted sin-

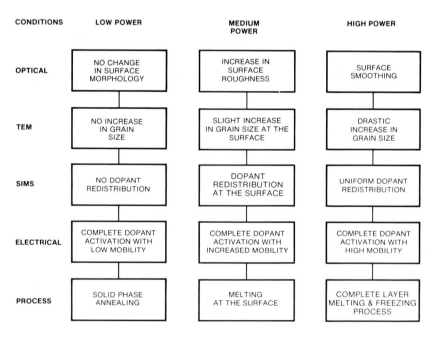

FIGURE 5.1. Summary of the effects of laser power on the annealing of polysilicon and the corresponding measurement methods.

gle-crystal silicon. A moderate increase in grain size occurs with dopant redistribution at somewhat higher irradiation intensities. In the medium power range, partial melting causes a growth of surface crystallites and significant dopant redistribution in the molten region. Annealing at the highest power levels results in complete melting of the (thin) polysilicon layer, with an attendant drastic increase in the grain size and uniformity of the doping distribution across the film. A sharp reduction in the electrical resistivity of the film is observed in this high power recrystallization stage. In what follows we provide experimental details on this latter process and also describe features that are observed where a single line scan is compared with an overlapped scan frame.

5.1.1 Sample Preparation

The polysilicon films that are typically used for these studies are 0.5-0.6 μm thick layers deposited by low pressure chemical vapor deposition (LPCVD) on a layer of thermally grown silicon dioxide In some cases the silicon dioxide layer is replaced with a silicon nitride layer. After deposition, the wafers may be doped by either ion implantation or thermal diffusion and then subjected to the beam processing operations.

For the laser processing experiments to be discussed below, a cw argon laser was operated in the multiline mode with a laser beam width on the sample surface of 40 μm. This circular spot was scanned across the surface at a rate of 12.5 cm/sec in all cases. The variables employed were substrate temperature, laser power, and the amount of overlap between adjacent scan lines. For reasons that are suggested in Chapter 2 it is advantageous to use subtantial substrate thermal bias (350°C) so that relatively small increments in laser power will produce substantial increments in the surface temperature of the irradiated film. For the experiments reported below, laser power was increased in 1 W increments from 5 W to 13 W, corresponding to a range of surface temperatures from 600° to 1500°C.

It should be emphasized that, while these conditions can produce very desirable changes in the properties of the films, they are certainly not the only set of conditions that would achieve these results; a different laser spot size, scan rate, laser power and substrate temperature can be chosen to yield very similar results.

5.1.2 Growth Mechanisms [5.5]

To study the mechanisms of crystallite growth, Gerzberg et al. [5.5] implanted arsenic and boron ions into separate samples to serve both as dopants and markers which could be used to investigate the recrystallization process. SEM was used to monitor changes in surface texture as a function of laser power (Fig. 5.2). For the arsenic implanted samples annealed at low and medium laser power levels (5 and 9 W respectively), no significant change in surface morphology was observed, as shown in Figs. 5.2(a) and (b). Above 10 W, however, distinct lines were apparent in the nonoverlapped sample with the microstructure inside the line pointing in the direction of the scan. Figure 5.2(c) is a SEM micrograph of an annealed line obtained in this high power region, showing the directionality in the line, the smoothing of the original fine surface roughness, and the formation of large surface structures.

The grain size obtained at the high power level was analyzed through transmission electron microscopy (TEM) and diffraction analysis. Samples were jet thinned from the back surface before examination. Figure 5.3 presents TEM micrographs of boron-implanted polysilicon samples irradiated with both single and overlapping lines. The long crystallites in the single line scan are characteristic of the high power region. These crystallites are found to be continuous, columnar structures extending from the SiO_2 interface to the surface of the film [5.6]. The difference between the single and overlapped scanning conditions will be explained below in terms of thermal conductivity variations.

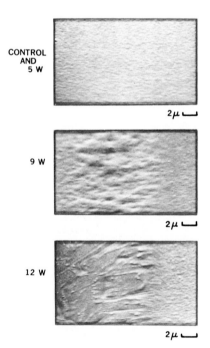

CONTROL AND 5 W

2μ

9 W

2μ

12 W

2μ

FIGURE 5.2. SEM photomicrographs of laser processed polysilicon at power levels of 5 W, 9 W and 12 W.

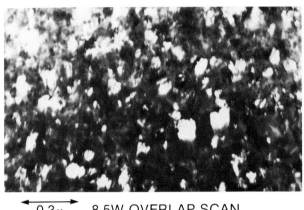

<--- 0.3μ --->   8.5W OVERLAP SCAN

<--- 3μ --->   8.0W SINGLE LINE

FIGURE 5.3. TEM photomicrographs of single and overlapping lines on boron-implanted polysilicon.

Figure 5.4 is a series of TEM photomicrographs of arsenic-implanted polysilicon, laser annealed at low-, medium-, and high-power levels. The as-deposited samples display fine grain structure with an average grain size of ~550 Å . Laser processing at 7 and 9 W produces grain growth but no significant change in surface morphology compared to the as-deposited samples. However, a sharp change is observed at power levels of 10 W and above, as evidenced by the 15 to 25 μm grain size in the sample annealed with 11 W laser power. The mechanism for this dramatic growth in crystallite size and its implications are discussed in Sec. 5.1.5.

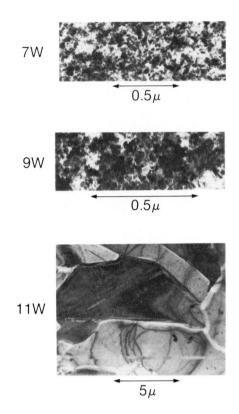

7W

0.5μ

9W

0.5μ

11W

5μ

FIGURE 5.4.   TEM photomicrographs of single and overlapping lines on arsenic–implanted polysilicon.

A global description of the recrystallization process can be gained from plots of the average grain size vs. laser power for the boron- and arsenic-implanted samples shown in Figs. 5.5 and 5.6. In Fig. 5.5, the grain size of the boron–implanted samples is seen to increase monotonically in the medium power range (6 to 8 W). The transition to the high power level occurs in the single line scan when the power is greater than 9 W. In a narrow power range from 8 to 9 W, grain size increases from 1 µm to 10 µm and, at a laser power level of 11 W, the grain size (length) increases to 20 µm.

Figure 5.6 is a plot of grain size for the arsenic-implanted polysilicon films after laser processing. As was the case for the boron-implanted samples, grain size increases monotonically with laser power from 6 to 9 W. The transition to the high power level occurs between 9 and 10 W, above which a 20 µm grain size is commonly observed.

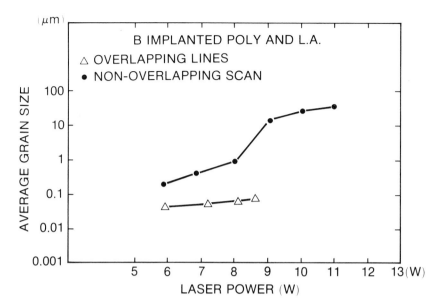

FIGURE 5.5. Average grain size of boron-implanted laser-annealed polysilicon as a function of annealing power levels for single and overlapping lines.

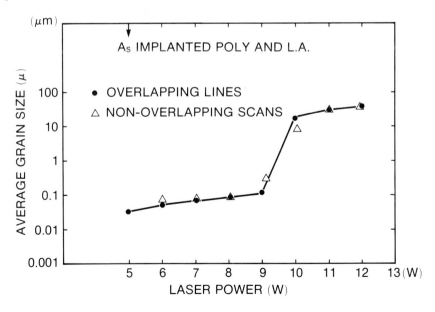

FIGURE 5.6. Average grain size of arsenic-implanted laser-annealed polysilicon as a function of annealing power levels for single and overlapping lines.

5.1.3 Dopant Distribution

Secondary ion-mass spectrometry (SIMS) was used to obtain dopant-profile measurements corresponding to the various power levels described in the previous section. Oxygen primary ion-bombardment and positive secondary ion-mass spectrometry (B^+ and As^+) were employed, using the CAMECA IMS-3f ion microanalyzer. The boron profiles were determined through standard depth profiling; however, sensitive measurements of the As profiles required secondary ion initial kinetic discrimination to reduce the amount of $(^{30}Si^{29}Si^{16}O)^+$ detected at the same nominal mass as $^{75}As^+$. This analysis was sufficient to provide an As detection limit of 3 to 5 $x10^{18}$ atoms/cm^3. As can be seen in Fig. 5.7, the high oxygen content of SiO_2 and minor surface-charging effects produce an anomolously high mass-75 intensity that misrepresent the true arsenic content in the oxide region. This problem, however, did not hinder the measurements in the polysilicon layer.

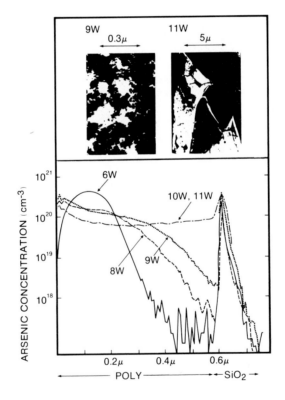

FIGURE 5.7. SIMS profile of arsenic-implanted polysilicon after laser annealing at different power levels. Inserts are the corresponding TEMS for 9 and 11 W.

As can be seen in Fig. 5.7, the dopants did not move from their as-implanted positions at a laser power of 6 W. The profile is approximately Gaussian and can be adequately predicted from Lindhard-Scharff-Schiott (LSS) range calculations [5.7]. Increased laser power results in deeper impurity penetration into the polysilicon.

When the power level was raised to 11 W, the dopant profiles became flat throughout the polysilicon layer. The inserts in Fig. 5.7 are the TEM micrographs measured at 9 and 11 W. It can be seen that, although partial melting occurred at 9 W, the grains are still small (~ 3 μm). Their growth is considerable (~ 20 μm) only when the melt front reaches the underlying oxide layer.

Figure 5.8 shows the boron concentration in the polysilicon layer after laser irradiation. This boron redistribution further substantiates the theory of surface melting. (The artificial boron intensity observed in the oxide could be the result of a change in the ionization yield of the boron atoms implanted in the oxide.) The boron distribution for a 5 W laser power level is identical to the as-implanted distribution and, again, it can be calculated from the LSS theory. As the power is increased, the region wherein the boron profile is flat also increases. At 8.5 W, substantial redistribution of the boron over 0.2 μm is observed but below this depth the boron distribution remains essentially unchanged.

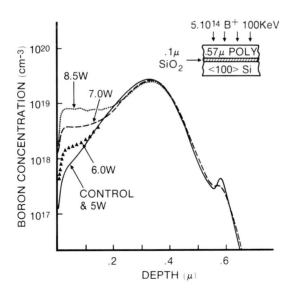

FIGURE 5.8.   SIMS profile of boron-implanted polysilicon after laser annealing with different power levels.

5.1.4 Electrical Properties

Sheet resistivity, carrier concentration, and Hall mobility were measured as a function of laser power by means of a Van der Pauw configuration. The sheet resistance, mobility, and active-carrier concentration (N_s) obtained from these measurements for the arsenic-implanted laser-annealed polysilicon are shown in Fig. 5.9. The most important observation is that the carrier concentration is fixed at an approximate level of 3.5×10^{15} cm^{-2} (the implantation dose was 5×10^{15} cm^{-2}), independent of laser power. This figure also shows that, even at the lower power level (5 to 6 W), for which the SIMS data indicate no diffusion, the dopants are almost all active. This is in agreement with work on cw laser and electron beam annealing of ion-implanted single crystal silicon, wherein the implanted dopant can be activated with no diffusion by a solid phase recrystallization process. At the medium-power level (6 to 9 W), sheet resistance decreases because of greater

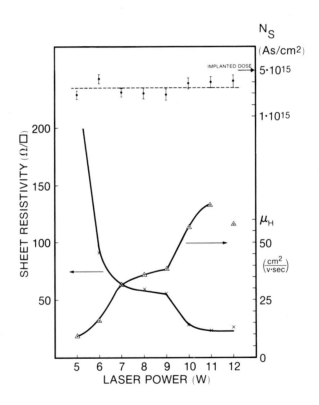

FIGURE 5.9. Sheet resistance, Hall mobility, and total active carriers of 5×10^{15} As$^+$ implanted in polysilicon with 170 keV and cw laser annealed as a function of laser power.

mobility, with no change in the number of active carriers. This increase in mobility drops at 9 W and, as the laser power is raised to 10 W, a sharp transition to the high-power region is accompanied by a rapid rise in mobility. In this region, as power is raised to 10 to 12 W, the trend toward greater mobility decreases, but with no effect on N_s.

In the initial experiments with boron-doped samples, the laser power could not be raised above 8.5 W without peeling the polysilicon from the oxide. As a result, the high-power region, where the mobility approaches that of single-crystal silicon, is missing in Fig. 5.10. It is important to note that, in an earlier experiment [5.6] where the polysilicon was deposited over Si_3N_4, implanted with the same boron dose and recrystallized at 11 W, a grain size of approximately 25 µm, resistivity of 269 $\Omega/\square$, and mobility of 44.5 $cm^2/V\text{-}sec$ were achieved. This means that, when polysilicon is deposited over nitride, it is possible to reach the high-power level without causing the film to peel. After this experiment was performed, the problem of peeling of the polysilicon from the oxide was solved by encapsulating the polysilicon with a thin film of SiO_2 or Si_3N_4.

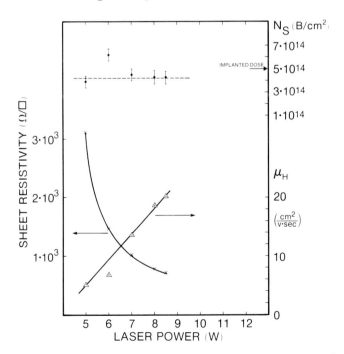

FIGURE 5.10.   Sheet resistance, Hall mobility, and total active carriers of 5 x 10^{14} B^+ implanted in polysilicon with 100 keV and cw laser annealed as a function of laser power.

These observations suggest the importance of capping the polysilicon film prior to recrystallization. For the experiments reported here, it was found that a thin Si_3N_4-related cap could be grown by carrying out a 1100°C N_2 anneal for 1 hour immediately prior to the laser annealing. The laser power level could then be increased without the occurrence of peeling [5.2]. Detailed studies of capping procedures and their effects on film integrity are described in Sec. 5.1.8.

5.1.5 Discussion of the Basic Results

The results obtained from the grain size analysis, dopant redistribution and electrical properties provide the basis for a proposed explanation for the effect of laser power on polysilicon annealing and recrystallization. To summarize, three laser power levels have been identified:

(a) Low Power. This range is characterized by no change in surface morphology, no increase in grain size or dopant redistribution and complete dopant activation with low carrier mobility. Laser heating without melting is responsible for these results, under conditions similar to CW laser annealing of ion-implanted single crystal silicon. A power level of 5 W is characteristic of this region.

(b) Medium Power. When the laser power is raised from 5 to 9 W, the electrical properties in the top layer are changed. Surface roughness increases as does grain size, and dopant diffusion occurs in the top 1000 to 3000 Å. This region is marked by greater mobility, reduced resistivity, and active implanted dopants. All of the data suggest that the polysilicon layer begins to melt from the surface downward. Because the actual annealing time is on the order of 1 msec, it is difficult to explain such phenomena, particularly the impurity redistribution, with simple solid-state diffusion.

It is important to note that, although melting occurs, no significant change in grain size is observed. This may be explained by a crystallization process that begins in the fine-grained polysilicon substrate that serves as the nucleating layer.

(c) High Power. All electrical properties change dramatically in the power range between 9 and 10 W. In this narrow region, there is a significant change in surface structure, a very large increase in grain size (from 1 to 20 μm), a steep rise in carrier mobility, and a uniform dopant distribution throughout the layer. Large grains develop when the layer is completely molten all the way to the underlying substrate.

Complete melting appears to be essential for the development of such grains. If samples are lightly doped and complete melting cannot occur without damage to the samples, then large grains will not develop.

5.1.6 Crystal Orientation of the Beam Recrystallized Films [5.8]

In this section we describe the structural changes produced by the laser recrystallization as they are measured by X-ray diffraction; and we further discuss the effect of the heat treatment and encapsulation techniques that are often required before laser irradiation can be employed. Data are presented for polysilicon films deposited on both silicon dioxide and silicon nitride since these are the most common films employed in present technology.

For these experiments films of low pressure CVD polysilicon were deposited at 625°C to a thickness of ~5500 Å. Samples of each type were then thermally annealed at 1100°C in flowing nitrogen. The samples were laser processed with the cw scanning argon laser under the melting conditions just described so as to form large grain structures.

A conventional X-ray diffractometer was used to analyze these samples. The results of the analysis are collected in Table 5.1. Measurements were taken with the samples mounted both parallel and perpendicular to the laser scan direction, but no anisotropy was seen. The results reported are the average of the two measurements. The intensities of the signals from the (111)-, (220)- (311)-, (400)-, and (331) peaks were measured and normalized to indicate the relative amounts of crystallites with (111)-, (110)-, (311)-, (100)-, and (331)orientation in each film. The measured signals were normalized to account for the signal strengths expected in a thick, randomly oriented sample, and also for the finite film thickness [5.9].

While the surface structure observed in polysilicon by optical microscopy does not always correlate with the grain structure, some general observations can be made from Table 5.1 and Fig. 5.11. The films which were not laser processed had a fine-grain appearance. Those laser processed, but not thermally annealed, (14-1, 16-1, 16-3) appeared to have a feather-like, long-grain structure, while those which were also thermally annealed (13-3, 15-1) indicated a large-grained appearance on the surface. One sample (13-1) indicated a less regular appearance, probably related to a mixed structure obtained because of less complete recrystallization.

The normalized X-ray results are also shown in Table 5.1. As expected [5.9], the (110) orientation is dominant in all the films

TABLE 5.1. Crystallographic Orientations in Laser Recrystallized Polysilicon.

Sample	Insulating Film	Thermal Anneal	Laser Processing	Laser Power (W)	Grains	Normalized X-Ray Texture				
						(111)	(110)	(311)	(100)	(331)
16-U	Si_3N_4	No	No	--	Fine	25	720	78	0	∿58
16-1	"	No	Yes	8	Long	335	133	363	∿156	168
16-3	"	No	Yes	10	Long	445	183	342	∿135	174
15-U	"	Yes	No	--	Fine	250	478	200	0	194
15-1	"	Yes	Yes	8	Large	395	68	308	208	219
14-U	SiO_2	No	No	--	Fine	50	615	62	0	∿65
14-1	"	No	Yes	8	Long	365	379	262	198	245
13-U	"	Yes	No	--	Fine	258	491	150	∿83	226
13-1	"	Yes	Yes	8	Mixed	193	595	249	167	168
13-3	"	Yes	Yes	11	Large	485	234	275	260	194

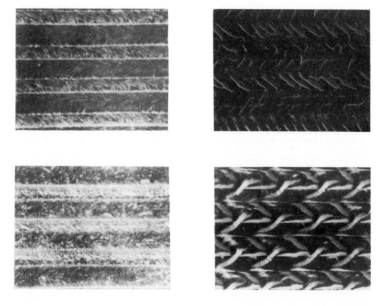

FIGURE. 5.11. Dark-field photomicrographs of laser-recrystallized polysilicon on Si_3N_4 [(a) and (b)] and on SiO_2 [(c) and (d)]; (b) and (d) show samples thermally annealed before laser recrystallization.

immediately after deposition, with the normalized intensity being about ten times larger than that from any other orientation. Thermal annealing of the samples at 1100°C decreases the preference for (110) texture, although this orientation is still dominant. The (111), (311), and (331) orientations all increase significantly, with little difference seen between the films deposited on silicon nitride and those deposited on silicon dioxide.

Laser recrystallization of the polysilicon films deposited on silicon nitride decreases the amount of (110) texture significantly; the crystallites with (111) and (311) orientation increase especially, with the (111) being generally dominant. Similar trends were seen both on samples which were thermally annealed and those which were not, with the thermally annealed samples showing an especially marked decrease in the (110) texture but little further increase in the (331) texture.

The polysilicon films deposited on silicon dioxide and not thermally annealed showed the same trends as those deposited on silicon nitride, with, perhaps, somewhat less pronounced changes; there was less decrease in the (110) texture and less increase in the (311) texture. This difference may be partially related to the different laser recrystallization conditions, which were optimized for each type of insulator. The thermally annealed sample which was laser recrystallized to produce a large-grained structure (13-3) also showed similar trends to those found on the silicon-nitride-containing sample. The other sample (13-1) was treated at a lower laser power so that the large-grained structure was not obtained; this film probably contained a mixture of large and small grains. The increase in the amount of (110) texture is probably related to some growth of the structure initially found in the as-deposited films. This mixed case is, of course, of less interest than the fully recrystallized films.

A significant increase in (100) texture is seen in many cases, but in no cases does (100) become the dominant orientation. In general, the laser-recrystallized samples exhibit less-strong preferred orientation than do the as-deposited films. The influence of this mixed orientation on the oxidation rate and interface charges of structures containing laser-recrystallized polysilicon will be discussed in Sec. 5.4.

5.1.7 Grain Boundary Diffusion

A conclusion drawn in Sec. 5.1.5 from the dopant distribution experiment for the arsenic implanted sample was that at 10 W, when the dopant profile became essentially flat throughout the film, the melt depth has extended all the way to the dielectric interface. Considering the diffusivity of impurities in the melt (about 10^{-4} cm^2/sec [5.10]) and the duration of the melt (which should be approximately the dwell time of the laser, or about 0.3 msec), the diffusion length of the dopant is about 2 μm. Hence the dopants would be quite uniformly distributed within the molten portion of the polysilicon film. While this is observed for the high power levels (10 W and 11 W), it is seen that the impurity profile for arsenic at the intermediate power levels (8 W, 9 W) exhibits a tail that is inconsistent with the above considerations. The diffusivity associated with this tail is on the order of 10^{-7} cm^2/sec. For comparison, the solid phase diffusivity of impurities in single crystal silicon is about 10^{-11} cm^2/sec near the melting temperature [5.11]. Hence the diffusivity in the tail cannot be explained by either a simple melting or solid phase diffusion within the single crystal grains. Furthermore, the presence of the tail cannot be associated with a temperature difference in the film, since from temperature calculations, the temperature difference across the film is expected to be very small. The tail thus seems to be associated in part with the occurrence of melting, since at 6 W where the temperature of the film is close to melting, no such broadening is found.

Note that while such an anomalous diffusion is observed in the arsenic-implanted sample, the boron-implanted sample is well-behaved, showing two regions of distinctly different diffusion. To eliminate any sample preparation difference between an arsenic-implanted sample and a boron-implanted sample, and to investigate such an anomalous diffusion, the following two experiments were carried out.

In the first set of experiments, As$^+$ and B$^+$ were implanted into the same sample. As$^+$ was implanted with a dose of 1×10^{15}/cm^2 at an energy of 60 keV, and B$^+$ was implanted with a dose of 1×10^{15}/cm^2 at an energy of 150 keV. The samples were then irradiated at various power levels as in 5.1.2. SIMS profiling was then carried out, the result of which is shown in Fig. 5.12.

It is seen that while at intermediate power levels a tail was again observed in the arsenic profile, the boron profile showed an abrupt change. There was no noticeable broadening of the boron profile deep within the film, and the boron profile is in agreement with a difference of diffusivity in the two different regions. Hence, partial melting of the film at intermediate power levels is established.

In the second experiment, the effect of grain boundaries on the anomalous diffusion of arsenic was investigated. The starting samples were similar to the samples above, except that the sample was deposited on nitride. (The nitride layer also eliminated the extraneous SIMS peak in the dielectric layer due to the presence of oxygen.) Prior to As$^+$ implant, a set of samples was first

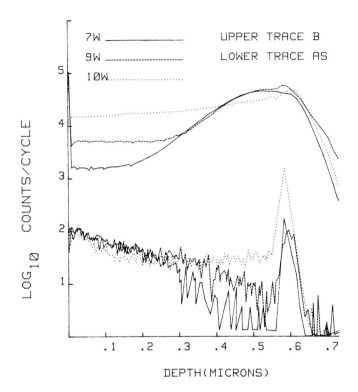

FIGURE 5.12. SIMS profile of sample implanted with arsenic and boron after laser irradiation at different power levels.

laser recrystallized at a power level such that long grains were formed. The two sets of samples were then implanted with As^+ with a dose of $1 \times 10^{15}/cm^2$ at an energy of 60 keV. The two sets of samples were then each irradiated at an intermediate power level where partial melting would occur. (Due to the use of a lens with a longer focal length (160 mm), and a broadening of the laser beam probably from a change in the output mode of the laser, P_{melt} increased to 8.5 W, so 11.5 W was used. On comparing with previous results, a value of $P = 1.4\ P_{melt}$ can be used as an approximation.)

The results are shown in Fig. 5.13. It is seen that the sample with no laser treatment prior to As implantation gives a post-laser annealing profile similar to those previously observed. However, the sample that had first been recrystallized into long grains showed a much sharper As profile after a subsequent laser processing step with a tail region that is reduced. Hence it is clear that the grain boundaries substantially enhance the diffusion of arsenic.

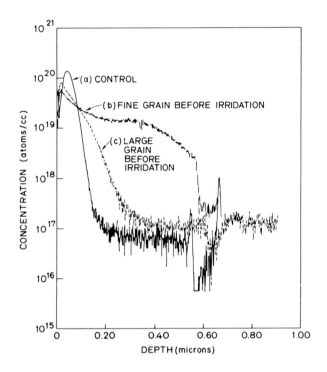

FIGURE 5.13. SIMS profile of samples. (a) Control (no laser irradiation); (b) Fine grain, implanted with As+ before laser irradiation at 11.5 W; and (c) Large grain, implanted with As+ before laser irradiation at 11.5 W.

Enhanced diffusion has also been observed in conventional fine-grain polysilicon, where a similar enhancement factor has been observed for both arsenic and boron at typical furnace temperatures (about 1000°C) [5.12]. The apparent difference between boron and arsenic in the present case is consistent with the following explanation. Due to the difference in the location of the peak of the implant, the high concentration region of arsenic was molten while the high concentration region of boron remained in solid phase. In the case of arsenic, then, any dopants lost in a local region within the melt due to diffusion into the grain boundaries of the solid phase are easily replenished from a neighboring molten region. In the case of boron, however, such dopants lost are not as easily replenished due to the difficulty of diffusion in the solid phase. Even if the grain boundaries act as dopant sinks in addition to being enhanced diffusion regions, the relatively low boron concentration at the surface is insufficient to cause any significant increase of boron concentration deep into the film. The location of the molten region with respect to the peak concentration region can thus explain the apparent difference in diffusion behavior. However, a possible difference in behavior between arsenic and boron near the melting temperature cannot be ruled out. An experiment with boron implanted near the surface of the film should clarify this point.

Thus it is concluded that the partial melting of the film that occurred at intermediate power levels, and diffusion along grain boundaries is responsible for the diffusion tail obtained at intermediate power levels.

5.1.8 Effects of Encapsulation Layers on the Properties of Recrystallized Films [5.13]

Experimental results to be described throughout this chapter show that many of the electrical properties of laser recrystallized polysilicon films are comparable to those of bulk silicon. However, surface ripples that can form during the melting and cooling processes can produce significant surface roughness in the recrystallized film. This in turn causes severe problems for applications in which the recrystallized films are to be used for silicon integrated circuit fabrication, where photolithographic operations require a high degree of surface flatness.

As a solution to this problem, several groups [5.14, 5.15] have reported on the use of an encapsulation layer on the polysilicon to maintain flatness during recrystallization. The most recent work is that of Ohkura et al. [5.13], who studied the encapsulation conditions necessary to achieve surface flatness of the recrystallized film in the range of 50 nm peak to valley height over a 4 inch (100) Si wafer.

For their studies, Ohkura et al. employed a 0.4 μm polysilicon film deposited by LPCVD on a (100) silicon wafer coated with 0.6 μm of thermally grown SiO_2. Encapsulation layers of SiO_2, Si_3N_4 and combinations of both were deposited by CVD on the polysilicon layers. The total encapsulation layer thickness was varied from 5 nm to 1 μm. An Ar laser beam focussed to 40 μm diameter was used to melt and recrystallize the polysilicon layer. Power and beam scanning speed ranged from 3 to 15 W and 1 to 100 cm/sec, respectively. The wafer was held at 500°C during the irradiation. The encapsulation layers were then eteched off and surface ripple measurements were made by a Talystep system.

The principal results obtained from these measurements were as follows. For both SiO_2 and Si_3N_4 encapsulation, surface roughness was found to fall off rapidly as the encapsulation thickness was increased. A surface roughness of about 35 nm can be obtained for 50 to 100 nm thick encapsulants for both cases. However, above a certain thickness, stripping of the polysilicon layer and/or void formation begins to occur. For SiO_2 encapsulation, both the SiO_2 and the polysilicon layer are stripped off when the encapsulating layer thickness exceeds 500 nm. For a Si_3N_4 encapsulant, voids begin to form in the polysilicon layer prior to stripping, due most likely to stress induced by movement of the molten Si.

These results suggest that the SiO_2 encapsulant is soft enough to absorb surface motion of the molten silicon, but also soft enough to be readily stripped. The Si_3N_4 encapsulant is, on the other hand, too stiff to absorb the surface motion of the molten silicon, and also difficult to strip. These observations suggest the use of a two layer (Si_3N_4 jon SiO_2) encapsulant, in which the SiO_2 is used to absorb surface motion and the Si_3N_4 layer is used to prevent the SiO_2 from being stripped.

Ohkura et al. found that an encapsulant of 20 nm Si_3N_4 on 50 nm SiO_2 gives a normalized window of laser power for recrystallization from 10 W to 17 W, compared to a 10 W to 12 W window in the capless case, with surface roughness in the encapsulated film of ~ 35 nm.

The desirable surface flatness achieved with an appropriate encapsulant does not come without a price, however. Drowley and Kamins [5.16], studying the effects of oxygen and nitrogen incorporation in laser recrystallized films, have shown that the encapsulating layers are partially dissolved during recrystallization. If an SiO_2 encapsulant is used, oxygen from this film is incorporated into the polysilicon during recrystallization to a level of $3-4 \times 10^{18}/cm^3$. If Si_3N_4 is used as the encapsulant, nitrogen is incorporated at a level of $2-4 \times 10^{17}/cm^3$, <u>and</u> oxygen is also incorporated at a level of $3-4 \times 10^{18}$. After subsequent processing, some of the included oxygen or nitrogen moves to the back interface (and other interfaces such as grain boundaries in the film), forming SiO_x and SiN_x. The SiN_x at the back surface produces poor electrical properties as will be discussed shortly.

It should be mentioned that the incorporation of oxygen and/or nitrogen in the films can have a very deleterious effect on both minority carrier lifetime and carrier generation lifetime, thus raising an as yet unanswered question about the suitability of these films for fabrication of bipolar transistors. The incorporation of oxygen and/or nitrogen will also affect the types of residual defect structures in the film and the nature of the recrystallization process itself.

5.2 RESISTIVITY REDUCTION IN BEAM–RECRYSTALLIZED POLYSILICON
FILMS [5.17]

     In the previous section it was shown that the sheet resist-
ivity of doped polysilicon films could be reduced by more than a
factor of two over that obtained by thermal processing. A similar effect has also been reported using pulsed laser processing
[5.17]. However, it has also been pointed out [5.17] that the
resistivity reduction obtained with pulsed laser annealing is
not thermally stable. Instead, the resistivity was found to
increase substantially above the initial values during a subsequent thermal–after–laser annealing treatment. An illustration
of these results for a phosphorus doped polysilicon film is
given in Fig. 5.14.

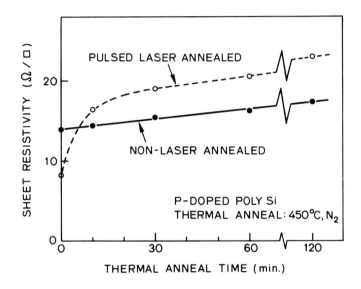

     FIGURE 5.14.   Sheet resistivity of laser-annealed (0) and
unannealed (●) poly Si as a function of subsequent thermal
annealing time (450°C); P-diffused poly Si.

     In this section we summarize experiments that were performed
to elucidate the mechanism of sheet resistivity reduction in
heavily doped polysilicon by laser annealing and its thermal
stability during post laser heat treatment. A summary of both
Q-switched Nd:YAG and cw argon laser annealed films will be
presented, leading to conclusions on the minimum resistivity
that can be achieved for phosphorus doped, laser recrystallized
polysilicon films.

5.2.1.   Electrical Properties

Figure 5.15 shows the resistivity changes that occur in ion implanted (3×10^{16} P^+ ions/cm^2), laser-recrystallized polysilicon films (1800 Å thick) during thermal annealing at 1000°C. Immediately following laser annealing, the cw laser-annealed poly Si (CA poly) exhibited a measured resistivity of 14 $\Omega/\square$, and the pulsed laser annealed poly Si (PA poly), 17 $\Omega/\square$. In each case, the resistivity is smaller than that of the thermal annealed control sample (31 $\Omega/\square$). However, it can be noted that the resistivity of both the CA poly and the PA poly increase rapidly in the first stage of thermal annealing. The stable resistivity of the PA poly is ~ 40 $\Omega/\square$, and is much higher than that of the thermally annealed poly. On the other hand, the CA poly rises to only 20 $\Omega/\square$ with the thermal treatment, and remains lower than the control sample. The reason for this large variation is related to dopant precipitation and has been thoroughly discussed by Shibata et al. [5.18].

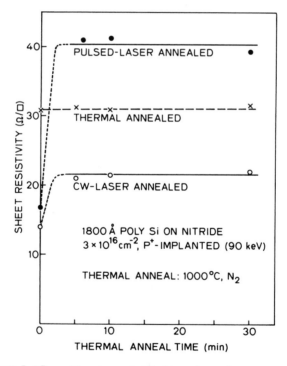

FIGURE 5.15.   Sheet resistivity of cw-laser-annealed ($\circ$) and pulsed-laser-annealed ($\bullet$) Si as a function of thermal annealing time. The samples are 1800 Å Si, 3×10^{16} cm^{-2} p^+ implanted with 90 keV. The thermal-annealed controls (X) are also shown. The thermal annealing was performed at 1000°C in N_2 ambient.

5.2.2 Limits of Resistivity Reduction

The dramatic decrease in sheet resistivity obtainable by cw beam processing is accompanied by a nearly 100% activation of the implanted impurities. The highest carrier concentration in as-laser-processed samples is in excess of $10^{21}/cm^3$, which exceeds the solid solubility limit of phosphorus in single crystal silicon ($4 \times 10^{20}/cm^3$ [5.19]). However, activation of impurities at levels in excess of the solid solubility produces a sheet resistance which is thermally unstable, leading to the precipitation of dopants during subsequent heat treatment. Precipitation then occurs at grain boundaries as well as in the grain crystallites. As a result of precipitation, the lowest resistivity obtainable in laser processed phosphorus doped polysilicon is achieved at an average doping level of $5 \times 10^{20}/cm^3$.

It is instructive to plot the mobility data as a function of the active impurity concentration. The results are shown in Fig. 5.16. The solid line indicates the mobility calculated from the Irvin curve assuming 100% activation of the dopant.

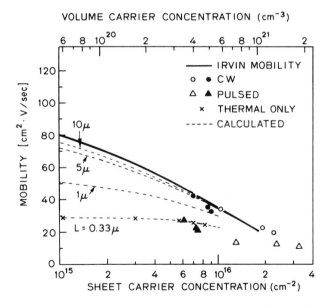

FIGURE 5.16. Hall mobility of Si films plotted as a function of sheet carrier concentration (cm^{-2}) and the volume carrier concentration (cm^{-3}). Open ($0,\Delta$) and solid ($\bullet,\Delta$) symbols refer to the values before and after the subsequent thermal anneal, respectively. The solid line represents single-crystal mobility calculated from the Irvin curve. Calculated mobilities of Si for a variety of grain sizes are also shown by dotted lines.

It is seen that the majority carrier mobility in the CA poly is
very nearly equal to the single crystal values obtained by Irvin,
while that in the PA poly or the TA poly is substantially smaller.
It is interesting to observe that mobility values for doping
concentrations that exceed the solid solubility are obtained by
a relatively straightforward extrapolation of the Irvin curve at
lower doping concentrations.

5.3 ELECTRONIC PROPERTIES OF LASER RECRYSTALLIZED FILMS INTENDED FOR MOS DEVICE FABRICATION

The possibility of fabricating devices directly in beam
recrystallized polysilicon films is of considerable interest in
integrated circuit technology. The characteristics of devices
fabricated in these films will be strongly dependent on both the
electronic properties in the film and the properties of the interfacial layers between the polysilicon and various insulators that
are grown during device fabrication. In conventional silicon-gate
MOS applications, the polysilicon is very heavily doped, similar
to that discussed in Section 5.2, so that the polysilicon film
is never depleted even at the insulator boundary. However, when
an active device is to be placed in a laser recrystallized polysilicon film, the film will be only moderately doped, so that
depletion and inversion layers can easily be formed by residual
charges that may reside in the insulator or at the polysilicon/
insulator interface. The possibility of using beam recrystallized polysilicon films for practical devices depends on keeping
these residual charges at sufficiently low levels that they do
not dominate the device behavior. The utility of the films for
integrated circuit fabrication also depends critically on the
carrier velocity-electric field characteristic of the films, and
particularly on whether this characteristic is affected by the
large number of grain boundaries that may exist within the film.
In this section we describe experiments that were undertaken to
explore these questions.

5.3.1 Velocity-Field Characteristics

To assess the potential of these films for MOSFET applications, it is useful to know at least empirically how grain
boundary scattering will affect the carrier velocity-vs-electric
field characteristics, especially in the high electric field
region. Measurements of electron velocity in both laser recrystallized and silicon-on-sapphire films made by Cook et al. [5.20]
are shown in Fig. 5.17 where we also include for comparison
measurements made on single crystal material by Norris and
Gibbons [5.21]. As can be seen, the electron velocity in

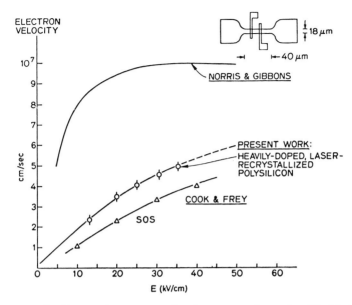

FIGURE 5.17.   Electron velocity vs. electric field in Si, SOS and laser-recrystallized-polysilicon.

heavily doped, laser recrystallized polysilicon films is actually larger than that in SOS films, even though the SOS films are more lightly doped and nominally "single crystal". Neither of the films gives the sharp saturation and high terminal velocity that are characteristic of single crystal material, though adequate velocity is obtained for device purposes.

5.3.2 Oxidation Characteristics [5.8]

Kamins et al. [5.8] have conducted a set of basic experiments to determine the oxide thickness grown on laser-recrystallized polysilicon under conditions which might be used in an integrated-circuit process. Only polysilicon films deposited onto silicon nitride with no subsequent nitrogen anneal were studied. After laser recrystallization, some of the wafers were doped by a 950°C, $POCl_3$ predeposition which produced a sheet resistance of 10 $\Omega/\square$ in single crystal silicon. Both undoped and doped films were then oxidized.

Two different oxidation cycles were used. A 125 min, 1000°C TCE/O_2 oxidation, which forms 1000 Å of oxide on (100)-oriented, single crystal silicon, was used in some cases. A 210 min, 800°C steam oxidation was also investigated in order to emphasize the effects of the silicon orientation and structure. The oxide thicknesses were measured on the polysilicon and single crystal control wafers with a UV spectrophotometer [5.22]. After the oxide thicknesses were measured, the oxidation cycles were repeated without stripping the oxide so that thicker oxides, which could be measured more easily, were grown.

The oxide thicknesses after the first oxidation cycle are shown in Table 5.3. The oxide thicknesses grown on the undoped polysilicon with the 1000°C TCE/O_2 oxidation fall between those on the (100)- and (111)oriented, single crystal silicon oxidized simultaneously, being about 6% greater than that on (100)-oriented silicon after the first oxidation. After the second oxidation the oxide thicknesses continued to fall between those of the two orientations of single crystal silicon. Oxidation under these conditions is influenced by diffusion of oxygen through the already-formed oxide, as well as by surface reaction, so the effect of the silicon structure and orientation is small.

TABLE 5.3. Oxide Thickness on Laser-Recrystallized Polysilicon

Doping	Oxidation Temperature (°C)	Oxidizing Ambient	Oxide Thickness (Å)				
			Poly-Silicon		Single-Crystal Silicon		
			Recrystallized	Fine-Grain	<100>	<111>	<110>
Undoped	1000	TCE/O_2	1050	1060	980	1100	~1120
Undoped	800	Steam	1930	1830	1140	1670	~1770
Phosphorus Doped	800	Steam	6190	4830	6030	6000	

The 800°C steam oxidation is controlled primarily by the surface reaction so the effects of orientation and structure should be more significant, as observed, with the oxide thickness on (111)-oriented, single crystal silicon 46% greater than that on (100)-silicon. The oxide thickness on the fine-grain poly-silicon appears to be slightly greater than that on (110)-silicon, with the oxide on the laser-annealed regions probably even slightly thicker. Thus, the oxide thickness grown on undoped, laser-recrystallized polysilicon under surface-reaction-controlled conditions differs only slightly from that on fine-grain polysilicon and is similar to that on the fast-oxidizing orientations of single crystal silicon.

The influence of the laser recrystallization on heavily phosphorus doped polysilicon is more dramatic. Under these conditions, the oxidation rate is controlled by the dopant concentration at the surface, and the effect of crystal orientation is less significant, as can be seen by comparing the oxide thicknesses on the two orientations of doped, single crystal silicon in Table 5.3. The oxide thickness on the doped, fine-grain polysilicon is markedly less than that on the single crystal silicon or on the laser-recrystallized polysilicon, while that on the laser-recrystallized polysilicon is close to that on the single crystal silicon.

Thinner oxides on heavily doped, fine-grained polysilicon than on equivalently doped single crystal silicon have been observed before and have been related to the lower active dopant concentration near the surface of the polysilicon [5.22]. During the doping cycle, the dopant can diffuse farther into the polysilicon than into the single crystal silicon so that the average surface concentration is lower, and the oxide subsequently grown is thinner. In addition, even if the dopant concentration were the same, the lower electrical activity in the polysilicon would keep the Fermi level closer to midgap so that the dopant-enhanced oxidation would not be as significant.

Similar reasoning can account for the thicker oxide grown on the laser-recrystallized polysilicon than on fine-grain polysilicon. The dopant appears to diffuse away from the surface of the laser-recrystallized polysilicon at about the same rate as in single crystal silicon and less rapidly than in fine-grained polysilicon since grain boundary diffusion is less important; consequently, the oxide grown is thicker than on fine-grain polysilicon.

We conclude from these results that, while fine-grained polysilicon films exhibit a strong (110) preferred orientation, the preference for a particular orientation is less pronounced in laser-recrystallized polysilicon, although there is some preference for (111) orientation. The oxide thickness grown on lightly doped laser-recrystallized polysilicon is not greatly different from that on fine-grained polysilicon and is probably dominated by the fast-oxidizing orientations under surface-reaction-controlled conditions. The oxide thickness on heavily doped, laser-recrystallized polysilicon is similar to that on single crystal silicon and is much greater than that on fine-grained polysilicon.

5.3.3 Charges at a Laser Recrystallized Polysilicon/Insulator Interface [5.23]

Kamins et al. [5.23] investigated the behavior of the oxide/polysilicon interfaces by fabricating capacitor structures and making capacitance-voltage measurements with the depletion region extending into both the film and in some cases the underlying substrate. Both thermally grown silicon dioxide and low pressure CVD silicon nitride were investigated as the insulating layer.

All of the structures studied in these experiments (Fig. 5.18) were fabricated on 0.01 Ω-cm n-type silicon wafers which subsequently served as the gate electrode. A 1000 Å thick, TCE/O_2 gate oxide was grown on some wafers at 1000°C, while others were covered with 1000 Å of low pressure CVD silicon nitride. A 5500 Å thick film of LPCVD polysilicon was then deposited at 625°C on all wafers. The wafers which contained an oxide layer were next annealed at 1100°C in N_2 for 1 hour to ease control of the subsequent laser recrystallization process. Wafers containing a nitride layer did not require this thermal annealing step. All polysilicon films were implanted with 10^{12} boron ions/cm^2 at 100 keV. Portions of each wafer were then recrystallized with a scanning cw argon laser, with a beam width of about 70 µm. The substrate temperature was held at 350°C, and a power level was chosen which resulted in the formation of the long columnar crystallites described in Section 5.2. Aluminum was then deposited and defined into squares 300 µm or 900 µm on a side, the aluminum serving as a contact to the polysilicon "substrate". The polysilicon was plasma etched using the metal as a mask. After a 450°C H_2 anneal, the high frequency (1 MHz) capacitance voltage characteristics were measured, with the depletion regions extending into the polysilicon. To facilitate discussion, all gate voltages mentioned below are those applied to the n^+ single crystal wafer which served as the gate electrode.

5.3.3.1   Recrystallized Films on SiO$_2$

The structures with silicon dioxide beneath the polysilicon will be considered first. Laser powers of 14–16 W produced overlapping scans and the desired large-grain structure in the polysilicon. In the laser-recrystallized areas well-defined accumulation and inversion regions were observed in the capacitance-voltage characteristics. The capacitance was very close to that expected for the 1000 Å target oxide thickness. When the opposite polarity voltage is applied, the capacitance decreases

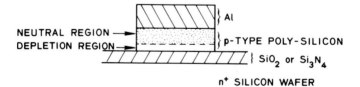

FIGURE 5.18.   Cross section of test capacitor structure, showing depletion region extending into the polysilicon when a the "gate" voltage is applied to the n^+ wafer. ©1980 IEEE

to a value indicating a depletion region approximately 2000 Å wide in the polysilicon. This polarity of voltage also tends to deplete the n^+ gate wafer, but the heavy doping in this wafer increases the voltage necessary for significant depletion and also limits the width of the depletion region so that the effect of the depletion into the n^+ wafer can be neglected in the following discussion.

The depletion region width in the polysilicon corresponds to a dopant concentration of approximately $(2.2 - 2.6) \times 10^{16}$ cm^{-3}, confirming that virtually all the dopant is activated by the laser recrystallization. The doping concentration is somewhat higher than the average dopant concentration of 1.8×10^{16} cm^{-3} expected if all the 1×10^{12} cm^{-2} boron atoms implanted were active and were uniformly distributed through the 5500 Å-thick polysilicon film. (Laser recrystallization does not appear to change the average film thickness.) This anomalously high apparent concentration may possibly be related to dopant segregation near the bottom of the polysilicon film as it recrystallizes or to residual defect states which limit the depletion-region width.

In the regions of the same wafer not affected by the laser, the minimum capacitance is considerably less than in the recrystallized regions, corresponding to a maximum depletion-region width of about 4500 Å, close to the thickness of the polysilicon film, suggesting that the high resistance of the unaffected polysilicon may cause the entire thickness of the film to act as a dielectric when attempts are made to deplete the surface. (Aluminum from the deposited electrode may have penetrated into the polysilicon slightly, causing the apparent maximum depletion-region width to be slightly smaller than the polysilicon film thickness.) Samples which received neither the 1100°C thermal anneal nor the laser processing showed very little dependence of the capacitance on the gate voltage.

After a negative bias-temperature stress to insure that any positive mobile ions present did not influence the results, the flatband voltage in the laser-recrystallized regions was found to be in the range -2.9 to -3.4 V. If ϕ_{MS} is taken to be -0.90 V, the fixed-charge density is calculated to be about 4×10^{11} cm^{-2}, which is somewhat higher than that expected in single crystal silicon but not unreasonable since the processing was not optimized for this unconventional structure. (X-ray measurements indicate a crystal structure containing grains of various orientations, with a weak preference for (111) orientation under the conditions employed here [5.8], so that the minimum fixed-charge density would be higher than that expected for (100) silicon.)

The voltage between the measured flatband and inversion points of the C-V characteristic was greater than calculated,

however, suggesting the presence of some fast states or lateral
nonuniformities. If this distortion in the curve were entirely
related to fast states, their density between flatband and inversion would be about 2×10^{11} cm^{-2}. A portion of the distortion
in the C-V characteristic may be related to nonuniformities since
a region treated under conditions which produced incompletely
overlapped recrystallized regions showed distorted C-V characteristics corresponding to a parallel combination of the recrystallized and unaffected curves.

In the regions which were not affected by the laser, the
magnitude of the flatband voltage varied significantly and was
generally greater than 10 V, suggesting an effective charge
density greater than 10^{12} cm^{-2}.

Thus, the properties of the silicon-dioxide/polysilicon
interface under a layer of laser-recrystallized polysilicon resem-
ble, but are inferior to, those at the interface between a thermally grown oxide and single-crystal silicon. This interface is,
however, of much higher quality than most semiconductor-insulator
interfaces not formed by thermal oxidation. For example, the
interface between single crystal silicon and a deposited oxide
layer is generally unstable and can, at best, be characterized
by very high fixed-charge and interface state densities.

A more recent detailed study by Le and Lam [5.24] of electrical characteristics at the interface between laser-recrystallized
polysilicon and an underlying thermally grown oxide has shown
that the lowest fixed-oxide charge density is obtained when the
oxide layer is thermally grown in an oxygen/HCl ambient. A double-
layered encapsulating structure obtained by first depositing a
10 nm plasma CVD oxide on the polysilicon film and then a 6 nm
LPCVD nitride film on the oxide prior to laser recrystallization
was also found to be effective in reducing charge at the back
interface. Oxide charge densities in the range of $3 \times 10^{11}/cm^2$ were
obtained, consistent with the earlier work of Kamins et al.

5.3.3.2 Polysilicon Films on Si_3N_4

Capacitors containing an insulating layer of silicon nitride
were also tested. The polysilicon deposited on the silicon
nitride contained a wide range of crystal structure after laser
recrystallization depending on the laser power used, which varied
from 14 to 17 W. No long grains were seen at the lowest power,
while the highest power produced totally overlapped recrystallized
regions containing long grains. The capacitance-voltage characteristics were also quite different in regions annealed at different powers.

In all cases, the maximum capacitance corresponded to a silicon nitride layer with a relative permittivity of 7.0 for the 1000 Å thickness deposited. The behavior of the capacitance-voltage characteristics in depletion was markedly different in the differently processed regions (Fig. 5.19). The minimum value of capacitance was lowest in the region recrystallized at the lowest power (approximately 44% of the nitride capacitance); the minimum capacitance increased with increasing laser power to about 60% of the nitride capacitance for the highest laser power. In addition, at large positive gate voltages (> 30 V, in the direction tending to invert the polysilicon), the capacitance again increased toward its maximum value, suggesting the presence of a source of minority carriers to charge the inversion layer rapidly. In the region processed at the lowest laser power, there was no flat portion of the C-V curve at the minimum capacitance. As the laser power and the minimum capacitance increased, a flat region developed in the C-V characteristic. (Rapid minority-carrier generation was also seen in samples with a silicon-dioxide insulator at high positive gate voltages).

This behavior of the minimum capacitance would not be con-sistent with defects within the polysilicon film, since a higher defect concentration at higher laser powers would have to be postulated, in contrast to the reduced defect density expected as the long grain structure becomes more fully developed.

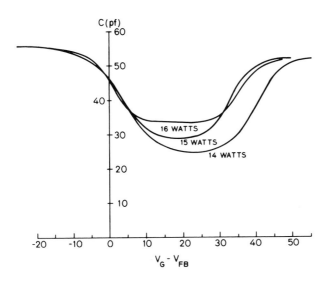

FIGURE 5.19. High frequency capacitance-voltage character-istics of structure containing a silicon nitride insulator under polysilicon films recrystallized at three different laser powers.

No meaningful value of the fixed charge or fast-state density can be extracted from the capacitance voltage measurements because of electron trapping in the insulator. Even before a bias temperature stress, the flatband voltage changes significantly when large voltage ramps are applied. (In contrast, no change was seen in samples containing a silicon-dioxide insulator when tested with similar, high-magnitude voltages.)

Limited tests indicated no significant improvement by laser annealing the nitride before the polysilicon was deposited, in addition to recrystallizing the polysilicon with the laser after the polysilicon deposition. The results obtained from the structures containing silicon nitride show that the polysilicon/silicon-nitride interface is not well behaved, a situation similar to that of a silicon-nitride/single crystal-silicon interface.

5.3.3.3 Summary of Interface Properties

Capacitance-voltage characteristics have been measured to determine the interface properties at the back surface of a layer of laser-recrystallized polysilicon. The interface between the recrystallized polysilicon and an underlying oxide layer can be characterized by an effective fixed-charge density and a fast-state density, both in the low-to-middle-10^{11} cm^{-2} range. The polysilicon/silicon nitride interface is not as well behaved. The minimum capacitance depends on the laser power used to recrystallized the polysilicon, and charge trapping at the interface precludes the determination of a meaningful value of interface charge.

5.3.4 Effect of Laser Recrystallization of Polysilicon on an Underlying Crystalline Silicon Substrate [5.25]

Gibbons and Lee [5.26] first showed that three-dimensional integrated circuits could be made by utilizing MOS transistors fabricated in both an underlying crystalline silicon substrate and (simultaneously) in layers of recrystallized polysilicon lying above the substrate. Since then a number of authors have developed other three-dimensional structures, a topic which we will discuss further in Sec. 5.4.

For structures of this type to be of maximum utility, the properties of the devices fabricated in the underlying single crystal silicon must not be degraded by the recrystallization of the polysilicon layer. Kamins [5.27] has shown that the threshold voltage and surface mobility of 59 randomly chosen MOS transistors fabricated in a 64 mm single crystal wafer were basically unchanged following recrystallization of an overlying polysilicon layer (separated from the substrate by a 0.1 mm oxide layer). Hence, the process is compatible with typical IC fabrication. Later, Kamins and Drowley [5.25] studied the effect of recrystallization on substrate minority carrier lifetime, which is a more exacting test of the process. These authors showed that the recrystallization process produces significant degradation of the generation lifetime in the underlying substrate as measured by MOS deep depletion techniques. However, a subsequent furnace treatment (30 min. at 950°C and 60 min. at 800°C in nitrogen) increased the generation lifetime to its original value except when the substrate was visibly damaged by melting. Measurements were also carried out on junction diode reverse leakage current for diodes made underneath polysilicon layers that were subjected to the recrystallization process, with similar results. The work of Kamins and Drowley, together with the earlier work of Kamins, thus suggests that the point defects in the substrate produced during laser recrystallization of an overlapping polysilicon film are substantially eliminated by subsequent heat treatment, and that recrystallization is therefore compatible with the fabrication of high quality devices in the substrate. Further evidence on this point is presented in Chapter 4.

5.3.5   Properties   of   Laser   Recrystallized   Films   on   Quartz Substrates

The recrystallization experiments and results described previously have been carried out exclusively with a silicon substrate. There are a number of applications, however, in which it would be desirable to recrystallize a polysilicon film on a nonsilicon substrate, especially a transparent substrate such as quartz (for imaging applications).

Suitable substrate materials must be compatible with silicon processing techniques and temperatures. In addition, the thermal expansion coefficient of the substrate should ideally match that of silicon.

Quartz and sapphire both satisfy the conditions of compatibility with silicon processing, and sapphire also has a thermal expansion coefficient near that of silicon. Indeed, the fabrication of recrystallized silicon films on sapphire both by beam-activated solid phase epitaxy and by laser recrystallization offer some opportunities that have not been explored. In contrast, several experiments have been performed to study the electronic properties of recrystallized films on quartz. Kamins and Pianetta [5.28] showed that device worthy films could be prepared despite the large mismatch in thermal expansion coefficients by etching the film into islands either prior to or following the recrystallization. Surface channel mobilities in these first experiments were found to be low, however, in the range of 250–300 cm^2/V–sec, and sometimes as low as 60–80 cm^2/V–sec.

In subsequent work, Tsaur et al. [5.29] showed that the stress in polysilicon films recrystallized on fused silicon substrates can in fact produce <u>larger</u> surface channel mobilities than are obtained in bulk silicon. Tsaur et al. correctly attributed this increase to increased electron population in high mobility valleys in the Brillouin zone, corresponding to tensile stress in (100)-oriented films.

Johnson et al. [5.30] obtained similar results using a CO_2 laser to recrystallize polysilicon films that were patterned into islands with an hour-glass shape to promote the growth of the preferred (100)-orientation while at the same time suppressing the formation of microcracks that arise on account of the thermal mismatch. Johnson et al. achieved surface channel mobilities in excess of 1000 cm^2/V-sec. Leakage effects were obtained from "back-channel" effects in these devices, suggesting that the recrystallization process at the quartz/polysilicon interface leads to material with surface-related defects. However, the leakage was not significantly different from that obtained when an oxidized silicon substrate is used as the base for polysilicon deposition and recrystallization. These experiments thus show that high quality recrystallized films can be prepared on appropriate nonsilicon substrates when precautions are taken to suppress microcrack formation arising from the thermal mismatch problem.

5.4 CHARACTERISTICS OF MOS DEVICES AND INTEGRATED CIRCUITS FABRICATED ON LASER RECRYSTALLIZED POLYSILICON FILMS

The electronic properties of laser annealed polycrystalline films have strongly suggested from the outset that these films can be used directly for the fabrication of active devices. A number of studies, some already referred to, have been performed to evaluate this possibility. Most of these evaluations have been performed using MOS field effect transistor structures, since these devices utilize majority carrier properties and should be minimally sensitive to lifetime and grain boundary effects that could dominate the characteristics of bipolar devices. Channel lengths for MOS transistors that are on the order of, or smaller than, the grain size in the laser recrystallized films may be expected to yield device performance that is comparable to that of devices fabricated in single crystal silicon, whereas the properties of grain boundaries have been found to dominate device behavior when devices are fabricated on fine grain polysilicon films [5.31].

In what follows we discuss the electrical characteristics of MOSFETs fabricated on such films under a variety of different conditions.

5.4.1 MOSFETs Fabricated on Polysilicon/Si₃N₄ Substrates

The first MOSFET devices fabricated in laser recrystallized polysilicon films were fabricated by Lee et al. [5.1]. The polysilicon samples used were 5500 Å thick, deposited by low-pressure chemical vapor deposition (LPCVD). The substrates were single crystal silicon onto which a 1000 Å layer of Si₃N₄ had been deposited. Phosphorus was implanted at an energy of 100 keV to a dose of 3 x 10^{12}/cm^2 to form the channel for the depletion-mode devices. Boron was implanted at an energy of 100 keV with a dose of 3 x 10^{11}/cm^2 for the enhancement mode devices. The wafers were then annealed by an argon cw scanning laser (circular beam) so that long grains were formed. The subsequent processing steps are described in Ref. 5.1. A photograph of the final depletion-mode device structure is shown in Fig. 5.20. Fabrication of the enhancement mode devices was similar except that steps related to the mesa formation were omitted since no isolation is required between devices.

Devices with channel lengths of 50 μm were fabricated; the channel widths were 250 μm for the depletion mode devices and 270 μm for the enhancement mode devices. Since the grains formed by laser annealing tend to align themselves with the laser scan directions, channels were fabricated both parallel and perpendicular to the laser scan direction to test the importance of such alignment.

The source-drain I-V characteristics of both the enhancement and depletion mode devices are shown in Fig. 5.21. Since the current in a depletion mode transistor flows through the entire thickness of the conducting layer, film properties can be calculated from the transconductance and drain current. In the linear region the transconductance is given by [5.32]:

FIGURE 5.20.   Photograph of a deep depletion-mode device.

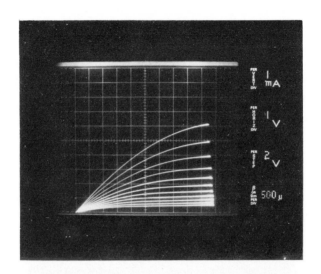

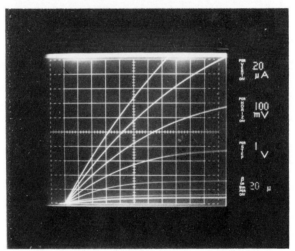

FIGURE 5.21.   (a) Source-drain I-V characteristics for a deep depletion-mode device (V_G = 0 to -12 V). (b) Source drain I-V characteristics for an enhancement-mode device.

$$g_m = \frac{I_D}{V_G} = C_{ox} \frac{W}{L} (V_D - V_S).$$

From the transistor channel geometry (L = 50 μm, W = 250 μm), the mobility is calculated to be 450 cm^2/V-sec, compared to a mobility of 750 cm^2/V sec in single crystal silicon at a dopant concentration of 6×10^{16} cm^{-3}, corresponding to the dose implanted into the polysilicon film. (Because the film melted during laser annealing, the phosphorus can be assumed to be uniformly distributed through the thickness of the film, and, as will be shown below, most of the implanted phosphorus contributed electrically active carriers. Therefore, the entire implanted dose can be used to calculate the average dopant concentration.) Because of the high oxidation temperature, the gate oxide thickness can be assumed to be the same as that on single crystal silicon (1000 Å).

The carrier concentration in the channel can also be calculated from the geometry of the device and the flatband voltage [5.32],

$$V_{FB} = \phi_{MS} - \frac{N_f}{C_{ox}} - \frac{1}{C_{ox}} \int_0^{x_{ox}} \rho(x)dx. \qquad (5.2)$$

If most of the dopant in the polysilicon is active, ϕ_{MS} is equal to that of single crystal silicon (-0.21 V), and the gate oxide thickness may again by assumed to be the same as that on single crystal silicon. Since N_f/q was found to be 1×10^{11} cm^{-2} on enhancement mode transistors fabricated on (100)-oriented single crystal wafers in the present experiment, and a value three times as high would be expected on (111)-oriented silicon, $N_f/q = 2 \times 10^{11}$ cm^{-2} is used in the calculation of V_{FB} for the polysilicon device to account for the random orientation of the crystallites. Assuming the third term to be negligible, a flatband voltage of -1.1 V is obtained with an uncertainty of ± 0.5 V because of possible deviations of N_f/q from the value used.

From the resistance at $V_G = V_{FB} = -1.1$ V, an average carrier concentration of 5×10^{16} cm^{-3} can be calculated from

$$N = L/q\mu W dR,$$

where $d = 0.5$ μm is the thickness of the film. Since this value is comparable to the average dopant concentration of 6×10^{16} cm^{-3} added by ion implantation, it is seen that a high percentage of the dopant is electrically active, and little has been lost during processing. The film properties observed in these laser-annealed polysilicon films are compared to the bulk properties of single crystal silicon and also to those of fine grained polysilicon at a dopant concentration of 6×10^{16} cm^{-3} in Table 5.4.

For the enhancement mode device the field effect mobility is similarly calculated from the source drain characteristics in the linear region and the device geometry (L = 50 μm, W = 270 μm) to

TABLE 5.4. Electrical Properties of Polysilicon and Single-Crystal Silicon at an Average Dopant Concentration of 6×10^{16} cm^{-3}.

	Laser-annealed Polysilicon	Single-crystal Silicon	Fine-Grained Polysilicon*
Carrier conc. (cm^{-3})	5×10^{16}	6×10^{16}	$\sim 1 \times 10^{12}$
Mobility (cm^2/Vsec)	450	750	~ 60
Resistivity (Ω cm)	0.28	0.14	$\sim 10^5$

*J. Y. W. Seto, J. Appl. Phys. 46, 5247 (1975), for holes.

be 340 cm^2/V sec. This value may be compared to the value of 630 cm^2/V sec expected for the field effect mobility in single crystal silicon of the same dopant concentration (50% of the bulk mobility [5.33]).

The threshold voltage (defined to be the gate voltage which induces a drain current of 1 μA) is measured to be +2.5 V. Again, assuming $N_f/q = 2 \times 10^{11}$ cm^{-2}, the threshold voltage is calculated to be −0.2 V in the absence of defect levels in the polysilicon. The difference between the measured and calculated threshold voltages may be attributed to the charging of defect levels before the surface can be inverted, as well as to uncertainties in the value of N_f/q used in the calculations. This difference represents a significant improvement over previous polysilicon MOSFETs where differences of about one order of magnitude larger were observed [5.31].

5.4.2 MOS Devices and Integrated Circuits on Polysilicon/SiO$_2$ Substrates

The experiments described in the previous section show that MOS devices can be fabricated on laser recrystallized films with electrical properties that are far superior to those obtained in as deposited material, and comparable to similar device characteristics obtained when single crystal silicon substrates are employed. However, as suggested in Sec. 5.3, the nitride layer is not an ideal choice as an insulating layer because of the high density of surface states at the nitride-silicon interface. For integrated circuit applications, a thick oxide layer is a far better choice as an insulating material. Hence most of the work done in this field since the original work of Lee et al. [5.1] has employed oxide layers.

5.4.2.1   MOSFETs [5.2]

The first MOSFET devices on SiO_2/Si substrates were those of Tasch et al. [5.2]. Enhancement mode and depletion mode transistors having W/L rations of 25/5, 25/10 and 25/25 ($\mu m/\mu m$) were fabricated on both annealed and unannealed areas of the wafers to provide a direct comparison of device type and annealed against unannealed polysilicon on the same wafer.

The I-V characteristics of enhancement (boron-implanted) and depletion (undoped polysilicon) devices, fabricated on laser annealed areas, are shown in Fig. 5.22. The threshold voltages (V_T)

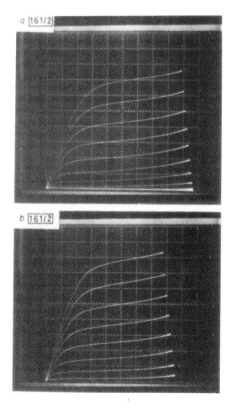

FIGURE 5.22.   Source-drain characteristics of transistors fabricated in polysilicon on oxide (W = 25 μm, L = 5 μm). The scales in both photographs are: 100 μA per vertical division, 1 V per horizontal division, 0.5 V per step, and g_m per division = 200 μS.   (a)   Enhancement mode (boron implant), V_G = 0–5V; (b) Depletion mode (undoped), V_G = 0–3.5V.

determined from measurements of I_{DS} against V_G were 0.35–0.45 V for the enhancement devices, and –0.5 to –0.7 V for the depletion devices. In the unannealed areas of the wafer, the measured threshold voltages of the polysilicon transistors ranged from 7 to 10 V, with little if any difference between implanted and unimplanted devices. The mobilities were determined from the measured slope of the drain–source current (I_{DS}) against gate voltage (V_G) curves at low drain voltage (V_D = 0.1 V) using the well known expression

$$g_m = \mu C_o \left(\frac{W}{L}\right) V_D \tag{5.3}$$

where μ is the surface mobility and C_o is the gate oxide capacitance per unit area. The measured surface mobilities were 170 and 215 cm^2/Vs, respectively, for the N–channel enhancement and depletion devices. These values compare favorably with 400 – 500 cm^2/Vs and 600 – 650 cm^2/Vs which are obtained in present N–channel SOS devices (using 0.5 μm silicon layers) and bulk NMOS devices, respectively. The measured surface mobilities on polysilicon devices (areas of wafer which did not receive laser annealing) ranged from 23 – 29 cm^2/Vs, which is in agreement with previous reported results [5.31].

Assuming the metal semiconductor work function difference ϕ_{MS} to be –0.8 to –0.9 V for a heavily doped N$^+$ silicon gate, and assuming the fixed charge at the SiO$_2$–Si interface to be Q_{ss}/q = 2.6 x 10^{10} cm^{-2} (typical of the process used to fabricate the devices in this study), then from the measured V_T of –0.5 to –0.7, a value of 10^{12} – 10^{13} cm^{-3} is inferred for the average doping level of the undoped laser annealed polysilicon, using the following equation for the threshold voltage [5.34]:

$$V_T = \phi_{MS} + 2\phi_F - \frac{(Q_B + N_f)}{C_o} \tag{5.4a}$$

where

$$Q_B = -q N_A t_{Si} \tag{5.4b}$$

In this equation, ϕ_F is the Fermi level in the silicon film, N_A is the average doping in the film and t_{Si} is the silicon-film thickness. The value of 10^{12} – 10^{13} cm^{-3} for the average doping of the undoped polysilicon is consistent with measured resistances of load resistors fabricated on undoped polysilicon deposited by the same process.

The above equation can also be applied to determine the average carrier concentration in the silicon film for the enhancement devices. Using ϕ_{MS} = –0.9 V, N_f/q = 4 x 10^{10} cm^{-2}, and the measured V_T value of +0.4 V, a concentration of 8×10^{15} cm^{-3} is is obtained. This is in good agreement with the average value of 1×10^{16} cm^{-3} calculated from the boron implantation dose and energy, when account is taken of partial diffusion of boron into the SiO$_2$ in subsequent processing. A more accurate prediction of the enhancement device threshold voltage, based on the implantation and subsequent process conditions, can be made with the aid of the SUPREM process modelling program [5.35]. For the ion implantation and fabrication conditions (temperatures and times) described earlier, SUPREM predicts a threshold voltage of 0.38 V. Also, the final calculated boron profile drops off rapidly beyond 300 nm and is down to 5×10^{13} cm^{-3} at 500 nm. Therefore, for the 500 nm silicon film on SiO$_2$, the SUPREM prediction is expected to be quite valid, since a negligible amount of diffusion of boron across the underlying SiO$_2$-Si interface is involved.

The leakage current between the source and drain in the off condition was examined, in view of the difficulty in achieving low leakage in SOS devices. Source-drain leakage currents were measured with the gate and source grounded and 5 V on the drain. The devices had channel widths of 25 µm. The results correspond to leakages of 1 - 6 pA per micrometer of channel width, and match the best reported values for SOS.

5.4.2.2   Ring Oscillators [5.4]

Ring oscillators have also been fabricated on laser recrystallized films deposited on thick oxide layers. The individual inverter stages in these ring oscillator circuits consisted of 6 µm channel enhancement mode driver devices with a deep depletion device as load. Each stage was constructed with a fan out of three, with a two stage source follower output on each ring oscillator as a buffer. Both seven and eleven stage ring oscillators were fabricated, both being functional and of very similar performance.

In Fig. 5.23, the delay per stage and the power delay product of an eleven stage ring oscillator are plotted as a function of the supply voltage (V_{DD}). These results were obtained for the ring oscillator operating in the large signal regime, namely, inverting between V_{DD} and ground. The minimum propagation delay obtainable was 44 nsec per stage at 10 volts V_{DD} and the minimum power delay product for sustaining oscillation was 4.1 pJ. At 5 volts V_{DD}, the propagation delay and the power delay product are 57.5 nsec and 7 pJ, respectively.

Isolated enhancement mode devices, which have the same geometry as the driver in the ring oscillator circuit, were also characterized. A surface electron mobility of about 300 cm^2/

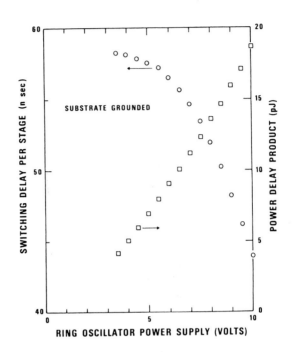

FIGURE 5.23.   Switching delay and product per state of an eleven stage ring oscillator with the substrate grounded.   The channel length of the driver transistors is 6 μm. ©1980 IEEE

V-sec (compared to 640 cm^2/V-sec measured in devices fabricated in bulk silicon slices), a threshold voltage of about -2 V and a sub-threshold leakage current of about 10 nA per micron channel width with -5 V gate voltages and 5 V supply voltage were measured. Isolated depletion mode devices having the same geometry as the load device in the ring oscillator exhibited a typical threshold voltage of -3.6 V and a drain current of typically 196 μA at 5 V supply and with the gate grounded. The subthreshold current decreased by about 3-1/2 decades per volt gate voltage. The shape of the subthreshold I-V characteristics and the fact that the leakage current did not scale with the channel width indicate that the edge-related leakage current dominates, a result not surprising since no channel stop implant was used. As a reference, the propagation delay in similar ring oscillators fabricated in bulk silicon, with a depletion width of 3 μm between the n^+ region and the substrate (and thus the same n^+ to substrate capacitance compared to that provided by the 1 μm thick oxide layer) and operating in the large signal mode is typically 36 ns at 5 V V_{DD}.

5.4.2.3   SOI/CMOS Circuits [5.36]

The most comprehensive circuit evaluation of beam recrystallized polysilicon films on oxidized silicon substrates published to date is that of Tsaur et al. [5.33]. These authors fabricated a CMOS test circuit chip containing six arrays of 360 to 533 parallel transistors, two 31stage ring oscillators and two inverter chains. The SOI structures consisted of a 0.5µm thick Si film deposited (CVD) on a 1µm thick layer of SiO_2 grown on a 2" <100> Si wafer. The film was capped with 2µm of CVD SiO_2 and 30 nm of sputtered Si_3N_4 before recrystallization. The recrystallization process employed a graphite strip heater in place of a cw scanning laser (details of this recrystallization process are described in Sec. 5.5). The yield of useable devices and circuits was in the range of 83%, with local metallization defects accounting for approximately 10% of the failures. The ring oscillators employed transistors with a gate length of 5µm and exhibited typical switching delay times of ~2ns and power-delay products of 0.2-0.3 pV at a supply voltage of 5V.

This work suggests that high quality circuits with yields comparable to those of similar circuits on bulk crystalline silicon can be fabricated on 2 inch SOI wafers. Current attempts to extend the wafer size to 100 and 125mm have led to recrystallized films that to date exceed the tolerances for wafer flatness that are acceptable for VLSI lithography.

5.4.2.4   Dynamic RAM Cells

A promising application of beam recrystallization for memory device fabrication has been proposed by Jolly et al. [5.37]. These authors have developed a dynamic RAM cell in beam recrystallized polysilicon that provides significant advantages over its single crystal counterpart. The structure is illustrated schematically in Fig. 5.24. By placing thin oxides both above and below the

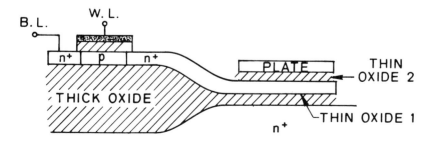

FIGURE 5.24.   Dynamic RAM cell. ©1983 IEEE

storage region, Jolly et al. were able to double the storage capacitance of the cell. Since the principal property of importance in such a cell is the ratio of storage capacitance to bit line capacitance, doubling the storage capacitance represents a significant improvement in circuit performance. In addition a thick oxide underneath the bit line reduces the bit line capacitance, representing a further improvement of the cell over its single-crystal counterpart. The complete isolation of the storage region by oxides also reduces the susceptibility of the cell to soft errors resulting from the collection of charges by junction leakage or alpha particle bombardment. The authors point out that long storage times are feasible, limited only by the leakage on the lower surface of the recrystallized film in which the access transistor is fabricated. Techniques for reducing this leakage are discussed in Sec. 5.4.5.

5.4.3   NMOS Logic Circuits in CO_2 Laser Recrystallized Silicon on Quartz [5.38]

In Sec. 5.3.5 we discussed the work of Johnson et al. [5.30], who showed that high quality single crystal islands could be grown on fused quartz substrates. High performance depletion mode and enhancement mode n-channel MOSFETs were fabricated in this material, with electron mobilities in excess of 900 cm^2/V-sec and leakage currents of less than 1 pA/μm of channel width. Building on these basic results, Chiang et al. [5.38] have optimized processing steps involving ion implantation and high temperature annealing cycles with the aid of the SUPREM simulation program [5.35] to achieve simultaneously low leakage currents and voltage thresholds appropriate for NMOS logic circuits. The threshold voltage variation achieved from this work was 0.3 volts and the yield of useable discrete devices was 98%. NMOS ring oscillators with 3 ns propagation delay/stage have been fabricated using this technology. Operational inverters and two-phase dynamic shift registers have also been demonstrated, leading to the prospect of integrating logic circuits and image-sensing arrays of thin film transistors on fused quartz substrates.

5.4.4   Minimum Feature Size Considerations [5.39]

All of the devices described to this point have employed geometries that were large enough to include many individual grains in the underlying film. Under these circumstances the properties of the devices are largely independent of the direction of current flow in the channel with respect to the direction of travel of the laser beam during recrystallization. However, this feature is not maintained as the device dimensions shrink.

The role of the grain boundaries in the laser recrystallized material on the performance of fine geometry MOSFETS was first addressed by Ng et al. [5.39]. These authors fabricated MOSFETS of various dimensions and correlated the electrical characteristics with channel length and channel width.

The output characteristics of MOSFETS with different channel lengths are shown in Fig. 5.25(a). The familiar "kink" effect is observed due to the floating substrate, a characteristic of silicon-on-insulator structures [5.40]. For channel lengths of less than 3 µm, there is a leakage current from source to drain which is independent of the gate bias. This leakage current is found to be weakly temperature dependent down to 77°K. In order to determine the origin of this leakage current, it was measured as a function of channel length for a fixed drain bias at a large negative gate voltage (-3V). The results are shown in Fig. 5.25 (b). A unique characteristic of these data is that between channel lengths of 2.3 and 3.3 µm, the leakage current drops by more than six orders of magnitude. The presence of this "threshold" channel length suggests that the leakage is not due to any mechanism that is proportional (or nearly so) to 1/L, such as leakage through grain boundaries or leakage through the bottom polysilicon /SiO_2 interface. For leakage current in short-channel devices, the possibility of punch-through effects is also eliminated due to its drain voltage dependence [5.41]. Ng et al. therefore interpret these results as arising from grain boundary diffusion of As from the source and drain regions into the channels during device fabrication. After the source and drain implantation, the wafers were subjected to a total of 90 min. at 900°C, which is a sufficient time-temperature cycle to diffuse As far as 1.5 µm into the grain boundaries [5.42]. This diffusion would explain the abrupt rise in leakage current below a channel length of ≈3 µm. Subsequent work by Ng et al. [5.43], in which source and drain As implants were annealed by a rapid thermal annealing process, enabled gate lengths to be reduced below 1 µm with no significant increase in leakage. The annealing cycle in this case consisted of a 10 second anneal at 1000°C, for which As diffusion along the grain boundaries would be negligible. Nineteen stage ring oscillators fabricated by this technique show a propagation delay of 118 ps/stage, which is the fastest reported to date for SOI structures. A photomicrograph of the device and the measured oscillation behavior is shown in Fig. 5.26.

In summary, characteristics of MOSFETS fabricated in laser crystallized polysilicon show an interesting dependence on channel length. A sharp increase in source-to-drain leakage current is observed for channel lengths below ≈3 µm when long thermal cycles are used to anneal the source and drain implants. The effective electron surface mobility is found to increase with decreasing channel length, and the highest values, compared to

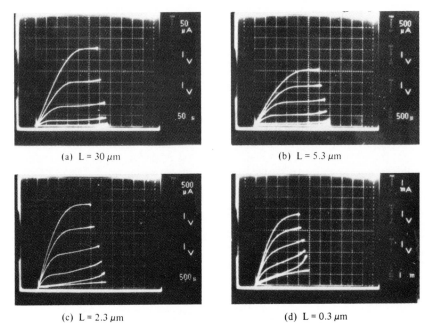

(a) L = 30 μm   (b) L = 5.3 μm

(c) L = 2.3 μm   (d) L = 0.3 μm

FIGURE 5.25(a). Output characteristics of MOSFETS with 30 μm channel width and varying channel lengths. ©1981 IEEE

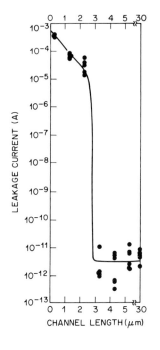

FIGURE 5.25(b). Source-to-drain leakage current as a function of channel width. Channel width = 120 μm. V_{ds} = 0.1 V. ©1981 IEEE

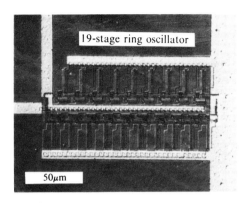

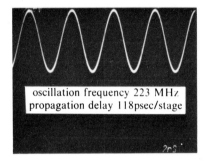

FIGURE 5.26.   MOSFET circuits in Si-on-insulator by laser crystallization and rapid thermal annealing. ©1984 Elsevier Science Publishing Co.

those from similar crystallization technique, are obtained. However, in order to capitalize on the increased mobility, enhanced diffusion through grain boundaries must be avoided, for example by using a rapid thermal annealing cycle for annealing.

5.4.5   Three-dimensional Devices on Recrystallized Polysilicon

The experiments described in the previous subsections show that MOSFETs and ring oscillators can be fabricated on the "free" surface of a recrystallized polysilicon film to obtain devices with characteristics very similar to those that would be obtained with single crystal silicon. The interface charge measurements discussed in Sec. 5.3.3 also indicate that N_f values at the $SiO_2/$ recrystallized polysilicon interface can be kept at or below the mid-$10^{11}/cm^2$ level. Additional measurements [5.27] indicate that N_f values at the crystalline silicon/SiO_2 interface can be kept at or below the mid $-10^{11}/cm^2$ level. These results suggest that devices can be made in which the bulk silicon is used for one device and the bottom of the recrystallized poly-silicon film is used for a second device. A structure that

would benefit directly from this possibility is a one-gate-wide
CMOS inverter in which the bulk silicon is used for the p channel
device and the recrystallized polysilicon film for its n channel
complement.

The basic device structure was first constructed by Gibbons
and Lee [5.26] and is shown in Fig. 5.27. The joint use of a
single gate to drive both the n- and p-channel devices led Kleitman [5.44] to suggest the term JMOS to describe this structure.
Fabrication details for this structure are described by Gibbons
and Lee [5.27]. In what follows we discuss briefly the fabrication and basic electrical characteristics of JMOS structures
that were made to explore the central idea. We then consider
subsequent investigations of the JMOS structure as a high packing
density form of CMOS, as well as other three dimensional device
structures that are related to it.

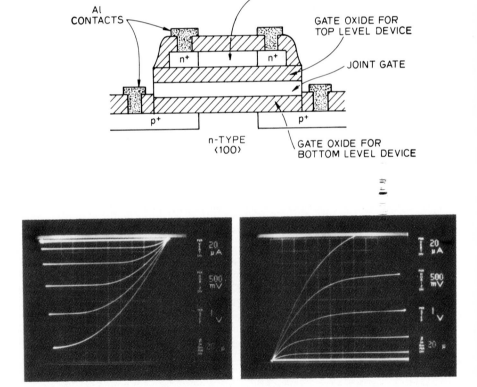

FIGURE 5.27. Schematic of a "high-rise" CMOS structure.

The drain characteristics of devices obtained from the fabrication schedule outlined [5.33] are shown in Figs. 5.27(b) and (c). An analysis of these characteristics shows threshold voltages and carrier surface mobilities comparable to those reported in the work previously described.

A similar three dimensional structure with the p channel device on top has been reported by Colinge et al. [5.45], and a self-aligned fabrication technique leading to reduction of overlap capacitance and an improved fabrication schedule has also been published by Goeloe et al. [5.46].

Alternatives to the JMOS form of three dimensional device integration have also been published by a number of authors. Kamins [5.47] has fabricated a CMOS structure with the p-channel devices in a layer of recrystallized polysilicon and the n-channel devices in adjacent, laterally displayed regions of the underlying single-crystal silicon. A schematic illustration of the transistor pair is shown in Fig. 5.28. This process proposed by Kamins allows the use of existing circuit layouts with only minor modification, which is an important feature.

Kawamura et al. [5.48] have fabricated a three dimensional CMOS integrated circuit in which one type of transistor is fabricated directly above a transistor of the opposite type, each transistor having its own gate. The structure is shown schematically in Fig. 5.29. Seven stage ring oscillators were fabricated with this device, with a propagation delay of 8.2 ns per stage. One of the most interesting features of the construction proposed by Kawamura et al. is that it can be repeated vertically to stack several transistors with separate gates and proper insulation between each device, using only minor modifications of existing technology.

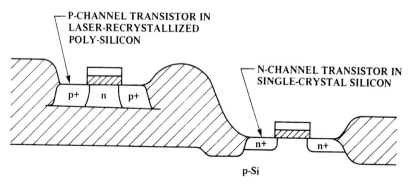

FIGURE 5.28.   Schematic cross section of transistor pair, showing a p-channel transistor in recrystallized polysilicon and an n-channel transistor in the single-crystal substrate. ©1982 IEEE

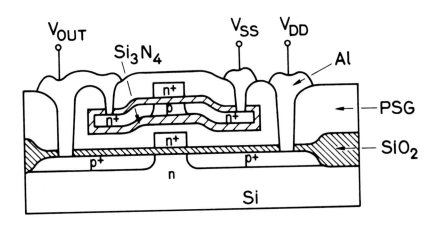

FIGURE 5.29.   Schematic cross section of the 3-D CMOS IC, showing an n-channel transistor in recrystallized silicon and a p-channel transistor in the single silicon substrate. ©1983 IEEE

5.4.6 Memory Applications of Three Dimensional Integration

In the preceding section we have described a representative set of the efforts that have been made to study the fabrication and properties of three dimensional devices. Studies of the speed of such configurations by SPICE computer circuit simulation have shown that, for the configurations so far envisaged at least, planar SOI devices are faster than their three dimensional counterparts (Gibbons et al. [5.49]). Hence circuit speed is not likely to be a feature of commercially important three dimensional circuits in the near future. However, the significant improvement in device packing density obtainable in three dimensional integration does offer promise of early application.

Two types of three dimensional memory circuits have been reported. A stacked CMOS static RAM cell has been described by Chen et al. [5.50]. For maximum packing density, the JMOS configuration of Colinge et al. referred to earlier was employed by these authors, in which the n channel devices are fabricated in the single crystal silicon. A schematic illustration of their device is shown in Fig. 5.30.

Interest in this circuit arises because it could provide a significant improvement in the noise immunity of the circuit compared to its n-MOS counterpart which uses high value polysilicon resistors as load elements.

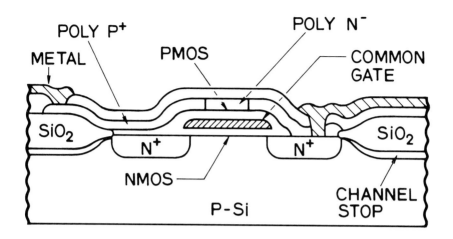

FIGURE 5.30.   Stacked CMOS structure. ©1983 IEEE

A second potential contribution of three dimensional device integration to the construction of memory circuits is that of Sturm et al. [5.51], who have modified the DRAM of Jolly et al. described earlier to obtain significant improvements in the leakage of the access transistor. A schematic of the structure is shown in Fig. 5.31, which can be obtained from the earlier circuit of Jolly et al. by simply folding the plate and a portion of the polysilicon film into the region of thick oxide underneath the access transistor.

The procedure for fabricating this structure is described by Sturm et al. [5.51] and is relatively straightforward. A large reduction of the leakage on the lower surface of the recrystallized film in which the access transistor is fabricated is obtained when sufficient bias is applied to the field plate. Storage times of several hundred seconds are obtained with this improvement, leading to device properties that rival those of DRAMS made in single crystal material.

5.4.7 Stacked MOSFETs in a Single Film of Laser-Recrystallized Polysilicon [5.52]

All of the devices envisaged in the work referenced above are based on fabricating only one device in each beam-recrystallized polysilicon film. However, one can imagine multiply-stacked structures in which <u>both</u> sides of a given recrystallized polysilicon film are driven simultaneously by independent gates. Such a structure, in which two transistors are fabricated in the same

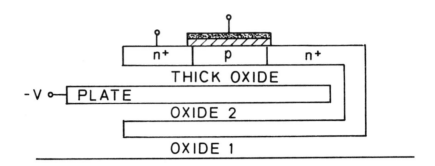

FIGURE 5.31.   Schematic cross section of the folded SOI
DRAM. ©1984 IEEE

laser-recrystallized polysilicon film was fabricated for explora-
tory purposes by Gibbons et al. [5.52]. For convenience in the
initial fabrication, the p-channel transistor employed a bulk
silicon substrate as a gate and has its channel on the lower
surface of the recrystallized polysilicon film. The n channel
device employed the upper surface of the laser recrystallized
film, which was modulated by an overlying polysilicon gate that
was fabricated conventionally. The source and drain areas of the
two devices were implanted in a cross-shaped island, as shown in
Fig. 5.32. The shape of this configuration suggests the name
cross-MOS for this type of 3D integration. While a structure
utilizing two transistors in a single device region has been
reported in bulk silicon [5.53], the present silicon-on-insulator
structure allows for the independent operation of the separate
gates.

5.5   IMPROVEMENTS IN THE RECRYSTALLIZATION PROCESS

    We now describe a number of experiments which have been per-
formed to improve the crystallographic quality of beam recrystallized films. These improvements include attempts to control the
nucleation and growth processes in order to (1) increase the
average grain size of the polycrystalline films, (2) form large
areas of single crystal material and/or (3) control the crystallographic orientation of the recrystallized material. To understand and appreciate the principles behind the experiments to be
described, it is first necessary to understand those factors that
influence the crystallographic features of the recrystallized
films. We therefore begin this section with a discussion of the
nature and origin of the features observed in continuous thin

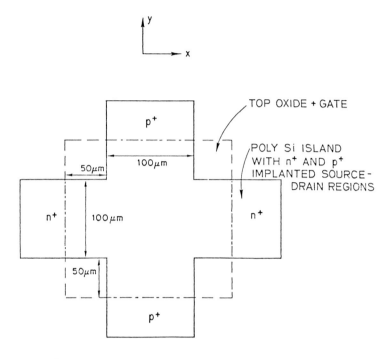

FIGURE 5.32.   Schematic top view of cross—MOS structure.
©1982 IEEE

films which have been recrystallized using a standard circular
laser beam. Subsequent sections will then describe modifications
to the film and/or recrystallization procedure which can lead to
significant improvements in the quality of the processed films.

5.5.1 Effect of Temperature Gradient

It is observed that the large crystallites which are formed
during laser recrystallization with a circular beam are developed
at an angle to the laser scan boundaries, producing a chevron
structure in the direction of the laser scan [5.6]. The direction
and curvature of these crystallites can be understood as follows:
Let "a" be the radius formed by the laser spot where the polysilicon film is completely molten, as in Fig. 5.33. "a" then
defines the radius of the region where long grains are formed. A
radial temperature gradient is produced. With the spot stationary
and the power of the laser abruptly reduced, the polysilicon film
will start cooling from the edge toward the center. The direction
of recrystallization, being in the direction of the temperature
gradient, will then result in the formation of radial grains.

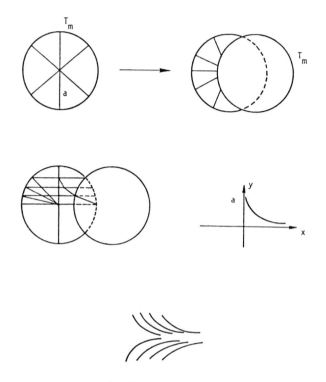

Resultant grain growth

FIGURE 5.33.   Schematic illustrating effect of temperature gradient on grain growth.   Scan direction is from left to right.

Consider now a scanning spot. The part of the spot that falls behind the scan cools off and recrystallizes. Since the isotherm is curved, the direction of the thermal gradient is not constant. As the spot moves away, recrystallization that starts out at a part of the recrystallization front close to the edge of the scan line continues at a lower part of the front close to the center of the line. A curvature in the recrystallization direction is thus produced.

One can compute the shape of the recrystallization direction if precise information about temperature profile and recrystallization velocity is known. For slow enough scan velocities, such that steady state temperature calculations apply (an estimate of the error in true temperature due to this assumption gives a value of about 5% at a scan speed of 12 cm/sec), the instantaneous temperature gradient is the same as that of a stationary beam. In particular, if the recrystallization front catches up with the molten spot fast enough, so that the super-cooled melt recrystallizes as soon as the spot is displaced infinitesimally, then the direction of the temperature gradient is given by the instantaneous radial vector. Mathematically, from Fig. 5.33, we can write

$$\frac{dy}{dx} = - \frac{y}{a^2 - y^2} \tag{5.5}$$

For the initial condition $y = a$, $x = 0$, the solution is

$$x = - a^2 - y^2 + a \ln \frac{a + a^2 - y^2}{y} \tag{5.6}$$

Initially, at the edge of the spot, the grains grow normal to the scan direction. As they continue to grow, they curve more towards the direction of the scan until at the center the recrystallization direction is along the scan direction. One can see that the recrystallization speed at various parts of the recrystallization front is related to the recrystallization speed at the center by a geometric factor. The recrystallization speed at the center must be equal to the scan speed.

5.5.2   Beam Shaping [5.54]

Since the shape of the trailing edge liquid-solid interface strongly influences the shape of the resulting grain structure, it is reasonable to assume that a modification of this interface would result in an altered structure. For instance, if the interface was concave with respect to the solidification front, crystallites would tend to propagate toward the outer edges of the laser scan. Consequently, grains which originate near the outer edges of the front would be consumed at the scan boundaries rather than growing inward in a competing manner as described for the circular case. In the same way, grains which originate near the center of the front would grow outwards towards the scan boundaries, resulting in wider grains. Experiments using this beam shape have been performed and have verified the anticipated behavior. The resulting grain structure was a reverse chevron pattern, fanning out in the direction of the scan.

When the liquid-solid interface is curved, as in both of the above cases, grain growth will not be unidirectional. For large area recrystallization, where it is necessary to have many overlapping laser scans, it would be beneficial if unidirectional grain growth were occurring, such that grains grown on a particular scan might be extended in length by subsequent scans. Using a slanted liquid-solid interface, it is possible to accomplish this type of grain growth behavior.

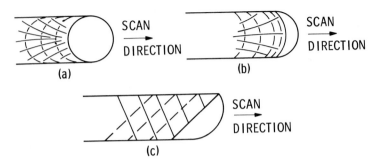

FIGURE 5.34.   Laser beam shapes and grain growth tendencies:
(a) circular, (b) crescent, and (c) half moon.

With a slanted interface, grains tend to grow from one side
of the scanned region to the other. This configuration results
in unidirectional grain growth with grains generally rectangular
in shape. Subsequent overlapping scans can then extend the length
of previously grown grains since the grain growth direction is
generally constant. In Fig. 5.34 we schematically summarize the
grain growth tendencies resulting from convex, concave and slanted
liquid-solid interfaces.

Experiments were performed by Stultz and Gibbons [5.54] to
study the possibilities of beam shaping. The laser used for
these experiments was a 18 W cw argon ion laser operated in the
multiline mode. The laser beam was expanded and collimated from
about 1 mm to 2 cm in diameter and then focused onto the sample
using a 180 mm focal length lens. A spatial filter was used to
improve the beam profile. Beam shaping was accomplished by
placing thin metal masks of appropriate shapes into the collimated portion of the beam. The samples for recrystallization
were mounted on a heated stage (350°C) which was translated
through the fixed laser beam at a rate of approximately 1 cm/s.
The scan overlap was nominally 20% of the scan width.

The samples used were 0.5 μm thick LPCVD silicon deposited
on quartz substrates. The deposition temperature was 625°C and
the deposition was performed using pure silane. The samples
intended for electrical characterization were implanted with
phosphorus to a dose of 1×10^{14} cm^{-2} at 100 keV. These samples
were then capped with 0.5 μm of CVD SiO$_2$ to prevent cracking.
Contact openings in the oxide were made and the contact areas
were implanted with phosphorus to 5×10^{15} cm^{-2}. Laser recrystallization was performed through the oxide and all electrical
measurements were made with the oxide cap in place.

The results of recrystallizing the silicon films using a
slanted interface were quite dramatic. Single scanned regions

comprised principally of rectangular grains were observed, with grain boundaries running at near right angles to the direction of the slanted liquid solid interface. The majority of the grains were as long as the scan width and about half as wide in the direction of the scan. Figure 5.35 is a photomicrograph of a single scanned region which has been Secco etched [5.55] to delineate the grain boundaries. The scan width is about 65 μm and the average grain size is about 65 μm x 25 μm. As shown, the slanted interface not only induced unidirectional grain growth but also significantly increased the average grain size of the recrystallized film, as compared to that achieved using a circular beam.

In order for a recrystallized film to be compatible with semiconductor processing techniques, the surface must be smooth and the film thickness uniform. The surface morphology of recrystallized poly using a circular beam generally degrades with increased beam diameter. For diameters greater than 100 μm, the chevron pattern begins to be less distinct and random grain growth becomes observable. To compare the effect of the slanted interface to the convex interface on the surface morphology, recrystallization was carried out by performing a series of scans using the masked portion of the beam as the trailing edge followed by a series of scans in the opposite direction. The scan width was expanded to about 150 μm. All other experimental conditions were held constant. The results are shown in Fig. 5.36. The photomicrograph using Nomarski optical microscopy shows that the surface is significantly smoother for the polysilicon recrystallized with the slanted interface.

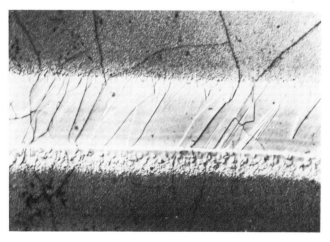

FIGURE 5.35. Photomicrograph of shaped beam laser recrystallized polysilicon using a slanted interface. The sample was Secco etched to delineate grain boundaries.

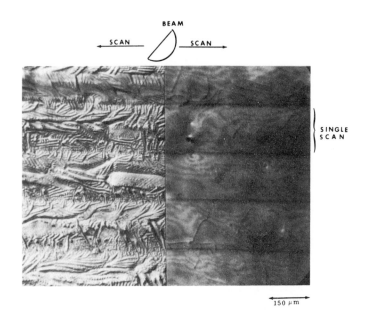

FIGURE 5.36.   Comparison of surface morphology of laser recrystallized polysilicon using slanted versus convex interface.

Previous studies on the structure of recrystallized polysilicon using a circular laser beam have shown that the grains had a mixture of crystal orientation (Sec. 5.1.6). Since films recrystallized using the slanted interface undergo unidirectional solidification, it was felt that a preferred orientation or textured structure might be present. Using a conventional X-ray diffractometer, as-deposited, circular beam recrystallized and shaped beam recrystallized films were analyzed for grain texture. In Fig. 5.37 we present the results of the diffractometer runs. All data has been corrected for finite film thickness and normalized to a common integrated intensity for direct comparison. A diffractometer trace constructed from powder diffraction data representing an ideal specimen with totally random grain orientation is also presented for comparison. As shown, the as-deposited and circular beam recrystallized films show no pronounced preferred orientation, while the shaped beam recrystallized film is clearly <100> oriented material.

Electrical characterization of the recrystallized films was made using van der Pauw-Hall measurements. Samples of 5 mm x 5 mm were prepared as described above. Circular and shaped beam recrystallization was performed through the oxide cap. The cap was used to suppress film cracking which results from the large difference between the thermal expansion coefficients of the

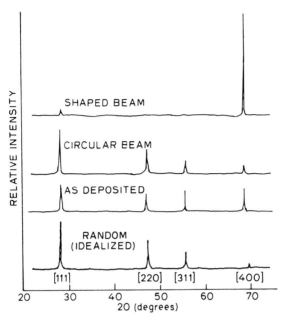

FIGURE 5.37.   Normalized diffractometer traces of shaped beam recrystallized, circular beam recrystallized, as-deposited and idealized random polycrystalline silicon.

silicon film and the underlying quartz substrate. Because of the shaped-beam-induced unidirectional crystal growth, the grain boundaries in the recrystallized film are nearly parallel to one another. In addition, each subsequent overlapping scan tends to extend the length of a grain grown by a previous scan. In this manner, grains with lengths in excess of 250 μm were grown. These grains, however, have large aspect ratios since the grain width is in general not increased by this process. As a result, the measured current-voltage relationship of a film taken parallel to the grain growth direction exhibits a much lower resistivity than that measured across the grain boundaries in the narrow direction. This results in large anisotropies in resistance when measuring the sheet resistivity using the van der Pauw-Hall method. In Table 5.5 we present electrical data obtained to compare shaped with circular beam recrystallized films. The measured carrier concentrations using the Hall technique are consistent with the nominal inpurity implant dose. The second column in Table 5.5 compares the resistance anisotropy observed using shaped versus circular beams. As expected, the shaped beam gives a higher anisotropy ratio resulting from the aligned grain boundaries. The remaining data in Table 5.5 should be considered for comparison only. It should be realized that because of the abrupt discontinuities in the recrystallized film due to grain

TABLE 5.5.   Electrical Characteristics of Shaped Beam and Circular Beam Recrystallized Silicon.

Recrystallization Conditions	Carrier Conc.(cm^{-2})	Resistance Anisotropy	Sheet Resistance $(\Omega/\square)$	Hall Mobility $(cm^2/V-s)$
Shaped Beam (6W)	6.5×10^{13}	4.1	311	307
Circular Beam (6W)	7.1×10^{13}	1.1:1	1258	69

boundaries and possible microcracks, a literal interpretation of the sheet resistance and Hall mobility cannot be made. However, it is worth noting that the mobility expected from crystalline material with the same impurity concentration is about 300 $cm^2/$ V-s, which is the value obtained using a shaped beam.

5.5.3 Effect of High Scan Speed [5.56]

The experiments described above were performed at a scan speed where a steady state approximation is valid. Schott [5.56] has examined the regime between pulsed laser annealing and conventional cw laser annealing which is achievable by increasing the scan speed of the laser. A brief discussion of this technique follows.

Thermal etching of the polysilicon films has been a problem from the beginning of laser recrystallization studies. For typical scan speeds in the 10-50 cm/s range, small laser instabilities or irregularities in the sample can initiate the stripping of the film off the insulator. The problem is less severe for nitride than for oxide due to the better wetting of nitride by molten silicon. Once initiated, thermal etching tends to propagate due to the increased thermal isolation of the remaining material. However, increasing the laser scan speed to 100 cm/s reduces the tendency to runaway etching [5.57], and increasing the scan speed to 250 cm/s virtually eliminates the problem. Decoration etching studies show no significant change in crystalline quality of the films resulting from the increased scan speed.

A related problem involves the use of a capping layer during recrystallization to prevent film movement or to influence crystalline texture. Here again there is a tendency of the molten film to de-wet, creating voids in the recrystallized film. Faster scan speeds (up to 500 cm/s) can minimize this problem.

The problems associated with the recrystallization of an MOS poly gate/interconnect level for reduced resistivity are compounded by the fact that the film is deposited over a patterned oxide. The variations in oxide thickness, and hence, thermal conductivity to the substrate, greatly increase the tendency to thermal etching, as shown in Fig. 5.38. By increasing the scan speed to 250 cm/s, etching can again be eliminated as shown in Fig. 5.39. Melting and large grain recrystallization is confined to the more thermally isolated region over the thicker field oxide and minimal effect is observed over the thin gate oxide. This is acceptable since the desired effect is in long interconnect lines over the field oxide. Even though discrete device performance is not affected, film movement during anneal can degrade circuit performance. If the film is patterned before anneal, however, and very fast scanning is used, the deposited energy density can be reduced to confine melting to small structures (e.g., narrow interconnect lines) on the field oxide, with minimal film movement and no problems in circuit behavior.

For vertical integration, one may wish to recrystallize a poly layer which is deposited over underlying device structures. Thermal pre-anneals are highly undesirable and the nonuniform recrystallization due to variations in thermal conductivity through the underlying substructure is a serious problem. Figure 5.40 shows that increasing the scan speed even further (to 500 cm/s) allows a reasonably uniform anneal over both thick and thin oxide layers simultaneously.

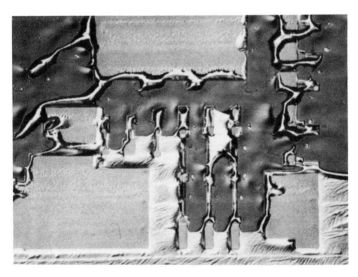

FIGURE 5.38.   Thermal etching of polysilicon over patterned oxide laser annealed at 50 cm/s (200X). ©1982 Elsevier Science Publishing Co.

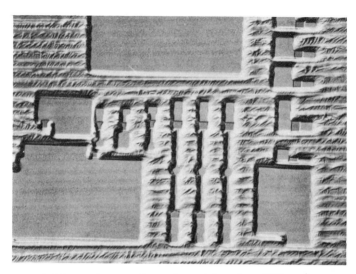

FIGURE 5.39.   Elimination of etching for 250 cm/s scan speed. Recrystallization of poly over thick (field) oxide but not over thin (gate) oxide (200X). ©1982 Elsevier Science Publishing Co.

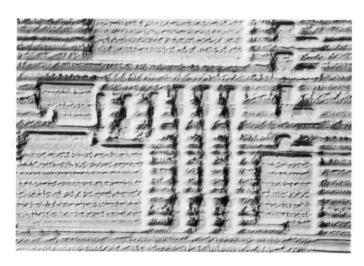

FIGURE 5.40.   Uniform recrystallization of poly over both thick and thin oxide at 500 cm/s (200X).   ©1982 Elsevier Science Publishing Co.

The fact that fast scanning reduces the sensitivity of recrystallization to variations in thermal isolation, has implications for local lateral seeding. Here one tries to obtain epitaxial regrowth of polysilicon in contact with the substrate which acts as a seed for the recrystallization of poly over oxide in adjacent areas (see Sec. 5.5.5 and Ref. [5.58]). Due to the difference in thermal conductivity to the substrate, it is difficult to get conditions of good epi growth in the contact and good recrystallization on oxide, simultaneously. If the oxide layer introduces a step over which the crystallization must proceed, it has been found that extended lateral overgrowth is even more difficult to achieve. Figure 5.41 shows a cross sectional view of an unannealed poly layer going up over a high oxide step of about 0.5 µm. Figure 5.42 shows a failure to induce single-crystal overgrowth for a scan speed of 50 cm/s. Note that there is also some thermal etching. Figure 5.43 shows that if the scan speed is increased to 500 cm/s, significant single-crystal lateral overgrowth occurs. The average extent of the overgrowth is 10 µm.

Recently explosive crystallization phenomena have been investigated in amorphous Si films on glass substrates [5.59]. Explosive crystallization was only observed above a certain threshold scan speed. Increasing the scan speed allowed the crystallization front to travel further and thereby increased the period of the characteristic arc-like features of explosive crystallization. Figure 5.44 shows a similar phenomenon occurring in the amorphous layer created by ion implantation. It is first observed (and then often sporadically) at 100 cm/s. As the scan speed is increased, no change in the period of the features (approximately 2 µm) is observed, however, up to a scan speed of 500 cm/s. The central portion of the track shows melting, while the outer portion does not.

FIGURE 5.41.   Unannealed poly layer over 0.5 µm field oxide step. (Marker equals 1 µm.)   ©1982 Elsevier Science Publishing Co.

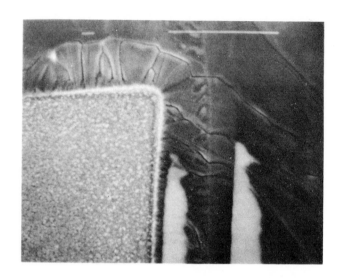

FIGURE 5.42.   No lateral overgrowth at 50 cm/s.   Laser scanned right to left, stepped top to bottom.   (Secco etched, long marker equals 10 μm.)   ©1982 Elsevier Science Publishing Co.

FIGURE 5.43.   Significant lateral overgrowth at 500 cm/s. Laser scanned top to bottom, stepped left to right.   (Secco etched, long marker equals 10 μm.)   ©1982 Elsevier Science Publishing Co.

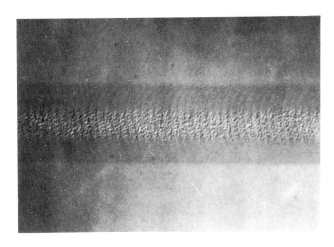

FIGURE 5.44. Explosive crystallization in implant amorphized layer in single crystal wafer for scan speeds greater than 100 cm/s (1000X). ©1982 Elsevier Science Publishing Co.

5.5.4 Beam Recrystallization of Patterned Silicon Films on Amorphous Substrates

We now describe experiments performed to study whether single orientation, single-crystal silicon could be prepared directly on an amorphous substrate (SiO$_2$ or Si3N$_4$) by patterning the poly-crystalline film prior to laser processing. This possibility is suggested by the work of Smith et al. [5.60], who found that single-crystal KCl could be grown by depositing the KCl on very finely engraved quartz substrates. It should be emphasized, however, that in the present experiments the substrates were not engraved or specially prepared in any way. Rather, the poly-crystalline film was simply etched to define islands of various sizes.

5.5.4.1 Polysilicon Islands on Si$_3$N$_4$/Si Substrates

The possibility that polysilicon islands on amorphous sub-strates could be recrystallized into single crystal material was first proven by Gibbons et al. [5.61]. The polysilicon samples used for this study were 0.55 μm thick, deposited by LPCVD on crystalline Si onto which had been deposited a 1000 Å layer of Si$_3$N$_4$ (LPCVD). Islands of polycrystalline Si were then formed by standard photolithographic techniques. The islands ranged in size from 2 x 20 μm to 20 x 160 μm. The 2x20 μm islands were arranged with the long dimension both parallel and perpendicular to the laser scan direction.

After formation of the islands, some of the wafers were multiply implanted with phosphorus according to the following schedule: $3 \times 10^{14}/cm^2$ at 50 keV, $6 \times 10^{14}/cm^2$ at 100 keV, $10^{15}/cm^2$ at 150 keV, and $6 \times 10^{15}/cm^2$ at 100 keV.

The samples were then irradiated with a scanning circular beam. The samples were held to a heated brass sample holder by a vacuum chuck during annealing. Substrate temperatures of 350°C were used for all experiments reported here. The laser was focused into a 40 μm spot. Two laser power/scan rate conditions were used: 9 W at ~0.15 cm/sec and 11 W at 12 cm/sec.

Nomarski optical observation of the annealed films revealed smooth featureless surfaces in all 2 x 20 μm islands, independent of their orientation with respect to the scanning direction. Featureless surfaces were also observed on 12 x 30 μm L shaped islands. However, the 20 x 160 μm islands show surface structure similar to that obtained on continuous films, independent of the orientation of the island with respect to the beam.

In Fig. 5.45(a) a representative scanning electron micrograph of a patterned polysilicon stripe (2 x 20 μm) array after scanning laser annealing is shown. No significant alterations in the morphology or geometric features of the stripe patterns were observed in any of the individual island structures examined after laser annealing for either of the laser power/scan rate conditions employed. In addition, the surfaces of the stripes shown in

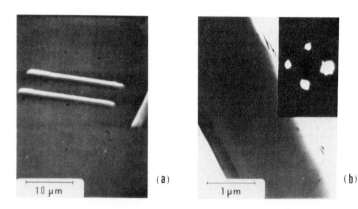

(a)   (b)

FIGURE 5.45.   Electron micrographs of laser-annealed island arrays.   (a) Scanning electron micrograph of laser-annealed islands in Si_3N_4 films; (b) bright-field transmission electron micrograph of laser-annealed island structure and selected area diffraction pattern (right inset) typical of region.

Fig. 5.45(a) appear smooth and structureless with an apparent absence of surface roughness typically noted in as-deposited polycrystalline films.

To examine the internal microstructure and crystallinity of the island structures, samples for TEM/TED analysis using a jet thinning technique were prepared. In all cases a thin (300 Å) nitride layer remained suspended across a central hole with the polysilicon islands available for observation in the electron microscope. The as-deposited island structures were composed of the polycrystalline grains ranging from 200 to 500 Å in average size. In all cases selected area diffraction patterns showed continuous ring patterns that are typically of (fine grained) polycrystalline films (see Fig. 5.45(b). In contrast, the ion implanted structures showed an absence of structure and the recorded diffraction plates exhibit the expected amorphous patterns.

After laser annealing, the 2 x 20 μm islands show an absence of fine grained or amorphous structure and are completely recrystallized to form single crystal stripes on the amorphous Si_3N_4 substrate [Fig. 5.45(b)]. Identical results were obtained in both the as-deposited (polycrystalline) and ion implanted (amorphous) stripes, suggesting that the formation of recrystallized single crystal islands is relatively independent of the initial crystalline state of the starting material for the experimental conditions used in this study.

Throughout the entire length of the stripe shown in Fig. 5.45(b) there are no microscopic defects, and selected area diffraction patterns (right inset) indicate that the island is single crystal and of (100) orientation. Of additional interest is the fact that the recrystallization is not dependent on the relative positioning of the island structure with respect to laser scan direction. Experiments conducted on samples in which the laser scan direction was parallel or perpendicular to the long axis of the strip exhibit identical results with no apparent differences in structure.

To further investigate the crystallinity of the island structures approximately one half of the thickness of the island structures was removed by chemical thinning, leaving the Si_3N_4 film intact as a support membrane. Selected area diffraction again showed only single crystal (100) patterns, confirming that recrystallization occurred throughout the entire thickness of the island.

Examination of the larger (20 μm x 160 μm) island structures and "L-shaped" arrays (12 μm x 30 μm) showed that laser annealing resulted in the formation of large polycrystalline grains of 10 μm maximum length, typically arranged in a chevron pattern

pointing in the direction of the laser scan. The occurrence of
stacking faults was also noted in some structures, independent
of island dimensions. The nucleation of stacking faults is not
thought to be essential for regrowth since smaller single crystal
island arrays and large columnar structures are observed containing no stacking faults or other microscopic defects.

From the data obtained we can conclude that defect free single crystal Si stripes of (100) orientation can be formed on
amorphous Si_3N_4 films by laser recrystallization.

5.5.4.2   Control of   Lattice Heat Flow [5.62]

A thorough investigation of the conditions necessary for producing crystalline or nearly crystalline islands on amorphous
substrates has been performed by Biegelsen et al. [5.62]. These
authors demonstrated how self-aligning techniques coupled with
choice of island shape and swept zone melting can lead to control
of nucleation and growth. In their experiments, the TEM_{00} output
of a cw argon ion laser was formed into a collimated elliptical
beam (by a cylindrical telescope) and focused to a spot ~30x90μm.
This spot was scanned (normal to its wide dimension) at 1 cm/sec
over the central axes of the 0.5-μm-thick chemical-vapor-
deposited (CVD) islands. The blackbody radiation from the heated
zone was imaged onto a vidicon either in transmission when transparent substrates (e.g. fused silica) were used, or in reflection,
when opaque substrates (e.g., silicon) were used. A microscope
lamp is also used to illuminate the nonradiating features.

Figure 5.46(a) shows a stationary molten zone dwelling in a
10-μm-wide island deposited on silica. The necking down of the
zone at the edges clearly indicates edge cooling. This is a
result of the fact that only the island is absorbing. The transparent substrate remains cold in the "sea" around the island and
acts as a lateral heat sink. (The darker region in the center of
the spot arises from a reduced collection efficiency caused by
surface fluctuations in the molten region.) An example of the
opposite situation is shown in Fig. 5.46(b). Here the islands
reside on predeposited layers of CVD polysilicon (1 μm thick)
and thermally grown oxide (88 nm). The oxide layer is maximal
antireflection thickness at room temperature. (Even at higher
temperatures, as the optical properties change, the absorption
in the silicon layer is always greater than that with no additional layer.) In this configuration more power is dielectric
absorbed by the sea than by the uncoated islands, so that a net
heat influx to the island results (or, more accurately, there is
less heat loss to the substrate at the island perimeter than at
its center). The resulting edge heating here may be undesirably
strong. To optimize the lateral thermal profile (and to use the

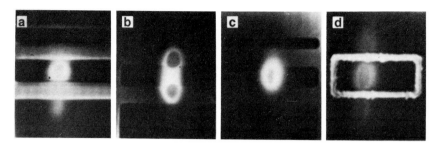

FIGURE 5.46.   Video monitor display of blackbody radiation
from silicon islands in various configurations: (a) deposited
on bare SiO$_2$, (b) continuous polysilicon layer and antireflection
dielectric layer interposed, (c) continuous polylayer and dual
layers interposed, (d) moat in continuous surround.   ©1981 Elsevier
Science Publishing Co.

minimum necessary heating of the substrate, thus minimizing
substrate damage) one can utilize other configurations as indicated in Figs. 5.46(c) and 5.46(d). Figure 5.46(c) is the same
as Fig. 5.46(b) with the addition of 64 nm of CVD silicon nitride
over the oxide below the islands. (However, this nitride can
also encapsulate the islands – see later.) This <u>dual</u> dielectric
coating is a maximally reflective coating, and the absorption in
the sea is intermediate between situations 5.46(a) and 5.46(b).
A configuration which is much simpler from a processing point of
view and which can be easily and spatially nonuniformly tailored
is shown in Fig. 5.46(d). This is the same as case 5.47(a) except
that instead of photolithographically stripping away the entire
sea, only a moat (here ~4 μm wide) is removed.

A demonstration of the results of heat flow control is provided by TEM. At relatively low laser powers, for the case of
edge cooling, large grains grow in the island centers and only
small grains at the edges. Conversely, for edge enhanced heating,
one can observe small-grain development at the island center. In
both cases at higher laser powers the small grains no longer
form; however, for the regions in which small grains would have
grown at low power, there still exists a higher probability of
competitive nucleation. This edge enhanced heating explains why
in the fortuitously chosen earliest embodiments of this technology
(i.e., polysilicon islands on silicon nitride coated crystal
silicon substrates) single-crystal islands could be grown, which
were larger than any grains grown under identical conditions
(i.e., on the same wafer, in the same laser scan) in continuous
polyregions [5.54].

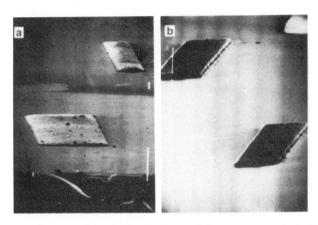

FIGURE 5.47.   SEM photographs of laser annealed islands on dual dielectric on crystal silicon substrates: (a) unencapsulated, (b) 20 nm Si_3N_4 encapsulation. ©1981 Elsevier Science Publishing Co.

For the slow scans rates used, the molten zone in a bare island is a convex, surface-tension-determined hill. Shock driven ripples quench into the solidifying topography and mass transport occurs to the trailing edge of the island. Figure 5.47(a) is a scanning electron microscopy photograph of two such islands viewed at a near glancing angle (~10°) to enhance topographic sensitivity. (The left island was scanned away from the viewer, the right toward the viewer.) It is found that ≤20 nm of an encapsulating layer, e.g. low-pressure CVD silicon nitride, reduces the above mentioned effects of surface tension on a free surface. Figure 5.47(b) shows an encapsulated island annealed under conditions identical to those of Fig. 5.47(a). In both cases complete melting has occurred. This is a worst case comparison. Encapsulation increases the optical absorption in the islands and therefore leads to a lower viscosity of the silicon. Therefore one would expect greater mass transport were it not for the physical encapsulation. Moreover, at no power did melted unencapsulated islands not show mass transport.) Flat islands can thus be achieved with negligible mass transport. A further considerable benefit of encapsulation arises from its suppression of the nucleation of ablation, i.e., surface fluctuations do not punch through and change the topology. One can thus irradiate with higher power and/or force the molten film to higher temperatures to improve crystal growth.

Another crucial step in the growth of single-crystal islands is the ability to force nucleation at a single point. The technique used here is island perimeter definition, i.e., a taper at the leading edge of the island. As the zone passes, the "point" cools first and acts as a seed for all subsequent growth on the island. (Other techniques, e.g. rectangular islands with tapered moat width might also provide the same result.) For illustration Fig. 5.48 shows TEM bright field images of two islands scanned simultaneously from left to right. The island with the leading taper is single crystalline (as shown by selected area diffraction) with a maximum width of 20 µm, whereas the island with a flat leading edge has multiply nucleated. The curving dark bands (labeled A in Fig. 5.48) are extinction and bend contours [5.63]. These are associated with single-diffraction directions and are continuous across the island. This example was chosen because it also illustrates the dominant structural defect in this growth method. The straight bordered dark areas (labeled B) are stacking faults. The lower island in Fig. 5.48 is single-crystalline in the sense that there are no grain boundaries, and selected area diffraction is everywhere the same. Frequently twins are also formed to relax the strains developed either in the growth process or arising from differential contraction relative to the substrate or encapsulant during cool down.

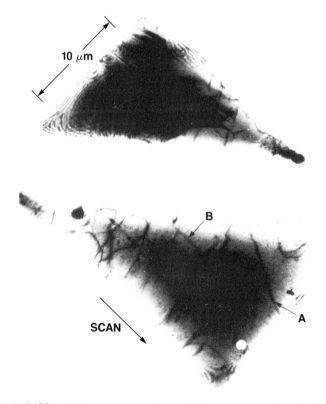

FIGURE 5.48. TEM photographs of laser annealed islands: un-encapsulated on dual dielectric and on crystal silicon substrate.

Islands 20 μm wide have been grown with no large defects and a
low density of small dislocation loops. However, control of
stress and defect formation remains as an important problem. No
attempt has been made yet to control the crystal axis normal to
the substrate.

5.5.4.3 Use of Antireflecting Stripes to Control Heat Flow

An interesting technique for controlling the heat flow based
on readily available silicon fabrication technology has been
published by Colinge et al. [5.64]. These authors use antireflecting (AR) stripes (deposited by LPCVD on the polysilicon
film prior to recrystallization and then patterned by standard
photolithographic techniques) to assure that more laser energy
is coupled into the film below the strip than in the open area
between stripes (see Fig. 5.49). The stripes are typically 10–
20 μm wide and are separated by 30–50 μm spaces. The laser
beam is scanned parallel to the stripes. Single crystal material
is propagated between the stripes (and parallel to them) near the
center of the beam. Grain boundaries tend to form at the edges
of the open area and to accumulate underneath the AR stripes
since this is the material that is hottest and recrystallizes
last.

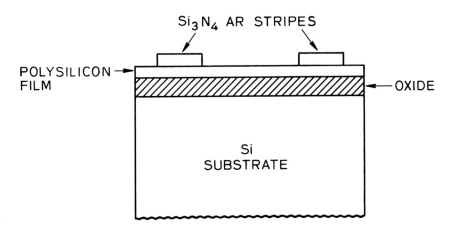

FIGURE 5.49. Use of antireflecting stripes to control heat
flow (after Colinge, et al. [5.64]).

5.5.5 Seeded Growth

5.5.5.1 First Experiment

In the previous sections of this chapter we have discussed a variety of approaches that have been used to recrystallize films into single (or "nearly-single") crystal material, in which the film is deposited directly on an amorphous substrate prior to recrystallization. However, if crystalline silicon substrates are employed, oriented single crystal silicon-on-insulator films can be grown. To accomplish this end, a polysilicon film is deposited on an oxided silicon substrate in which a certain region of the oxide has been etched away to expose the single crystal substrate. By scanning a focused cw argon laser beam onto the area where the polycrystalline silicon is deposited directly on the exposed silicon substrate, the polycrystalline silicon is converted into an epitaxial layer by a liquid phase process. By scanning the laser beam from the epitaxial region to the region where the polycrystalline silicon is deposited on the silicon dioxide layer, the polycrystalline silicon is converted into a single crystal through a a seeded growth process, where the previously formed epitaxial layer is used as the seed. This process is named "<u>lateral seeding</u>" [5.58] and is shown schematically in Fig. 5.50. Lam et al. [5.58] first explored this technique as a method for obtaining large area, controlled orientation, single crystal silicon films.

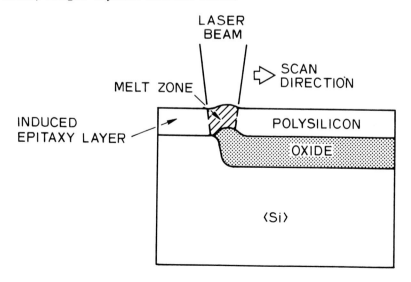

FIGURE 5.50.   Schematic representation of the lateral seeding process.
Reprinted by permission of the publisher, the Electrochemical Society, Inc.

Single crystal silicon wafers of {100} orientation were used
by Lam et al. as the substrate. A 140 nm thick layer of CVD
silicon nitride was deposited onto the surface of the wafers.
Photoresist was applied to the surface of the nitride layer and
rectangular patterns of various sizes were defined photolithographically. After developing and baking the photoresist, the
exposed nitride regions were removed by a plasma etching technique. A 0.55 μm thick layer of the then exposed silicon was
removed by an anisotropic plasma etch. After removing the photoresist, the wafers were placed in an oxidation furnace set at
1000°C with a steam ambient, timed to grow a one micron thick
layer of oxide, so that the oxide surface was almost coplanar
with the original silicon surface. After etching the wafers in a
10% HF solution for 90 sec, the nitride layer was removed by etching in hot phosphoric acid. The wafers were then etched in a 10%
HF solution for 30 sec, dried and loaded immediately into a low
pressure CVD reactor, where a 0.5 μm thick layer of polysilicon
was deposited at 620°C. A thermal anneal in nitrogen at 1100°C
for 1 hour completed the preparation. It was shown that the
1100°C anneal significantly reduces the "etching" (removing the
polysilicon from the surface of the insulator) of the polysilicon
during the laser recrystallization process [5.2]. Without this
thermal annealing step, the proper beam power window for the
recrystallization of the polysilicon-on-oxide region is typically
±0.5W. Furthermore, the beam power for proper liquid phase
regrowth of the polysilicon on-silicon region is higher than the
beam power for the recrystallization of the polysilicon-on-oxide
region due to the higher thermal conductivity of the polysilicon-
on-silicon region. Consequently, a beam power level chosen to
induce epitaxial regrowth in the polysilicon-on-silicon region
will be excessive for the recrystallization in the polysilicon-
on-oxide region and will usually result in the removal of the polysilicon from the oxide surface. The thermal annealing widens the
window of beam power for proper recrystallization of the polysilicon-on-oxide area to ±3W. The exact mechanism is not presently understood. This thermal annealing step allows the utili-
zation of a single beam power level to be used for both regions
simultaneously. The finished structure is shown in Fig. 5.51.

The laser-induced recrystallization was performed with a
scanning cw argon ion laser system. Two different lens systems,
a single 190 mm focal length spherical lens and a combination of
a 300 and a 100 mm focal length cylindrical lenses, were used.
Most of the experiments discussed were performed with the spherical lens. The raster line scan speed was 10.4 cm per sec and
the line-to-line step size varied from 5 to 20 μm. The wafers
were mounted such that the lines were scanned parallel and perpendicular to the oxide and single crystal substrate interface.
The wafers were held with a vacuum chuck and were heated to a
temperature as high as 500°C.

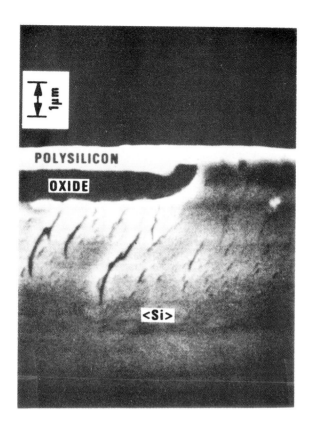

FIGURE 5.51.   Cross-sectional electron micrograph of the
finished structure.   The oxide layer is 0.96 µm thick.   The
polysilicon layer is 0.5 µm thick.   The lines below the oxide
region are artifacts of the sample preparation.
Reprinted by permission of the publisher,
the Electrochemical Society, Inc.

The wafers were studied by optical microscopy, using a Secco
etch [5.14], to decorate the grain boundaries and defects. Electron channeling was used to study the crystallographic orientation
of the laser recrystallized silicon-on-oxide material.

An optical interference contrast micrograph of a typical
laterally seeded silicon-on-oxide area is shown in Fig. 5.52. It
is generally observed that when the silicon-on-oxide area is
small, such as the smaller squares at the bottom of the picture,
no grain boundaries are observed, indicating that the silicon
layer is a single crystal. For the silicon-on-oxide areas that
are larger in size, large single crystal regions that are free
of grain boundaries are obtained immediately adjacent to the
epitaxial regions, but the entire silicon-on-oxide region is not
of one orientation.

FIGURE 5.52.   An optical interference contrast micrograph
of a lateral seeding area after etching in a Secco etch for 10
sec. The two large square silicon-on-oxide areas are 70 x 70 μm^2
in size. The 13 W laser beam scanned from left to right and the
lines stepped from top to bottom. Reprinted by permission of the
publisher, the Electrochemical Society, Inc.

The laser scan lines are also evident in Fig. 5.52.   It is
interesting to note that the single crystal region extended over
three successive scan lines, indicating that the single crystal
silicon-on-oxide area formed during the first scan served as the
seed for the growth during the subsequent scans. Therefore, the
single crystal area indeed propagated laterally, using the previously formed single crystal as the seed.

5.5.5.2   Seeded Growth with a Patterned Anti-Reflection Coat-
           ing [5.65].

We discussed in Sec. 5.5.4.3 a technique for employing a
patterned anti-reflection coating to manage the lateral heat
flow problem in a recrystallizing film. Drowley et al. [5.65]
have combined this idea with the idea of seeded growth to produce
recrystallized material that is completely free of grain boundaries except underneath the AR coating. The AR coating pattern
used by Drowley et al. consisted of a series of parallel stripes
terminating in seeding windows, as shown in Fig. 5.53. A 50 μm
x 250 μm elliptical laser beam was used for recrystallization.
The beam was scanned <u>perpendicular</u> to the stripes, with the long
axis of the beam <u>parallel</u> to the scan direction. The beam was
stepped by 1-2 μm between successive scans to promote crystal
growth. The grain boundaries were completely confined to the
region beneath the AR stripes.

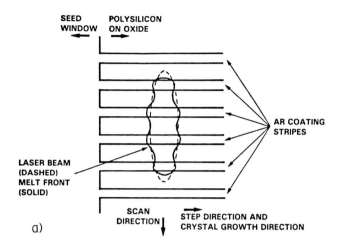

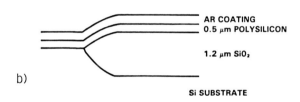

FIGURE 5.53.   (a) The recrystallization scheme.   (b) Cross-section through the seeded SOI structure. ©1984 Elsevier Science Publishing Co.

Propagation distances of up to several millimeters can be obtained for a structure containing 10 µm wide AR stripes and 10 µm wide spaces. With spaces greater than 20 µm between the AR stripes, single crystal propagation is limited to 30–50 µm from the seed edge; at that point new (typically low angle) grain boundaries nucleate and propagate in the space, similar to the behavior seen in lateral laser epitaxy without the patterned AR coating.

5.5.5.3   Experiments in Seeded Island Growth

Trimble et al. [5.66] have demonstrated lateral seeding in two types of structures. For the first structure, continuous SiO_2 sheets with narrow "via" holes filled with crystalline Si were fabricated by local oxidation of (100) Si wafers. Polycrystalline Si films were then deposited and patterned into rectangular pads by a second local oxidation process. The two patterns were

matched so that the vias were within the rectangular Si islands
and near one of their edges. A schematic cross-section of a
typical island is shown in Fig. 5.54. The first local oxidation
process ensured that there was no topological or thermal discontinuity of the deposited Si film in the via region. The second
local oxidation sequence effectively provided SiO_2 crucibles
or tubs surrounding Si islands from all but one (top) side.

The second structure was similar to that investigated by Lam
et al. [5.58] and consisted of rectangular SiO_2 pads recessed into
crystalline (100) Si wafers and then covered with a continuous
sheet of deposited polycrystalline Si (Fig. 5.53).

In both structures 1 μm of thermal, steam grown SiO_2 insu-
lated the deposited Si from the bulk Si substrate. Si films of
≈0.6 μm thickness were prepared by low pressure chemical vapor
deposition (LPCVD) at ≈650°C. Their microstructure was that of
fine grained Si with a large amorphous component.

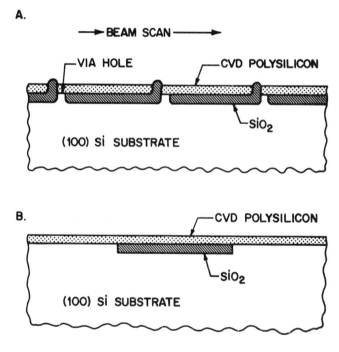

FIGURE 5.54.   Schematic cross-sections of the two structures
studied. (a) Si islands recessed into SiO_2 "tubs", with via
links to the substrate; (b) continuous Si deposited over recessed
SiO_2 pads. ©1982 Elsevier Science Publishing Co.

The crystallization and epitaxial overgrowth of the deposited
Si films were performed with a multiline cw Ar^+ laser. An intense,
10-20W, laser beam was usually focused to ≈50 μm spot on the
Si samples. The samples were vacuum clamped to a resistively
heated stage at 400°C, which was moved in a desired fashion by a
computer controlled x-y table. In some cases, the beam was shaped
with 2 cylindrical lenses into an elongated ellipse, and scanning
mirrors could be used to superimpose motion of the beam over the
table movement.

Before describing the results of recrystallization experiments, the temperature profiles expected for uniform irradiation
of different semiconductor structures will be briefly discussed.
Fastow et al. [5.67] pointed out that Si pads on oxidized Si are
hotter near the edges, contrary to the conventional crystal
growth experience and to the case of Si islands on transparent
substrates. The two characteristic situations are presented in
Fig. 5.55(a) and 5.55(b) with the corresponding temperature
profiles. The second case, with the lowest temperature at the
center, is highly advantageous for transforming the whole island
into a single crystallite, as the nucleation would start at the
single point in the center.

HEAT FLOW

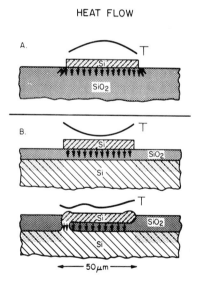

FIGURE 5.55. Temperature profiles of Si-on-insulator island
structures during uniform, argon laser irradiation. The direction
of heat flow is shown with arrows, and the lightly dotted regions
are directly heated by irradiation.   ©1982 Elsevier Science
Publishing Co.

The structure of Fig. 5.54(a) is similar to the lower part of Fig. 5.55(b). Here, a via hole link between the deposited and bulk Si acts always as a heat sink since the thermal conductivity of Si is ~100 times that of SiO_2. Most early attempts to induce epitaxial overgrowth with laser beams failed, due to large thermal mismatch between the seeding and the overgrowth regions. The problem was usually compounded by the presence of a ~1 µm step where the Si film crossed the edge of the SiO_2 layer. The local oxidation procedure, utilized in both structures shown in Fig. 5.54, planarized the surface before Si deposition, thus eliminating the topographic discontinuity. Vias were made intentionally very narrow, $\leq$ 5 µm in width, to increase the thermal impedance. Their placement at the edges of Si islands, which would otherwise be the hottest areas, reduced the ratio of temperatures between the seeding and the overgrowth regions to acceptable levels. In Fig. 5.56, we show optical micrographs of two 20 µm x 100 µm Si islands after a single traverse of a laser beam from left to right. The surface of the islands was etched to delineate grain boundaries. It is well known that unless the Si film is melted through its whole thickness, the regrowth is seeded from the remaining crystallites and submicron grains result. This allows us to estimate the temperature profile in the upper island in Fig. 5.56 which was swept with a 11W beam. It is clear that incomplete melting occurred at the vias on both ends of the rectangular pad, and also in the center, in good agreement with the analysis of Fig. 5.54(b). The lower island, which was swept with a 13W laser

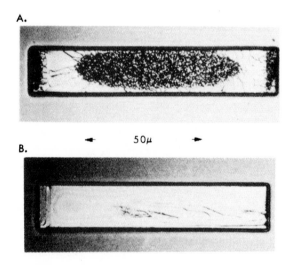

FIGURE 5.56.   Optical micrographs of laser recrystallized via islands, after chemical delineation of grain boundaries. (a) 11W, and (b) 13W laser beam was swept from left to right. ©1982 Elsevier Science Publishing Co.

laser beam, was completely melted and recrystallized from the via at the left edge. Selected area electron channeling confirmed that the crystallographic orientation was the same as the substrate; i.e., (100) plane with <100> axes aligned with the edges of the island.

A cross section of a typical recrystallized island is shown in Fig. 5.57. The oxide has been etched away to enhance contrast, followed by light etching of the Si to show the grain boundaries. The recrystallized Si is smooth, but some thickness variation is noticeable, which is likely caused by mass redistribution in the wake of the laser beam. A "bird's beak" distortion from planarity of the film near the via is characteristic of the local oxidation process. The irregularity of the surface did not, however, affect the crystallization process.

A closer inspection of Fig. 5.56 reveals that the single crystalline region breaks up into smaller grains at a distance of ≈ 30 μm from the seeding point. Others have reported similar problems in different lateral overgrowth structures [5.58]. The nature of this limitation of the seeded overgrowth is not fully understood. It was suggested by Lam et al. [5.58] that the stress buildup in the recrystallizing film, caused by the differing thermal expansion coefficients of Si and SiO_2, produces a progressively increasing defect density, with the eventual breakup of the crystalline lattice by feather shaped regions filled with dislocations, and then grain boundaries are formed. Lam et al. have also noticed that the extent of a successful overgrowth increases with the increasing substrate temperature, which is compatible with the excessive stress hypothesis. A different explanation was recently proposed by Leamy [5.59]. He argues that

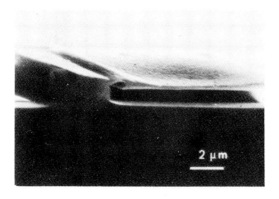

2 μm

FIGURE 5.57. Cross-sectional SEM micrograph of an island with a via. SiO_2 was etched away to improve contrast (after Ref. 5.57). ©1982 Elsevier Science Publishing Co.

zone refining of impurities contaminating the deposited film would lead to a cellular structure of crystalline Si, surrounded by a supersaturated solution of the impurities. The stress effects associated with the cell structure would then cause formation of low angle grain boundaries.

By modifying the laser irradiation procedure, the problem of limited overgrowth was eliminated. These results do not permit, however, determination of the predominant factor responsible for the limited extent of single crystalline growth in the conventional configuration.

The first successful modification of the crystallization process involved rapidly scanning the laser beam in 2 orthogonal directions such that a crescent shaped Lissajous figure resulted instead of a conventional circular molten zone. Sweeping the altered molten zone along the longitudinal axis of the rectangular islands produced seeded single crystalline regions as large as 50 x 100 μm, shown in Fig. 5.58.

In Fig. 5.59(a), an alternative recrystallization procedure is shown schematically. Here the laser beam was scanned across the islands with an incremental motion in the overgrowth direction of only 10 μm per scan. Each time the beam crosses the island, it remelts most of the Si recrystallized in the preceding scan. Since the center of the island is cooler than the edges, new crystalline growth always starts there. There is of course competing crystal growth from the side of the molten zone remote from the vias. This random growth is eliminated, however, by

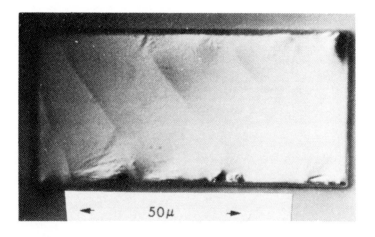

FIGURE 5.58.   Optical micrograph of single-crystalline via island produced by crescent-shaped beam (after Ref. 5.57). ©1982 Elsevier Science Publishing Co.

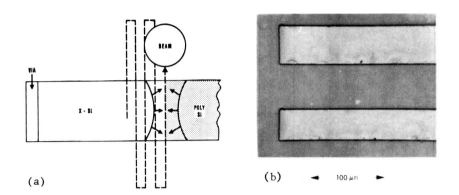

FIGURE 5.59. Transverse scanning procedure. (a) Schematic; (b) optical micrograph of transversly scanned, single crystalline via islands (after Ref. 5.57). ©1982 Elsevier Science Publishing Co.

the subsequent scans. In Fig. 5.59(b), results of transverse laser scans are presented. Single crystalline areas as large as 50 x 500µm have been formed in this manner.

5.5.5.4 Seeded Lateral Epitaxy of Thick Films

The work reviewed in the previous section of this chapter has been concerned entirely with the recrystallization of <u>thin</u> films on insulating substrates (with or without seeding), in which a scanning laser (or electron) beam is used as a heat source. With appropriate attention to heat flow considerations, the material produced in this way is single crystal (100) silicon with electrical properties that approach those of bulk silicon. However, the material contains low angle grain boundaries (sub-boundaries) that could be important in limiting the applicability of the films for bipolar transistor fabrication.

Celler et al. [5.68] have recently described a recrystallization process based on uniform melting of thick (e.g., 10-15 µm) silicon layers deposited over amorphous insulators, such as SiO_2, with an extended radiative heat source. Periodic openings in the insulator serve both as seeding regions for epitaxial recrystallization and to impose the lateral temperature gradients necessary for propagation of the solid-liquid interface. Typically Si over an entire 3-inch wafer is melted simultaneously in a few seconds, after which recrystallization proceeds over a 10-100 second period. The resultant films are single crystalline, free of any grain boundaries and contain only moderate densities of dislocations and stacking faults.

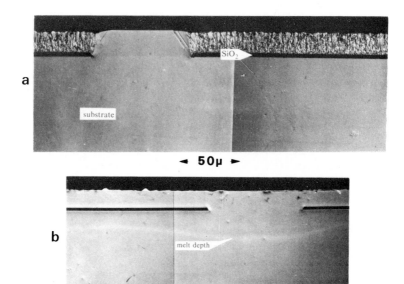

FIGURE 5.60. Optical micrographs of two wafer cross-sections after 10 sec Schimmel etching; (a) as deposited sample, (b) after lamp melting and recrystallization. ©1983 Elsevier Science Publishing Co.

The sample geometry employed by Celler et al. is shown in Fig. 5.60. Sample preparation involves oxidation, photolithography and wet etching, Si deposition (APCVD) and encapsulation with a second oxide film (2 µm of LPCVD oxide).

The system used for rapid heating is shown in Fig. 5.61. It consists of two rectangular chambers with lateral dimensions 10 x 12.5 inches, positioned one above the other. The upper chamber contains a bank of air-cooled tungsten halogen lamps suspended below a gold plated reflector. Two quartz windows separate the lamps from the wafer chamber. The wafers are placed on quartz pins ~0.5 inch above a water cooled aluminum base. The lamps provide uniform heating of the top surface of the wafers. Equally uniform radiative cooling of the back surface causes a vertical temperature gradient of ~100°C/cm, which allows controlled melting of the films deposited on the Si wafers.

At ~78 W/cm^2 incident power level, the entire deposited film and some of the underlying substrate is melted. Crystallization of the deposited silicon begins at the openings in the oxide and proceeds laterally over the oxide. Films with a thickness of 10-15 µm and surface dimensions of ~1 mm on a side can be recrystallized in this manner, making the process competitive with standard dielectric isolation techniques for fabrication of bipolar transistors with high collector breakdown voltages.

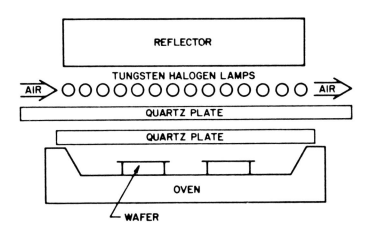

FIGURE 5.61.   Schematic cross-section of the RTA apparatus.

5.5.6   Large Area Recrystallization

One of the principal limitations evident in the early work on laser recrystallization is the small processing area which results from the use of the highly focused beam. This small area limits the area and quality of the recrystallized films, as discussed earlier. As a result, large area recrystallization techniques have been developed whereby a scanning narrow molten zone spanning the entire width of the sample, in some cases an entire wafer, has been used to process the material. This process is known as zone recrystallization and was first successfully demonstrated in silicon by Fan et al. [5.69] using a graphite strip heater. Geis et al. [5.70] have published a review of zone recrystallization efforts as they have developed since the early work of Fan et al.

In the early work of Fan, the sample was placed on a flat graphite heater which was used to raise the temperature of the entire sample to near the melting point. A narrow graphite strip located above the sample was then resistively heated and used to create a narrow molten zone in the thin polycrystalline film. The molten zone was then scanned across the film by translating the narrow strip across the sample as shown schematically

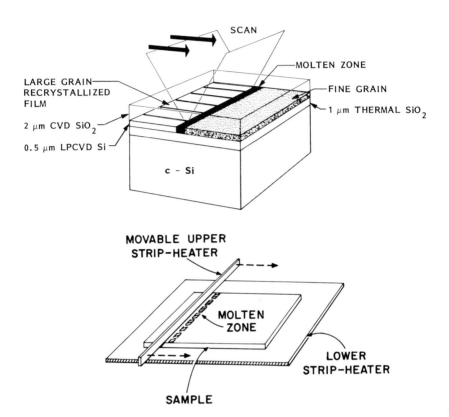

FIGURE 5.62.   (a) Schematic drawing of sample used for zone recrystallization experiments and (b) schematic diagram showing zone recrystallization using a graphite strip heater.

in Fig. 5.62(a). This technique gave large grain films (single crystal films when seeding was used) with excellent electronic properties. Subsequent to this work, other radiant sources such as focused high pressure mercury arc lamps [5.71] have been used in the zone recrystallization process. In the following sections the structure of zone recrystallized films and the origin of the characteristic low angle grain boundaries observed in these films will be discussed.

5.5.6.1 Sample Preparation and Experimental Conditions

The samples used for the zone recrystallization experiments are silicon thin films which have been deposited on thermally oxidized single crystal silicon wafers. The silicon films are deposited by low pressure chemical vapor deposition (LPCVD). The thermal oxide and LPCVD silicon are nominally 1 micron and 0.5

microns thick, respectively. Before recrystallization, the silicon layer is encapsulated with 2 microns of CVD silicon dioxide. This encapsulating layer is found to be necessary to prevent agglomeration of the molten silicon during the recrystallization process. Further, although encapsulating layers thinner than 2 microns were found to prevent agglomeration, the thicker encapsulant provided smoother recrystallized surfaces and induced a predominantly <100> texture in the recrystallized film. A schematic illustration of the sample structure used in this work is shown in Fig. 5.62(a), in which a focussed heat source is suggested as an alternative to the graphite strip geometry.

The deposited surface layer can be visualized as performing two functions; (1) reducing the total surface free energy of the molten layer by providing a second surface for the silicon to wet and (2) mechanically restraining the film from balling up or agglomerating. However, it is also found that the encapsulating layer improves the surface smoothness and induces a <100> texture in the recrystallized layer. The efficacy of the encapsulant on these two additional features was also found to improve with increasing thickness. These two effects will be discussed in more detail in later sections.

The process of zone recrystallization as it is performed with a scanning arc lamp system is as follows. The sample is placed on the substrate heater located in the reaction chamber. The chamber is evacuated and backfilled with argon gas. The substrate heater is then used to raise the sample temperature to near its melting point, between 1000°C and 1350°C. The sample temperature is monitored using an optical pyrometer. After the sample reaches the desired temperature, the arc lamp or strip heater is turned on and scanned across the film. This creates a narrow molten zone, 0.2 cm - 0.3 cm in width, which is translated through the film at rates typically less than 0.5 cm/s. Although a variety of initial sample temperatures, scan rates and power combinations have been used to create a stable molten zone, in general the best results are obtained using a sample temperature of approximately 1250°C and a scan rate of 0.2 cm/s. Unless otherwise stated, the results presented in the following sections were obtained using these conditions.

5.5.6.2 Structure of Zone Recrystallized Films

Although the zone recrystallized silicon films are in general very smooth and virtually featureless when viewed under conventional optical microscopy, some linear features or striations are faintly visible using Nomarski differential interference microscopy. To delineate the linear features, samples are etched using the Secco defect etchant [5.55]. After etching, the samples

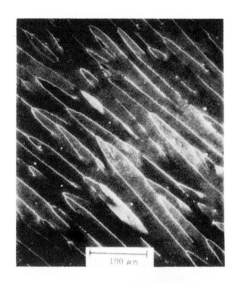

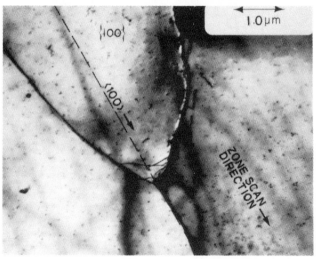

FIGURE 5.63. (a) SEM photomicrograph of Secco etched zone recrystallized film showing "hair pin" shaped low angle boundaries. (b) TEM bright field micrograph of hairpin feature.

exhibited a regular array of hairpin-like features which tended to be aligned in the zone scan direction, as shown in Fig. 5.63. The spacing of these features is essentially uniform over a given sample. However, from sample to sample, the spacing ranges from 25 microns to over 100 microns and is found to be dependent on the recrystallization conditions, film thickness and impurity concentrations of the film. The dependence of the spacing on these various parameters will be discussed below.

Transmission electron microscopy (TEM) was used to characterize the nature of the linear defects. Figure 5.63(b) is a TEM brightfield photomicrograph of one of the hairpin features. As shown, the linear features run nearly parallel to the <100> direction and the material between them has an in-plane orientation of of {100}. Further microscopic analysis revealed that the linear features were in fact low angle grain boundaries and the material between them subgrains, all in general having a azimuthal orientation near <100> and a texture of <100> in the zone scan direction. The subgrains were further analyzed using a scanning electron microscope (SEM) in the channeling contrast mode. In this mode, contrast is obtained due to variations in the crystallographic alignment of the material being irradiated [5.72]. Thus variations in the subgrain azimuthal orientation will lead to contrast under these conditions. Figure 5.64 is a SEM photomicrograph showing such contrast and thus indicating that there is indeed such a variation between subgrains in the film. To quantify the subgrain orientation variation, TEM Kikuchi pattern analyses were performed. By measuring the shift of the Kikuchi patterns obtained on either side of the low angle boundary, relative to the diffraction pattern, a very accurate determination

FIGURE 5.64. SEM photomicrograph of zone recrystallized film taken under channeling contrast conditions showing variation in azimuthal subgrain orientation.

of the degree of misorientation can be made [5.72]. A TEM photomicrograph of a typical low angle boundary and the Kikuchi patterns obtained from either side is shown in Fig. 5.65. Analyses
of these type figures indicated that the misorientations ranged
from a few tenths of a degree to a couple of degrees throughout
the film.

Low angle boundaries are generally just arrays of polygonized
dislocations [5.73]. To further characterize the low angle boundaries, TEM analyses of the dislocation networks forming these
boundaries were made. It was determined that the dislocations ran
in the ⟨110⟩ direction at 60 degrees to their Burger's vector, 1/2
⟨110⟩. This type of dislocation is typical in covalent crystals
with the diamond structure such as germanium and silicon [5.74].
A TEM photomicrograph of a low angle boundary dislocation array
and a stereographic projection showing the relative orientations
of the boundary, the dislocations and the Burger's vector are
shown in Fig. 5.66.

Although microscopic analyses of the recrystallized film
provide some very precise information concerning the nature of
the small defects in the material, it is very difficult to obtain

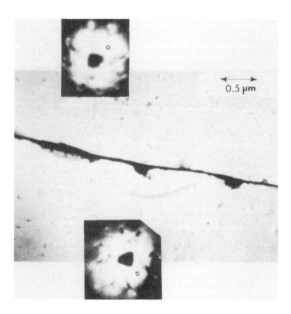

FIGURE 5.65.   TEM photomicrograph of low angle boundary with
inset Kikuchi patterns from either side. Subgrains have tilt misorientations ranging from a few tenths of a degree to a couple of
degrees.

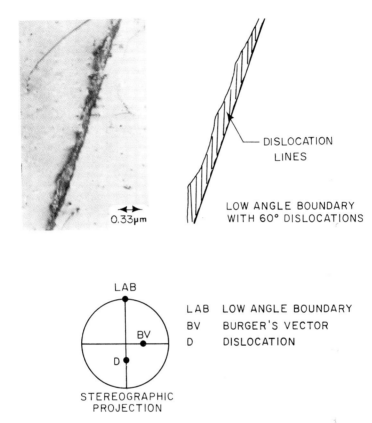

FIGURE 5.66.   TEM photomicrograph of dislocation array forming low angle boundary. Also shown are a model of dislocation array and a stereographic projection of low angle boundary, dislocation and Burger's vector.

information about the macroscopic features of the film using these techniques. This is especially true when the features of interest are much larger than the probe or beam used to evaluate the material. There is however, an ingenious technique proposed by Smith for evaluating the crystallographic orientations in large grain polycrystalline films using an array of etch pits [5.75]. This technique provides a means of determining texture, azimuthal orientation and the location of large angle grain boundaries in large grain polycrystalline silicon. With this technique, the silicon film to be analyzed is coated with a SiO_2 film of 200 nm or more. (In these experiments, the 2 micron thick SiO_2 encapsulant was used.) The SiO_2 is then coated with photoresist and a grid array of circular holes is patterned into

the resist using standard photolithographic techniques. The
holes are nominally 25 microns in diameter with a spacing of
50 microns between openings. Following the development of the
photoresist, the pattern of holes is transferred to the SiO_2
using a buffered HF oxide etchant. Finally, the silicon film
which has been exposed by the oxide etching is etched in a 40%
by weight solution of KOH in DI H_2O at 80°C for 6–8 minutes.
KOH is a very anisotropic etchant in silicon, etching much more
rapidly in the <100> direction than in the <111> [5.76]. Consequently the symmetry and alignment of the etch pits can be
related to the crystallographic nature and orientation of the
silicon being etched. Figure 5.67 schematically illustrates the
features of an etch pit which would be formed in a thin film
of silicon having {100} orientation parallel to the plane of
the film. As shown, the pits are truncated pyramids with {111}
facets or sides and 4–fold rotational symmetry. The diagonals
of the pits are in the <100> directions. Similarly, if the pits
were formed in a film with a {111} orientation they would exhibit 3–fold or 6–fold rotational symmetry. Clearly the symmetry
and diagonal alignment of the pits can be used to determine the
in-plane orientation and texture of the film being etched. Figure
5.68 is an optical photomicrograph of a zone recrystallized film
which has been etched as described above. The 4–fold rotational
symmetry and diagonal alignment of the etch pits indicate a
{100} in-plane orientation and <100> texture in the zone scan
direction, respectively. Although these data corroborate the TEM
analyses described above, the real utility of this technique is
that this information has been obtained over a very large film
area, orders of magnitude larger than the area sampled using the
TEM. Thus using a standard optical microscope, relatively expan-
sive areas of the film can be quickly analysed for this type of
crystallographic information.

A second important feature of the etch pit technique is that
the location of large angle grain boundaries is very straightforward. As shown in the photomicrograph in Fig. 5.69 a grain
boundary will result in a misalignment of the diagonals of the
etch pits in the grains on either side of the boundary, corresponding to the crystallographic misorientation of the grains
themselves. Further, as also shown, etch pits which happen to
fall on the grain boundary itself will have distorted features.

Using the etch pit technique described above, the zone
recrystallized films were found to be composed of large grains,
typically 0.05 cm to 0.2 cm in width and up to several centimeters
long. For the most part, the grains exhibited {100} texture and
alignment as discussed earlier; however, in certain cases {111}
orientations were observed. The films which had the {111} grains
were films which had been recrystallized under very rapid scan
conditions, > 0.4 cm/s. Under these conditions, the freezing

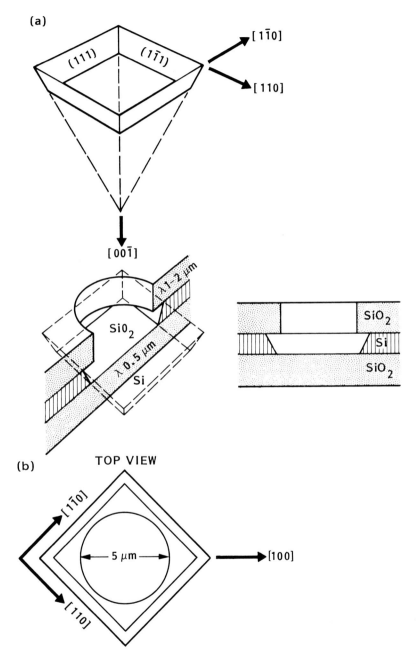

FIGURE 5.67.   Drawing showing features of anisotropic etch pit formed in a thin silicon film with {100} orientation.

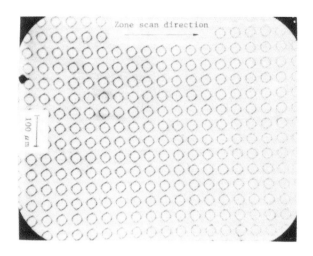

FIGURE 5.68. Photomicrograph of zone recrystallized film which has been etched using etch pit technique. Sample contains no grain boundaries.

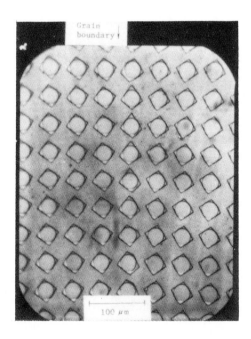

FIGURE 5.69. Photomicrograph of zone recrystallized sample which has been etched using etch pit technique. Sample has a grain boundary running down center of figure.

rate was slower than the scan rate. Consequently the solidification front would lag far behind the actual lamp position, creating a large extended molten zone. In time, the molten zone would become supercooled and unconstrained dendritic growth would occur. This type of growth leads to a very faceted film with random crystallite orientations. An example of a film grown under these conditions is shown in Fig. 5.70.

As stated earlier, films grown with thin encapsulating layers, e.g., less than 0.5 microns, tended to have faceted surfaces, even in the absence of dendritic growth. A film with such a faceted surface is shown in Fig. 5.71.

5.5.6.3   Origin of Structure in Zone Recrystallized Films

Crystallization is simply the long range ordering of atoms in a periodic solid phase structure or lattice. The final configuration of the crystalline (or polycrystalline) state is determined by a compromise between the thermodynamic driving

FIGURE 5.70.   Photomicrograph of sample which was zone recrystallized under very rapid scan conditions leading to dendritic growth.

forces and the kinetics of the reaction, and is the lowest free
energy state which can be achieved under the constraints of the
kinetics involved in the process. In the zone recrystallization
of thin films, minimization of the free energy, associated with
the crystalline bonding configuration, the grain boundary density
and the interfacial area is the primary thermodynamic driving
force which influences this final state. The kinetic limitations
are principally atomic mobility, nucleation, and adatom attach-
ment rates. In zone recrystallization, linear growth rates of
several millimeters per second are achieved. This is extremely
rapid when compared to other crystal growing techniques such as
Czochralski or Bridgman, where linear rates of tenths of millimeter per minutes are common [5.77]. Consequently it is likely
that the nucleation rate is the dominant growth rate limiting
mechanism in the zone recrystallization process.

Although there are several comprehensive theoretical treatments on the nucleation and growth of thin films [5.78-5.80],
quantitative evaluation of thin film zone recrystallization process is impeded by the lack of consistent and experimentally
verified materials data, even in the well studied silicon-silicon
dioxide system. For example, the possible role of oxygen as an
impurity which induces constitutional supercooling during the
recrystallization process is of considerable interest, as will

FIGURE 5.71.   Photomicrograph of sample with a faceted surface resulting from zone recrystallizing with a thin encapsulating
layer.

be discussed below. However, the necessary parameters for making a quantitative assessment of this role are either not available or the published values vary sufficiently that virtually any argument can be reasonably supported. As a consequence only qualitative and/or semi-quantitative arguments can be forwarded to explain the various observed phenomena. The following discussions on the observed features are presented with this in mind.

There are three striking features of the zone recrystallized film which are of interest: (1) the prevalence of the <100> azimuthal orientation, (2) the <100> growth parallel to the zone scan direction and (3) the origin and spacing of the low angle boundaries.

The predominance of the {100} in-plane orientation is a feature which is extremely fortuitous but not well understood. It is fortunate in that fabrication of MOS integrated circuits is best performed in {100} silicon material. Unfortunately there is no well-founded explanation for this growth behavior. It has been argued that the interfacial free energy between silicon and silicon dioxide is minimum for the {100} planes [5.81]. However, there exists to date no data on the interfacial free energy versus orientation for this system. Another possible explanation takes into account the anisotropic stress-strain characteristics of silicon. Since the silicon film is sandwiched between two layers of silicon dioxide, it will experience stress from differential thermal expansion as the sample is heated up. In addition there will be a contribution to the stress field resulting from volumetric contraction of the silicon upon melting. Consequently grains of different orientation will experience different densities of elastic strain energy. If these grains are in fact the seed grains for the film growth, then a textured growth from the grain with a preferred orientation might occur [5.82-5.84].

Another mechanism which is known to affect grain growth tendencies is the formation and mobility of dislocations [5.85]. Since this is indeed a polycrystalline system composed of many low angle boundaries as described above, growth accommodation by dislocation motion may play an important role in the final crystal orientation.

A final explanation for the {100} orientation has to do with the match between the planar atomic densities of {100} silicon and amorphous (or microcrystalline) silicon dioxide [5.86]. In the cited reference, an analysis of the orientation dependent oxidation rate of silicon based on planar atomic densities is given. It has been suggested that an extension of this analysis might account for the observed {100} texture [5.87]. Specifically, if the {100} planes in silicon offer the best atomic match-up with the SiO_2 surface, then growth with this orientation would result

in the least number of dangling and/or strained bonds and thus
would be energetically favorable. Unfortunately some experimentally unverified assumptions on the nature and composition of
the SiO_2 surface must be made to support this argument. These
assumptions have to do with the possible existence of microcrystalline phases of β-crystoballite and the nature of the bonding configuration in the transition region between the SiO_2 and
silicon regions.

One of the experimentally observed features in the zone recrystallized films which is not necessarily in accord with the
above explanations is the fact that films which have been recrystallized on Si_3N_4 also have the {100} orientation. Thus for
example, unless the bonding configuration and surface free energy
relations between the silicon crystal faces and both silicon
nitride and silicon dioxide are very similar, this added observation is in conflict with the interface controlled models. It is
clear from these remarks that more work must be done in this area
to obtain a satisfactory explanation for the in-plane {100} film
orientation.

As described earlier, the zone recrystallized films tend to
grow with the ⟨100⟩ direction parallel to the scan direction. It
has also been observed that the solidification front during the
zone processing is faceted rather than planar [5.84]. Lemons
[5.88] has argued that this growth is a result of constitutional
supercooling with oxygen as the most likely impurity leading to
this effect. Although constitutional supercooling can lead to
cellular and dendritic growth, the basic experimentally observed
trends in the zone recrystallized films are not entirely in accord
with this mechanism. Further, the materials parameters used to
support this argument vary sufficiently in the literature that
an equally strong argument against constitutional supercooling
can be made.

Constitutional supercooling is a situation where interface
instability can result from the existence of a zone of "constitutionally supercooled" material ahead of the liquid solid interface. This zone is termed supercooled because its equilibrium
melting temperature is depressed below the nominal melting temperature of the material due to an increase in the concentration
of an impurity in that region. This increase in concentration
is a result of the segregation of the impurity during the solidification process. (An excellent review of this phenomena is
given in Ref. 5.85, Chapter 9.) It has been shown that the
existence of a zone of constitutional supercooling ahead of a
smooth planar interface is given by the following condition

$$\frac{G}{V} < \frac{m_L \, C_s(0)(1-k_o)}{k_o \, D_L} \qquad\qquad (5.7)$$

where G is the temperature gradient in the liquid at the interface, V is the freezing velocity, m_L is the liquidus slope of
the phase diagram at the impurity concentration of interest, $C_s(0)$
is impurity concentration in the solid at the interface, D_L is
liquid diffusion coefficient and k_o is equilibrium segregation
coefficient.

To determine under what growth conditions constitutional
supercooling will exist, it is necessary to obtain values for
k_o, m_L, $C_s(0)$ and D_L. The published values of the segregation
coefficient for oxygen in silicon range from 0.5 [5.89] to 1.25
[5.90]. Thus it is not even clear whether oxygen is rejected
from or incorporated into the liquid zone. Published values for
the solid solubility limit of oxygen in silicon range from 10^{-5}
to 10^{-3} atomic fraction and the liquid diffusion coefficient
ranges from 10^{-5} cm^2/s to 10^{-4} cm^2/s. Finally, the liquidus
slope can be calculated from thermodynamics if the heat of fusion
and segregation coefficient are known.

If Eq. (5.7) is evaluated using $k_o = 0.5$, $m_L = 225$, $D_L = 10^{-5}$,
$C_s(0) = 10^{-3}$, and $V = 0.2$ cm/s, it is found that a thermal gradient
greater than 4500°C/cm is required to prevent constitutional
supercooling, which is a very unlikely situation under these
experimental conditions. Thus one could conclude with Lemons that
the film was indeed constitutionally supercooled. On the other
hand, if Eq. (5.7) is evaluated using $k_o = 1.25$, $m_L = 100$, $C_L = 10^{-4}$, $C_s(0) = 10^{-5}$ and $V = 0.2$ cm/s, then a thermal gradient of
only 0.5°C/cm would prevent constitutional supercooling. Clearly
the issue of whether or not the film is constitutionally supercooled during recrystallization cannot be made on the basis of
available data in the literature. In the following, an alternative model based on absolute supercooling and nucleation limited
kinetics is presented. This model adequately explains the faceted
growth behavior and is consistent with the observed trends in low
angle boundary spacing.

It is well known that "pure" materials with moderately high
entropies of fusion and anisotropic growth rates tend to grow with
a faceted interface. Silicon falls into this category, and thus
even in the absence of impurities, might be expected to grow in
this manner. However, another more intuitive way of viewing this
is that if the nucleation and growth rates on different crystallographic planes vary significantly, then a growing crystal would
be expected to be bounded by the slowest growing planes. In
silicon, the {111} planes are the slowest, as evidenced by its
habit form as well as by experimental observations during the
growth of crystals [5.92]. Indeed, consistent with this description, the interface facets observed during the zone recrystallization of silicon thin films have been shown to be closely
aligned with the ⟨111⟩ direction [5.84].

Now, for nucleation of new plane to occur, a certain amount of thermodynamic driving force in the form of undercooling from the equilibrium melt temperature, Tm, must be present. If we call the nucleation temperature Tn, then there will exist a region in the molten zone bounded by Tm and Tn where liquid and solid can coexist. The width of this region is set by the thermal gradient in the melt. Under steady state conditions, the shape of the solidification front will be one which permits the fastest growth rate while at the same time being bounded by the {111} planes. A planar interface aligned with the Tn isotherm offers the minimum surface area for nucleation and growth while at the same time leaving the zone of undercooled melt in front of it unoccupied by solid. Thus this is an unlikely configuration for it to assume. On the other hand, a faceted interface offers the maxiumum area for nucleation while at the same time having liquid and solid co-existing in the undercooled region. The angles between the facets are set by the crystallographic relation between the planes form-ing the facets. The length of the facets is determined by the thermal gradient in the melt; a steeper gradient reducing the distance between Tm and Tn and thus reducing the penetration depth and length of the facet.

The growth of a film with a faceted interface actually has two growth directions; one microscopic and one macroscopic. As-suming that the nucleation and growth rates on each of the facet planes are approximately equal, and imposing the restriction that the in-plane orientation must be {100} , as has been observed, then by symmetry the macroscopic growth direction must be in the <100> direction. The microscopic growth direction would be the projection of the <111> directions onto the {100} film plane, which are vectors pointing at 45 degrees to the macroscopic <100> growth direction. A schematic of this growth structure is shown in Fig. 5.72.

Assuming a growth behavior as described above, the origin of the low angle boundaries is easily understood. As the {111} facets grow and converge upon one another, any microscopic mis-orientation in any crystallographic direction will necessarily lead to a mismatch at the point of convergence. This mismatch will be accommodated by the generation of a dislocation. As growth continues, rows of dislocations will be formed along a line running parallel with the macroscopic growth direction, <100>. These rows of dislocations are the observed low angle grain boundaries.

Thus far a model has been described which explains the faceted growth and the origin of the low angle boundaries based on the anisotropic nucleation and growth rates in silicon. This model suggests that layer can occur. The facet size and thus low angle boundary spacing is dependent on the thermal gradient

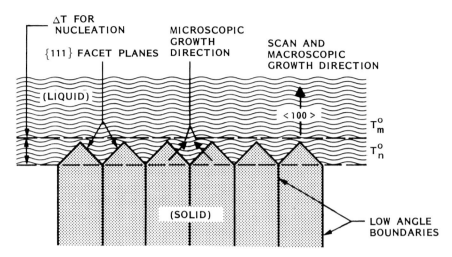

FIGURE 5.72.   Schematic drawing of faceted liquid-solid interface showing origin of low angle boundaries. T_m^o is the melting temperature of pure silicon and T_n^o is the temperature at which nucleation occurs.

in the molten zone and not on the impurity concentration. Experiments conducted to investigate this conclusion are described below.

5.5.6.4   Effect of Scan Rate

In the first series of experiments, the sub-boundary spacing was determined as a function of the zone scan rate. With all other parameters held constant, an increase in scan rate will produce a decrease in the thermal gradient. Thus an increase in scan rate should result in an increase in low angle boundary spacing. Over the limited range of scan rates under which the zone recrystallization process could be conducted without introducing other growth mechanisms; e.g. dendritic, this indeed was found to be the case. Scan rates from 0.03 cm/s to 0.35 cm/s produced boundary spacings from 25 microns to over 100 microns. Although variations of up to a factor of 2 in boundary spacing were found on a given sample, the overall trend clearly indicated an increase in spacing with increasing speed. However, these results do not conclusively rule out constitutional supercooling. In the classical case, the cell spacing in constitutionally supercooled material is inversely proportional to the product of the growth rate and the thermal gradient [5.93]. Since an independent control of the growth rate and temperature gradient is not available in the recrystallization system, it cannot be determined whether the increase in the scan rate was larger or smaller than the decrease in the temperature gradient.

5.5.6.5   Effect of Oxygen

A second experiment was performed in light of the suggestion
of Lemons [5.88] that oxygen would be the most likely impurity
creating the constitutionally supercooled condition. A series of
silicon films were doped with oxygen to 2.5×10^{17} cm^{-3}, 2.5×10^{18}
cm^{-3} and 2.5×10^{19} cm^{-3} prior to zone recrystallization. These
impurity concentrations correspond to oxygen atomic concentrations
of approximately 0.001%, 0.01%, and 0.1%, respectively. After
recrystallizing the films, they were Secco etched to delineate
the grain boundaries [5.55]. Two interesting features were
observed in the recrystallized and etched films. First, the sub-
boundary spacing was found to increase with increasing oxygen
concentration. Second, whereas the film with 0.001% oxygen
appeared similar to undoped films which had been previously
recrystallized and etched, the other series of samples were covered with small etch pits. Further, the sample with 0.1% oxygen
was uniformly covered with these pits whereas the pits on the
sample with 0.01% were located in bands running parallel to the
zone scan direction. The bands of etch pits on that sample
were uniformly spaced with a characteristic spacing approximately
3 times larger than the sub-boundary spacing. Photomicrographs
of these samples are shown in Fig. 5.73. The etch pits are interpreted as resulting from oxygen induced defects and/or oxygen
precipitates which etch more rapidly than crystalline silicon.

(a)

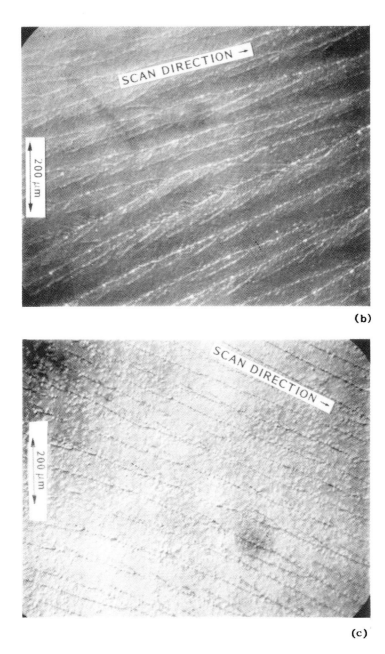

(b)

(c)

FIGURE 5.73.   Photomicrographs of zone recrystallized silicon containing (a) .001% oxygen, (b) .01% oxygen and (c) .1% oxygen. Samples have been Secco etched.

The banding is interpreted as an indication of the onset of true constitutional supercooling. In this case the interface has broken down into a cellular structure and the oxygen has segregated to the cell boundaries. However, superimposed on this structure is the faceted interface which still generates the low angle boundaries. This structure and interface is schematically shown in Fig. 5.74. The absence of banding in the 0.1% sample can be interpreted as a case where the oxygen concentration might be so high that the background concentration in the film after segregation is still very large and consequently pitting will occur in all areas of the film.

Although these experiments and observations still do not conclusively demonstrate whether the faceted growth is a result of constitutional supercooling or simply from anisotropic growth and nucleation limited kinetics, they do indicate the direction that future experiments must take to help unravel this question.

5.5.6.6   Other Defects in Zone Recrystallized Films

In Sections 5.5.6.2 and 5.5.6.3, the nature and origin of crystallographic defects which are present in zone recrystallized films were discussed. These defects are observed in otherwise high quality films. However, one also observes gross imperfections which are related to either the zone recrystallization conditions or the properties of the deposited film and/or encapsulant.

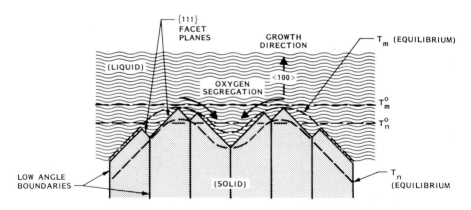

FIGURE 5.74.   Schematic drawing of model for constitutionally supercooled interface with superimposed faceted growth.

The principal defects observed in the zone recrystallized films other than the ones discussed previously are: (1) substrate slip and warpage, (2) substrate melt-through, (3) protrusions along sub-boundaries, (4) thinned out regions or film voids, and (5) film agglomeration or balling up.

Substrate slip and warp damage may or may not affect the electronic quality of the recrystallized film; however the non-planarity of the resulting sample precludes its use in modern integrated circuit processing equipment such as contact mask aligners. This defect is a result of the large thermal stresses the sample can experience during the recrystallization process. It can be minimized by insuring that there are no lateral thermal gradients introduced by the substrate heater and by using relatively slow thermal cycle rates during the preheat and cool down steps.

Substrate-melt through is a situation when the molten zone (or an area of it) becomes too hot and melting of the underlying substrate occurs. Under this condition the thermal oxide between the two molten areas can rupture and allow the film and substrate to fuse together. The resulting melt-through pit has a symmetry characteristic of the underlying substrate.

Sub-boundary protrusions are usually a result of film impurities. During the solidification process these impurities will segregate to the boundaries and locally depress the equilibrium melting temperature of the film. Consequently that region of the film will solidify after the material surrounding it. Since silicon undergoes a volumetric expansion during solidification, these small molten regions which are surrounded by solid silicon must expand upward upon freezing. Clean materials, steep thermal gradients and reduced scan velocities can effectively prevent this defect.

Regions where the recrystallized film has thinned out even to the point of exposing the underlying thermal oxide have also been observed in the recrystallized films. These nonuniformities in the film thickness can obviously pose a serious problem in device processing. Although it is believed that these "voids" may be a result of localized impurities or defects at the lower interface, no experimental evidence of this fact has been obtained. It has, however, been observed that the density of these defects decreases with increasing substrate bias temperature. This is reasonable since the surface free energy of a molten layer generally decreases with increasing temperature and thus the overall wetting properties of the film will be improved.

Film agglomeration or balling up is perhaps the most serious problem encountered during the zone recrystallization process.

It results in large areas of agglomerated silicon surrounded by the exposed thermal oxide. Although the mechanism involved in this defect is easily understood, the reliable control of it is not, at least as of this writing. The agglomeration is simply a result of trying to minimize the total free energy of the film by reducing the surface to volume ratio. The obvious solution to the problem is to reduce the interface free energy between silicon and silicon dioxide, i.e. improve the wetting characteristics of the film. The conventional approach to improving wetting characteristics is to introduce an impurity at the interface or change substrates. However, impurities are in general undesirable in semiconductor device grade silicon and the silicon-silicon dioxide system has electronic and processing properties which are very attractive. Consequently, an alternative approach is needed. As stated, encapsulating layers have been shown to be effective in preventing film agglomeration. It is generally believed that the role of the encapsulant is structural in that it mechanically suppresses film balling up. It also plays a role in reducing the surface free energy of the film by providing a second surface for the film to wet. However, since the silicon-silicon dioxide wetting characteristics are rather poor, this is only a small contribution. It has been observed that silicon wets silicon nitride better than silicon dioxide; however, the electronic properties of the silicon-silicon nitride interface are inferior to the silicon-silicon dioxide interface. Many investigators have found that a dual layer encapsulant is more effective than the single layer of SiO2 [5.13, 5.94]. However, it has also been observed by the same investigators that the efficacy of the second layer is highly dependent on the deposition technique and parameters.

Alternatively, it has been observed that the tendency for agglomeration is affected by the silicon film deposition technique and conditions. In fact it has been found that films which are not prone to agglomeration are relatively insensitive to the encapsulant. With these films the encapsulant type, deposition technique and thickness serve only to modify surface features of the recrystallized film. For example, silicon dioxide layers less than 0.5 microns have been successfully used in the recrystallization process. However, as shown earlier, the resulting films had very faceted surfaces and in some cases had a large number of grains which were not {100} oriented.

Attempts to recrystallize silicon films which have been deposited by sputtering and atmospheric pressure chemical vapor deposition (APCVD) have also been made. However the only films which were successfully zone recrystallized without agglomeration were deposited using LPCVD. Unfortunately LPCVD is not the simple answer, since it has been found that whereas films deposited in one LPCVD reactor gave good results, films deposited in another reactor may not. Chemical analyses of the films using

Auger and Secondary Ion Mass Spectrometry did not reveal any systematic differences between the deposited films. A variety of encapsulating layers was used on the films which severely agglomerated. However for those films, none of the encapsulants were successful in preventing agglomeration, including the dual layer cap described above.

Clearly this problem is the most perplexing and demanding, and its understanding and solution must be obtained before the thin film zone recrystallization process can become a universally useful technique.

5.5.6.7 Geometrical Techniques for Improving the Crystallinity of Zone Recrystallized Films

Thus far our development of the zone recrystallization process has concentrated entirely on attempts to crystallize thin, large area silicon films deposited directly on SiO_2/Si substrates. In the earlier work on laser recrystallization, both the seeding process and the introduction of diamond-shaped islands were found to improve the crystallinity and lead in some cases to material that was free of both grain and subgrain boundaries. These same techniques have been employed in the hopes of improving the quality of zone recrystallized films. In particular, Lam et al. [5.95] have used a seeding technique in which the seed was an annular ring of thickness ~1 mm on the outer edge of a 3 inch silicon wafer, the center of which was covered with a thermal oxide. High quality regrowth has been obtained with this geometry using a graphite strip heater, though surface flatness is still somewhat too large for the most demanding (VLSI) applications. The seeded films also display the hairpin structures that are characteristic of the unseeded films.

An interesting structure has also been developed by Atwater et al. [5.96] to limit and perhaps ultimately eliminate the formation of subgrain boundaries. The structure is similar to the use of the pointed island structures of Biegelsen et al. in that the film is patterned to provide a series of periodic, planar constrictions placed in such a relation to each other that only one grain can propagate through the pattern. A schematic illustration of the concept is shown in Figure 5.75. Atwater et al. have succeeded in recrystallizing films without sub-boundaries using this technique, but to date the area of the films is limited to a few square centimeters.

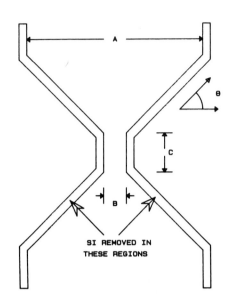

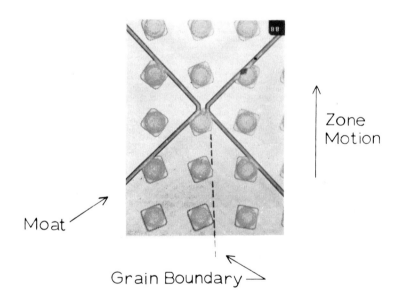

FIGURE 5.75.   (a) Hourglass shaped pattern for elimination
of sub-boundaries.   (b) Selection of a single orientation from
two initial grains by recrystallization through the neck of an
hourglass structure.

REFERENCES

5.1 Lee, K. F., Gibbons, J. F., Saraswat, K. C., and Kamins, T. I., Appl. Phys. Lett. 35, 173 (1979).

5.2 Tasch, A. F., Jr., Holloway, T. C., Lee, K. F., and Gibbons, J. F., Elec. Letts. 15, 435 (1979).

5.3 Kamins, T. I., Lee, K. F., Gibbons, J. F., and Saraswat, K. C., IEEE Trans. Elec. Devices ED-27, 290 (1980).

5.4 Lam, H. W., Tasch, A. F., Jr., Holloway, Lee, K. F., and Gibbons, J. F., IEEE Elec. Dev. Lett. EDL-1, 99 (1980).

5.5 Gerzberg, L., Gat, A., Lee, K. F., Gibbons, J. F., Peng, J., Magee, T. J., Deline, V. R., and Evans, C. A., Jr., J. Electrochem. Soc., Extended Abstracts, 80-2, 1053 (1980).

5.6 Gat, A., Gerzberg, L., Gibbons, J. F., Magee, T. J., Peng, J., and Hong, J. D., Appl. Phys. Lett. 33, 775 (1978).

5.7 Gibbons, J. F., Johnson, W. S., and Mylroie, S. W., "Projected Range Statistics in Semiconductors". Dowden, Hutchinson, and Ross (1975).

5.8 Kamins, T. I., Lee, K. F., and Gibbons, J. F., Appl. Phys. Lett. 36, 7 (1980).

5.9 Kamins, T. I., Mandurah, M. M., and Saraswat, K. C., J. Electrochem. Soc. 125, 927 (1978).

5.10 Kodera, Hiroshi, Jap. J. Appl. Phys. 2, 212 (1963).

5.11 Sze, S. M., "Physics of Semiconductor Devices", p. 68. Wiley (1981).

5.12 Kamins, T. I., Monoliu, J., and Tucker, R. N., J. Appl. Phys. 43, 83 (1972); also Tsukamoto, K. et al., J. Appl. Phys. 48, 1815 (1977).

5.13 Ohkura, M., Kusukawa, K., Yoshida, I., Miyao, M. and Tokuyama, T., Extended Abstracts of 15th Conf. on Solid State Devices and Materials, Japan Society of Applied Physics (August 1983), pp. 43-46.

5.14 Kamins, T. I., J. Electrochem. Soc., 128, 1824 (1981).

5.15 Lasky, J. B., J. App. Phys. 53, 9038 (1982).

5.16 Drowley, C. and Kamins, T. I. in "Laser-Solid Interactions and Transient Thermal Processing of Materials" (J. Narayan, W. L. Brown and R. A. Lemons, eds.), 13, 511-516, North Holland (1983).

5.17 Shibata, T., Izuka, H., Kohyama, S., and Gibbons, J. F., Appl. Phys. Lett. 35, 21 (1979).

5.18 Shibata, T., Lee, K. F., Gibbons, J. F., Magee, T. J., Peng, J., and Hong, J. D., J. Appl. Phys. 52, 3625 (1981).

5.19 Trumbore, F. A., Bell Syst. Tech. J. 39, 205 (1960).

5.20 Cook, R. K., Frey, J., Lee, K. F., and Gibbons, J. F., unpublished.

5.21 Norris, C. B., Jr., and Gibbons, J.F., IEEE Trans. on Elec. Dev. ED-14, 1, 38 (1967).

5.22 Kamins, T. I., J. Electrochem. Soc. 126, 838 (1979).

5.23 Kamins, T. I., Lee, K. F., and Gibbons, J. F., IEEE Elec. Dev. Lett. EDL-1, 5 (1980).

5.24   Le, H. P. and Lam, H. W., IEEE Elec. Dev. Lett. EDL-3, 6 (1982).

5.25   Kamins, T. I. and Drowley, C. I., IEEE Elec. Dev. Lett. EDL-3, 12 (1982).

5.26   Gibbons, J. F., and Lee, K. F., IEEE Elec. Dev. Lett. EDL-1, 117 (1980).

5.27   Kamins, T. I., IEEE Elec. Dev. Lett. EDL-3, 11 (1982).

5.28   Kamins, T. I. and Pianetta, P. A., IEEE Elec. Dev. Lett. EDL-1, 10 (1980).

5.29   Tsaur, B. Y., Fan, J. C. C. and Geis, M. W., Appl. Phys. Lett. 40, 322 (1982).

5.30   Johnson, N. M., Biegelsen, D. K., Tuan, H. C., Moyer, M. D. and Fennell, L. E., IEEE Elec. Dev. Lett. EDL-3, 12 (1982).

5.31   Kamins, T. I., Solid St. Electr. 15, 789 (1972).

5.32   Muller, R. S., and Kamins, T. I., "Device Electronics for Integrated Circuits". Wiley (1977).

5.33   Leistiko, O., Grove, A. S., and Sah, C. T., IEEE Trans. Elec. Dev. ED-12, 248 (1965).

5.34   Grove, A. S., "Physics and Technology of Semiconductor Devices", Chap. II. Wiley (1967).

5.35   Antoniadis, D. A., Hansen, S. E., and Dutton, R. W., "SUPREM II - A Program for IC Process Modelling and Simulation", Technical Report 5019-2, Stanford Electronics Labs (1978).

5.36   Tsaur, B. Y., Fan, J. C. C., Chapman, R. L., Geis, M. W., Silversmith, D. J. and Mountain, R. W., IEEE Elec. Dev. Lett. EDL-3, 12, 398 (1982).

5.37   Jolly, R.D., Kamins, T. I. and McCharles, R. H., IEEE Elec. Dev. Lett. EDL-4, 1, 8 (1983).

5.38   Chiang, A., Zarzycki, M. H., Meuli, W. P. and Johnson, N. M. in "Laser-Solid Interactions and Transient Thermal Processing of Materials" (J. C. C. Fan and N. M. Johnson, eds.), North Holland (1984) in press.

5.39   Ng, K. K., Celler, G. K., Povilonis, E. I., Frye, R. C., Leamy, H. J. and Sze, S. Z., IEEE Elec. Dev. Lett. EDL-2, 316 (1981).

5.40   Lepselter, M. P., IEDM Tech. Digest, 42 (1980).

5.41   Tihanyi, J. and Schlotterer, H., IEEE Trans. Elec. Dev. ED-22, 11, 1017 (1975).

5.42   Sze, S. M., "Physics of Semiconductor Devices", 2nd ed. Wiley (1981).

5.43   Ng, K. K., Aller, G. K., Povilonis, E. I., Trimble, L. E. and Sze, S. M., in "Laser-Solid Interactions and Transient Thermal Processing of Materials" (J. C. C. Fan and N. M. Johnson, eds.) North Holland (1984) in press.

5.44   Kleitman, D., private communication.

5.45   Colinge, J. P. and Demoulin, E., IEDM Technical Digest, 557 (1981).

5.46   Goeloe, G. T., Maby, E. W., Silversmith, D. J., Mountain, R. W. and Antoniadis, D. A., IEDM Technical Digest, 554, (1981).

5.47 Kamins, T. I., IEEE Elec. Dev. Lett. EDL-3, 11, 341 (1982).

5.48 Kawamura, S., Sasaki, N. Iwai, T., Nakano, M. and Takagi, M., IEEE Elec. Dev. Lett. EDL-4, 10, 366 (1983).

5.49 Gibbons, J. F., Giles, M. D., Lee, K. F. and Walker, J. F., SPIE Conference, Los Angeles (1983).

5.50 Chen, C.-E., Lam, H. W., Malhi, S. D. S. and Pinizzotto, R. F., IEEE Elec. Dev. Lett. EDL-4, 8, 272 (1983).

5.51 Sturm, J., Giles, M. D. and Gibbons, J. F., IEEE Elec. Dev. Lett. EDL-5, 5, (1984).

5.52 Gibbons, J. F., Lee, K. F., Wu, F. C., and Eggermont, G. E. J., IEEE Elec. Dev. Lett. 3, 8 (1982).

5.53 Hamdy, E. Z., Elmasry, M. I., and El-Mansy, Y. A., IEEE Trans. Elec. Dev. ED-28, 322 (1981).

5.54 Stultz, T. J. and Gibbons, J. F., Appl. Phys. Lett. 39, 499 (1981).

5.55 d'Argona, F. Secco, J., J. Electrochem. Soc. 119, 948 (1972).

5.56 Schott, J. T., in "Laser and Electron-Beam Interactions with Solids", p. 517. Elsevier (1982).

5.57 Gat, A. and Moore, J., private communication.

5.58 Lam, H. W., Pinizzotto, R. F. and Tasch Jr., A. F., J. Electrochem. Soc. 128, 1981 (1981).

5.59 Auvert, G., Bensahel, A. Georges, Nguyen, V. T., Henoc, P. Morin, F., and Coissard, P., Appl. Phys. Lett. 38, 613 (1981).

5.60 Gibbons, J. F. Lee, K. F., Magee, T. J., Peng, J. and Ormond, R., Appl. Phys. Lett. 34, 12, 831 (1979).

5.61 Smith, H. I. and Flanders, D. C., Appl. Phys. Lett. 32, 349 (1978).

5.62 Biegelsen, D. K., Johnson, N. M., Bartelink, D. J. and Moyer, M. D., in "Laser and Electron-Beam Solid Interactions and Materials Processing" (J. F. Gibbons, L. D. Hess and T. W. Sigmon, eds.), pg. 487. North Holland (1981).

5.63 Thomas, G., "Transmission Electron Microscopy of Metals". Wiley (1966).

5.64 Colinge, J. P., Demoulin, E., Bensahel, D., and Auvert, G. Appl. Phys. Lett. 41, 346 (1982).

5.65 Drowley, C. I., Zorabedian, P. and Kamins, T. I., "Laser-Solid Interactions and Transient Thermal Processing" (J. C. C. Fan and N. M. Johnson, eds.), North Holland (1984) in press.

5.66 Trimble, L. E., Celler, G. K., Ng, K. K., Baumgart, H. and Leamy, H. J., in "Laser and Electron-Beam Interactions with Solids, (B. R. Appleton and G. K. Celler, eds.), pp. 505-510. North Holland (1982).

5.67 Fastow, R., Leamy, H. J., Celler, G. K., Wong, Y. H. and Doherty, C. J., in "Laser and Electron Beam Solid Interactions and Laser Processing", (J. F. Gibbons, L. D. Hess and T. W. Sigmon, eds.), p. 495. North Holland (1981).

5.68 Celler, G. K. Robinson, McD., Lischner, D. J. and Sheng, T. T., in "Laser-Solid Interactions and Transient Thermal Processing of Materials", (J. Narayan, W. L. Brown and R. A. Lemons, eds.), pp. 575-580. North Holland (1983).

5.69 Fan, J. C. C., Geis, M. W. and Tsaur, B-Y., Appl. Phys. Lett. 38, 365 (1981).

5.70 Geis, M. W., Smith, H. I., Tsaur, B-Y., Fan, J. C. C., Silversmith, D. J., Mountain, R. W. and Chapman, R. L., in "Laser-Solid Interactions and Transient Thermal Processing of Materials, (J. Narayan, W. L. Brown and R. A. Lemons, eds.), pp. 477-489. North Holland (1983).

5.71 Stultz, T. J. and Gibbons, J. F., Appl. Phys. Lett., 41, 824 (1982).

5.72 Bowen, D. K. and Hall, C. R., "Microscopy of Materials." Cambridge Press (1973).

5.73 Kelly, A. and Groves, G. W., "Crystallography and Crystal Defect,", pp. 345-356. Addison-Wesley (1970).

5.74 Ibid, pp. 254-255.

5.75 Bezjian, K. A., Smith, H. I., Carter, J. M. and Geis, M. W., J. Electrochem. Soc. 129, 1848 (1982).

5.76 Kendall, D. L., Annual Review of Materials Science, 9, 373 (Huggins, Bube, Vermilyea, eds.), (1979).

5.77 Brandle, C. C., "Crystal Growth," (Pamplin, B. R., ed.), pp. 275-301. Pergamon Press (1980).

5.78 Sato, H., Annual Review of Materials Science 2, 217 (1972).

5.79 Joyce, B. A., Rep. Prog. Phys. 37, 363 (1974).

5.80 Chapera, K. L., "Thin Film Phenomena," Chap. 4. McGraw-Hill (1969).

5.81 Leamy, H. J., "Laser and Electron Beam Interactions with Solids", (B. R. Appleton and G. K. Celler eds.). North Holland (1982).

5.82 Vook, R. W. and Witt, F., J. Appl. Phys. 36, 2169 (1965).

5.83 Vook, R. W. and Witt, F., J. Vac Sci. Tech. 2, 243 (1965).

5.84 Geis, M. W., Smith, H. I., Tsaur, B-Y, Fan, J. C. C., Silversmith, D. J. and Mountain, R. W., J. Electrochem. Soc. 129, 2812 (1982).

5.85 Cahn, R. W., "Physical Metallurgy" (Cahn, ed.), Chap. 19. North-Holland (1970).

5.86 Tiller, W. A., J. Electrochem. Soc. 128, 689 (1981).

5.87 Tiller, W. A., private communication.

5.88 Lemons, R., "Laser-Solid Interactions and Transient Thermal Processing of Materials", (J. Narayan, W. L. Brown and R. A. Lemons, eds.). North-Holland (1983).

5.89 Trumbore, F. A., BSTI 39, 205 (1960).

5.90 Yatsurugi, Y., Akiyama, N. and Endo, Y., J. Electrochem. Soc. 120, 975 (1973).

5.91 Jackson, K. A., "Crystal Growth", (Bardsley, Hurle and Mullin, eds.), Chap. 5. North-Holland (1979).

5.92 Abe, T., Kikuchi, K. Shirai, S. and Muraoka, S., "Semiconductor Silicon", Electrochem. Soc. (1981).

5.93   Flemings, M. C., "Solidification Processing", Chap. 3. McGraw-Hill (1974).

5.94   Maby, E., Geis, M., LeCoz, Y., Silversmith, D., Mountain, R. and Antoniadis, D., IEEE Elect. Dev. Lett. EDL-2, 214 (1981).

5.95   Lam, H. W., personal communication.

5.96   Atwater, H., Smith, H. I. and Geis, M. W., Appl. Phys. Lett. 41, 747 (1982).

CHAPTER 6

Metal–Silicon Reactions and Silicide Formation

T. Shibata*, A. Wakita, T. W. Sigmon, and James F. Gibbons

STANFORD ELECTRONICS LABORATORIES
STANFORD UNIVERSITY
STANFORD, CALIFORNIA

6.1 INTRODUCTION

The reaction of thin metal films with both single crystal and amorphous silicon to form silicide compounds is a subject that is currently receiving considerable attention. This interest arises from the fact that future trends in silicon integrated circuit technology, particularly the push toward higher device packing density and decreased dimensions, will require low resistance, thin film materials with high temperature properties that are compatible with other silicon processing operations. Such films will find important applications as gates for MOS transistors, device interconnects and ohmic contacts [6.1-6.3].

Presently there are a number of silicide compounds that look promising for fulfilling these needs. An excellent review of the thermal equilibrium interdiffusion and reaction of thin metal films on silicon, including silicide formation, is given in Ref. [6.4]. Here both the deposition and furnace annealing processes are discussed in detail, along with useful measurement techniques for determining reacted film composition and quality.

Within the last two years, the use of laser, electron and ion beams to react metal/silicon films have also been actively investigated [6.5-6.12]. New and unique features provided by these forms of beam processing include shorter reaction times, localized heat treatment and rapid heating and cooling rates. As in the

---

*Present address: VLSI Research Center, Toshiba Research and Development Center, Kawasaki City, Kanagawa, 201 Japan.

case of beam annealing of ion-implanted semiconductors, both the basic mechanisms and the experimental results depend on whether pulsed or scanned cw beams are used to promote the reactions. For both cases experimental results are found in many cases to differ sharply from those obtained with standard furnace processing.

Since the use of pulsed laser and electron beams to form silicides is thoroughly reviewed elsewhere [6.13], only a brief description of the most important results will be given here for comparison with the cw case. Following this brief review, we will concentrate on the experimental results obtained when scanned cw laser and electron beams are employed to produce reactions in a metal/silicon system.

6.1.1 Pulsed Laser Silicide Formation

The most important results obtained when a laser beam is used to react a thin metal film with a silicon substrate are summarized in Table 6.1. As of this writing there is no published data on the use of pulsed e-beams for silicide formation. Since the principal results obtained with the cw-scanned e-beam [6.8] are similar to those obtained with the cw laser, they are not included in Table 6.1; however, important differences are discussed in Sec. 6.4 of this chapter.

It can be seen in Table 6.1 that the essential difference between the pulsed and cw scanned lasers is the length of time during which the laser energy is deposited in the film (20–200 ns as

TABLE 6.1   Summary of Silicide Formation Results for Pulsed and Scanned CW-Laser Processes.

	Pulsed Laser	CW Laser
Laser Irradiation Time	20–200 nsec	msec
Mechanism of silicide formation	Melting, mixing in a molten phase and rapid solidification	Solid phase diffusion (basically) Nucleation controlled reaction and eutectic melting can also be involved.
Results:		
Silicide phase formed	Mixed phase with Si precipitation	Single phase (generally) with no Si precipitation.
Composition of the film	Continuously changes by increasing powers	Selected by power levels.
Structure of the film	Cellular structure	Homogeneous layer
Silicide/silicon interface	Deep metal penetration into Si substrate	Well-defined interface
Application to refractory metals on Si	Unable to react thicker films	Possible with multiple laser scans
	Formation of bubble structure	

opposed to ~ 1 ms). It will become clear that this exposure time and the associated heating and cooling rates play a decisive role in the physical processes that occur during silicide formation.

As is the case for the pulsed laser annealing of ion-implanted, elemental semiconductors, melting and rapid resolidification occurs in metal/silicon films that are reacted using pulsed lasers [6.5]. The rapid melting is typically followed by the nucleation and growth of a number of metal/silicon species in the molten phase, after which a very rapid resolidification occurs due to the extremely high cooling rates involved [6.14]. The principal result is that the reacted film has frozen in a mixed phase compound that consists of several silicide compounds and also silicon and metal precipitation [6.15]. The composition of the reacted layer can be changed by adjustment of the laser power level. It is also possible to obtain compositions inaccessible by conventional furnace annealing. However, for the formation of <u>uniform</u>, large area, <u>single phase</u> compounds, pulsed laser processing is not presently useful.

The structure of the films, as revealed by transmission electron microscopy, indicates that they consist of small cellular silicide clusters on the order of 0.1 μm in diameter, interspersed with silicon precipitates that are separated from each other by metal-rich walls [6.16]. The simultaneous formation of large cells, on the order of 1 micron in diameter is also observed; and deep penetration of the metal into the silicon substrate has been verified [6.16].

The formation of this mixed-phase cellular structure would be expected from the rapid melting-resolidification mechanism for the reaction. These results are in sharp contrast to those obtained with the scanned cw laser, where uniform, large area, single-phase silicide compounds are produced.

6.1.2 Scanned cw-Laser and E-Beam Silicide Formation

Recent work in the application of scanned cw-lasers and e-beams for the formation of silicide compounds has led to a process that appears to be superior in many respects to that obtained with pulsed lasers. As in the annealing of ion-implanted amorphous layers in single crystal, elemental semiconductors, a <u>solid phase</u> process is responsible for the metal-silicon reactions obtained when the scanning cw beam is used at sufficiently low power levels. In the remainder of this chapter we will discuss the mechanisms by which single phase silicide compounds are formed when a scanned cw laser or electron beam is used. A brief overview of the data to be presented is provided below for convenient comparison with the pulsed case.

Basically it is found that single phase silicide compounds with little or no Si precipitation are obtained when a scanned cw beam is used to promote the metal-silicon reaction. The phase of the film (e.g., Pd_2Si or PdSi) can be selected by appropriate choice of the power level in the scanning beam. In those special cases where more than one phase appears, the ratio of the phases can be determined in some cases from the power/(beam radius) factor of the laser. The films are generally found to be homogeneous with a sharply defined silicon/silicide interface [6.17]. The formation of refractory metal silicides has also been accomplished, with the successful formation of $MoSi_2$, $NbSi_2$ and WSi_2 films reported in the literature [6.18]. Finally, it is possible to establish reaction kinetics by multiple scanning in a manner to be described below.

Although the majority of metal/silicon reactions studied depend on a solid phase interdiffusion mechanism, certain anomalies are observed. For the special case of PdSi formation, a nucleation-controlled reaction is believed to dominate [6.7], while for PtSi evidence of eutectic melting and rapid resolidification is obtained [6.19].

6.2   EXPERIMENTAL TECHNIQUES

In this section we first discuss the sample structures that are normally used to study the formation of silicides by beam processing. We then describe the principal analytical methods that are used to study the physics of the beam annealing process. These include Rutherford backscattering, the X-ray Read camera, and optical microscopy.

6.2.1   Sample Preparation

In Fig. 6.1 we show a schematic illustration of the sample structures utilized in these experiments. To obtain well-controlled reactions, the metal films were deposited by electron beam evaporation (onto p-type <100> silicon substrates) in a high vacuum, oil-free system. Sputter deposition of the metal films was avoided since it normally leads to sufficient Ar inclusion to inhibit the the desired reaction.

As suggested in Fig. 6.1(a), samples used for e-beam processing consisted only of the silicon single crystal substrate and the metal film. When the cw laser was employed to promote silicide formation, it was necessary to add a thin (200 Å) layer of silicon on the metal surface to provide an antireflection coating Fig. 6.1(b)]. Without this silicon overcoating, a high laser

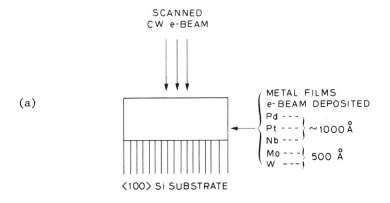

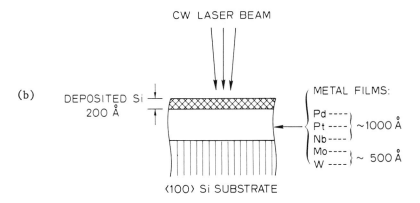

FIGURE 6.1.   Schematic of sample structure and film thick-nesses utilized in the (a) electron-beam and (b) cw laser reacted silicide experiments.

power is required to compensate for the substantial amount of laser power that is reflected from the metal surface. Then, once the silicide begins to form, the lower reflectance of the silicide compound results in an increased power absorption in the film during the reaction, leading to difficulty in controlling the reaction. This problem is easily avoided by use of the thin silicon overcoat. Since the reflectance does not change significantly for the electron beam, this overcoat is not required for the scanning electron beam experiments.

In Table 6.2 we list the principal metal/silicon systems to be discussed in this chapter. Also listed are some of the basic properties that characterize each sample, such as the metal layer thickness and whether or not a silicon overcoat is used.

TABLE 6.2   Description of Samples used for Scanned cw-Laser and Electron-Beam Silicide Formation.

Beam Process	Si-Overcoating	Metal Layer	Substrate
Scanned-Laser Beam	Si(200 Å)	Pd(1300 Å) Pt(1000 Å) Mo(530 Å) W (440 Å) Nb(1100 Å)	P-type <100> Si
Scanned-Electron Beam	None	Pd(1400 Å) Pt(1100 Å) Nb(1250 Å)	

6.2.2   Typical Laser/Electron Beam Annealing Conditions

The laser processing system used for all experiments reported here is the cw argon system described in Chapter 1. The laser was operated in the multi-line mode. The beam was focussed by a 135 mm lens to obtain a spot on the sample surface approximately 50 μm in diameter. The spot was scanned across the sample surface at a rate of approximately 12 cm/sec. The substrate was held at a fixed temperature during the laser irradiation. Substrate temperatures used for the various reactions studied are listed in Table 6.3.

The laser beam irradiation produces a temperature distribution that is Gaussian-like. The peak temperature in the center of the beam is given by the formula derived in Chapter 2 [Eq. 2.46]:

$$T_{max} = T_k + (T_0 - T_k) \exp [(1 - R)P/(4\pi)^{1/2}wA] \qquad (6.1)$$

where T_0 is the substrate temperature and T_k and A are constants in the empirical expression for the temperature-dependent thermal conductivity. w is the beam radius, P is the laser output power and R is the reflectivity of the sample. To determine the laser power required to achieve a specific temperature, it is necessary to know R and w. R can be experimentally measured before each scan, and w is determined from Eq. (6.1) by measuring the power at which the surface of a clean silicon wafer starts to melt (T = 1412°C). Since the depth distribution of temperature is constant within a few micrometers from the surface, the entire thickness of the metal film (0.1-0.5 μm) can be considered to be heated to a uniform temperature during laser annealing.

The error in calculating T_{max} that arises by using the power required to melt crystalline silicon as a calibration is estimated to be ± 10% for silicon [6.20]. Use of this calculated temperature to predict growth rates for silicides will be shown later in this chapter to provide very satisfactory agreement with theory. However, complications can occur when the sample reflectivity changes during annealing. The uncertainty of the annealing temperature can then be substantially larger than the estimation based on the silicon melting point.

For the scanning electron beam experiments, a Hamilton Standard electron beam welder (Model EBW 7.5) was employed at a working vacuum of 10^{-4} Torr. The samples were mounted on a copper heat sink with Dow Corning 340 heat sink compound. The heat sink was maintained near room temperature during the reaction. Typical electron beam parameters used for the reaction were: a substrate temperature of $\leqslant 50°C$; electron beam energy 31 keV; beam current, 0.1 – 0.3 mA; beam spot size (as measured by the reacted path) 100 μm; scan rate 13 cm/sec; and a step size of less than 100 μm.

6.2.3 Analytical Methods: Rutherford Backscattering

Analysis of both the reaction kinetics and the average film composition of the silicides produced by scanned laser and electron beam processing can be conveniently carried out with Rutherford backscattering (RBS) techniques. When combined with a structural analysis technique, such as glancing angle X-ray diffraction in a Read camera, Rutherford backscattering can provide a rapid, highly quantitative analysis of most thin film metal–silicon reactions. RBS proves to be particularly useful for analysis of beam-reacted films, as they tend to be uniform and of reasonably large area so that analysis by mm^2 size $^4He^+$ probing beams is easily accomplished.

The application of Rutherford backscattering to the measurement of thin film properties has been discussed elsewhere [6.4] and will not be dealt with in detail here. However, a brief description will be useful to the reader who is unfamiliar with this technique. The review given below will be based on the analysis of a thin metal film on a silicon substrate as it appears immediately after deposition and after a silicide-forming reaction.

To analyze material properties using MeV helium particles, the surface of the sample to be analyzed is irradiated with a well collimated monoenergetic beam of helium atoms in the MeV energy range. As the probing beam penetrates the material, it loses energy by two processes: interactions with the electrons in the solid and interactions with the host atom in nuclear scattering events. Only those particles that are scattered by close range nuclear encounters in the material yield the desired information on the material properties.

A ^4He ion that experiences a close range collision with a target atom will have its momentum essentially reversed and can thus be <u>backscattered</u> into a solid state detector, as shown in Fig. 6.2(a). Such a particle will create in the detector a number of electron-hole pairs that is proportional to its energy.

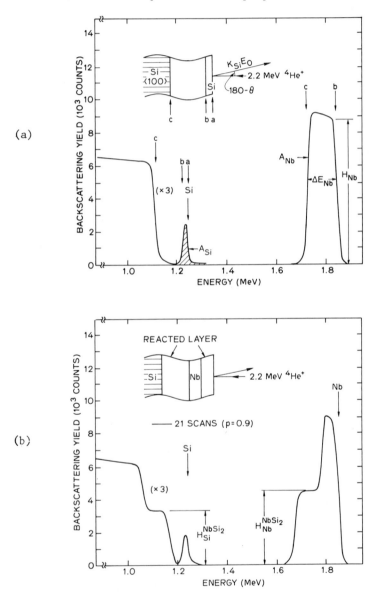

FIGURE 6.2. Rutherford backscattering spectra taken using 2.2 MeV ^{4}He particles for (a) as deposited Si/Nb/Si<100> sample and (b) identical sample with cw laser reaction (21 scans at p = 0.9).

This in turn creates a current that is proportional to the particle energy. This current is then used to measure the collected particle's energy. The function of the electronics following the detector is to convert this current (and energy) into a digital signal and store a count in the system memory at a location corresponding to the measured energy.

Since we are primarily interested in depth profiles and compositional analysis of reacted and unreacted thin metal films on Si substrates, we shall utilize an experimental backscattering spectrum to illustrate the RBS analysis technique. In Fig. 6.2 we show backscattering data taken on unreacted [Fig. 6.2(a)] and reacted [Fig. 6.2(b)] Nb/Si films. The initial sample structure was chosen for laser processing and consists of a Si(200 Å)/Nb (1100Å)/Si⟨100⟩ sandwich.

Although the large majority of ^{4}He particles incident on such a sample will penetrate many microns into the substrate before stopping, a small fraction (approximately one in 10^6) will experience close range (i.e., Rutherford) collisions and be backscattered out of the sample. A particle scattered from a Si atom on the surface of the sample of Fig. 6.2(a) will have an energy given by

$$E = K_M E_o \qquad (6.2a)$$

where K_M, the kinematic factor, is given by [6.21]:

$$K_M = \left[\frac{m \cos \Theta + \sqrt{M^2 - m^2 \sin^2 \Theta}}{m + M} \right]^2 \qquad (6.2b)$$

In this equation, m is the mass of the ^{4}He atom (4), M is the mass of the target atom, and Θ is the scattering angle in laboratory coordinates. For the geometry used in the RBS system used to take the data of Fig. 6.2, $\Theta = 170°$. Using Eq. (6.2b), we find $K_{Si} = 0.5647$ for Si and $K_{Nb} = 0.8429$ for Nb. Hence the resultant energy detected for a 2.2 MeV ^{4}He particle scattered through 170° by a collision with a Si surface atom will be 1.242 MeV. This energy is referred to as the "Si edge" energy and is indicated by an arrow at this energy in Fig. 6.2. The atomic composition at the surface of a sample can then (in principle) be measured by analysis of the associated backscattered energies.

Helium particles that are not backscattered from the target surface will enter the film, where they lose energy to electronic interactions. The stopping cross section ε is defined as [6.21]

$$\varepsilon \equiv \frac{1}{N} \frac{dE}{dx} \qquad (6.3)$$

where dE/dx is the energy loss rate and N is the atomic density of the material. As a 2.2 MeV [4]He particle penetrates the thin Si overlayer it loses energy at a rate of about 23.5 eV/Å. Thus a [4]He particle that penetrates through the 200 Å Si overlayer will have an energy of

$$E(x_1) = E_o - \frac{1}{|\cos \Theta|} \int_o^{x_1} \frac{dE}{dx} \, dx \tag{6.4}$$

at the Si/Nb interface. In the "near surface" approximation [6.21] the term dE/dx is considered a constant, having a value calculated from the initial beam energy, i.e.,

$$E(x_1) = E_o - \frac{1}{|\cos \Theta|} \left. \frac{dE}{dx} \right|_{E_o} x_1 \quad . \tag{6.5}$$

For the case under consideration

$$\left. \frac{dE}{dx} \right|_{E_o} = 23.5 \text{ eV/ Å}, \; x_1 \cong 200 \text{ Å, so}$$

$$E(x_1) = 2.2 \text{ MeV} - 4.7 \text{ keV} = 2.195 \text{ MeV}.$$

If the [4]He particle suffers a backscattering collision at x_1, it will arrive at the detector with energy E given by

$$E = K_i \; E(x_1) - \frac{1}{|\cos \Theta|} \left. \frac{dE}{dx} \right|_E x_1 \tag{6.6}$$

Here the subscript i has been used since a particle at the Si/Nb interface may backscatter from either a Si or Nb atom.

The difference between the energies of the incident and emergent beams can be written in terms of the energy loss [S] or stopping cross section factor [ε], and is

$$\Delta E = \Delta x[S] = N\Delta x[\varepsilon] \tag{6.7}$$

or $\quad [\varepsilon_o] = K\varepsilon(E_o) + \dfrac{1}{|\cos \Theta|} \, \varepsilon(E) \tag{6.8}$

or $\quad [S_o] = K \dfrac{dE}{dx} (E_o) + \dfrac{1}{\cos \Theta} \dfrac{dE}{dx} (E) \tag{6.9}$

using the near surface approximation. The values of $[\varepsilon_o]$ and $[S_o]$ are tabulated as a function of E_o and Θ with the energy E approximated as $K_m E_o$ [6.21]. For scattering from a Si atom at x_1, ΔE is

$$\Delta E = (200 \text{ Å})(47 \text{ eV/Å}) = 9.4 \text{ keV}. \tag{6.10}$$

Here the tabulated value for $[S_0]$ at 2.2 MeV has been used.

By extension of this analysis we find that 2.2 MeV ^{4}He particles backscattered from Si atoms in the Si overlayer will arrive at the detector with energies in the range 1.242 to 1.233 MeV. Under ideal conditions these particles would produce a rectangular backscattering spectrum, though as seen in Fig. 6.2(a), for very thin films the system response will broaden the rectangle into a bell-shaped curve.

The area under the backscattering spectrum associated with the thin silicon layer is

$$A = \sum_{i=1}^{n} C_i = H \frac{\Delta E}{\delta \epsilon} \tag{6.11}$$

where n represents the number of channels, C_i is the number of counts in the ith channel, and $\delta \epsilon$ is the system resolution in KeV/channel. This area can be related to the physical parameters of the measurement through

$$A = N \Delta x \frac{d\sigma}{d\Omega} \quad Q \Delta \Omega \tag{6.12}$$

where Q is the total charge of ^{4}He particles incident on the substrate, $d\sigma/d\Omega$ is the differential Rutherford cross-section and $\Delta \Omega$ is the solid angle subtended by the detector.

For a ^{4}He particle that reaches the Si/Nb interface and is scattered from a Nb atom the energy of the ^{4}He particle arriving at the detector will be

$$E_{Nb} = K_{Nb} \left[E_o - \frac{dE}{dx} \bigg|_{E_o}^{x_1} \right] - \frac{dE}{dx} \bigg|_{E_{Nb}}^{x_1} \tag{6.13a}$$

$$E_{Nb} = 0.8429[2200 - 4.7] - 5.2 = 1.845 \text{ MeV}. \tag{6.13b}$$

On the spectrum, Fig. 6.2(a), E_{Nb} is the energy at which the Nb signal is at "half-height". Although it is close to the "Nb edge" (1.854 MeV), defined as the energy of the ^{4}He backscattered from a surface Nb atom, it does not define the Nb edge since the Nb in this case is not located at the sample surface. Particles can scatter throughout the Nb film, producing the signal shown in Fig. 6.2(a). The area of this signal is proportional to the number of Nb atoms/cm^2 in the film. One can calculate the energy corresponding to the lower half height of the Nb signal from

$$\Delta E_{Nb} = [S_o]_{Nb} \; \Delta x_{Nb}. \tag{6.7}$$

To perform this calculation we must assume a value for the density of the Nb film. By providing an independent measurement of the Nb film thickness, and measuring ΔE_{Nb} experimentally an estimate of $[S_o]$ and a corresponding estimate of film density can be obtained.

By a similar analysis one can find that ^{4}He particles back-scattered from the silicon atoms at the interface c between the niobium and the silicon substrate will reach the detector with an energy around 1.1 MeV; an exact calculation requiring a knowledge of the density of the Nb film. Particles that penetrate the silicon substrate before being backscattered will give rise to the low energy portion of the spectrum ($<$ 1.10 MeV) shown in Fig. 6.2(a).

Let us now consider the changes in the RBS spectrum that occur when the silicon-metal-silicon sample undergoes a cw laser-induced silicide-forming reaction. In Fig. 6.2(b) we plot the backscattering yield vs particle energy for the Nb/Si couple after cw laser reaction. The surface Si film reacts with the underlying Nb, but the energy resolution of the backscattering system ($\approx$ 17 keV) is insufficient to detect a large change in the backscattered particle signal over that of the unreacted case. However, close examination of the Nb signal shows that the leading edge has moved up in energy to 1.854 MeV corresponding to backscattering of particles from Nb atoms at the sample surface. There is also a large change in the spectrum at the Nb/Si substrate interface. Here both the Si and Nb signals have spread out, with a step appearing in both signals. The observed increase in width of the Nb and Si signals is due to Si atoms in the Nb film causing an increase in ^{4}He particle energy loss and vice versa.

The composition of the silicide layer formed in this region ($\sim$ 1.05 $\to$ 1.18 MeV; 1.65 $\to$ 1.8 MeV) can be determined from the ratio of the spectrum heights from Eqs. 6.11 and 6.12.

$$\frac{N_{Nb}}{N_{Si}} = \frac{H_{Nb}}{H_{Si}} \frac{\left(\dfrac{d\sigma}{d\Omega}\right)_{Si}}{\left(\dfrac{d\sigma}{d\Omega}\right)_{Nb}} \frac{[\epsilon_{Nb}]}{[\epsilon_{Si}]} \tag{6.14}$$

In the near-surface, thin-film approximation, this equation can be written as

$$\frac{N_{Nb}}{N_{Si}} = \frac{H_{Nb}}{H_{Si}} \left(\frac{Z_{Si}}{Z_{Nb}} \right)^2 \tag{6.15}$$

where the relationships $(d\sigma/d\Omega)_M \alpha (Z_M)^2$, and $[\varepsilon_{Nb}] \approx [\varepsilon_{Si}]$ have been assumed. For the 21 laser scan reaction of Fig. 6.2(b) this ratio is found to be ≈ 0.5. Continued laser scanning of this film would result in complete mixing of the Nb and Si atoms with the Si signal extending from the substrate to 1.24 MeV. Such a reaction is shown in Fig. 6.8 for the case of Si/Pd/Si couples. For film thicknesses on the order of $0.1 \rightarrow 0.2$ μm, these approximations allow measurements to be made with an accuracy of $5 \rightarrow 10\%$.

The composition of the film can be determined not only from the height of the spectrum components, but also (and more accurately) by integration of the areas underneath the metal and silicon peaks corresponding to the intermixed layer. By assuming an atomic density for either the metal or the silicide film, a depth scale can also be generated which then allows measurement of the rate at which the reaction front proceeds through the film.

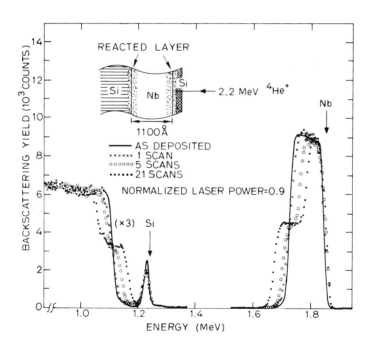

FIGURE 6.3. MeV ^{4}He ion backscattering spectra for Si(200Å)/ Nb (1100 Å)/Si<100> samples with multiple laser scans. Laser irradiation was done at p = 0.9, where P is defined in Sec. 6.3.1.

Reprinted by permission of the publisher, the Electrochemical Society, Inc. ©1981 Elsevier Science Publishing Co.

Data on the formation kinetics of the reaction can also be obtained from the RBS spectra. For this purpose we observe from the backscattering spectrum in Fig. 6.3 that the reaction clearly did not go to completion during the first scan. However, as the number of scans of the laser is increased from 1 to 21, the thickness of the reacted film is seen to expand into the pure niobium metal. In Fig. 6.4 we plot the thickness of the niobium silicide formed by the laser as a function of the number of laser scans. We will return to a thermodynamic analysis of this thickness relation below.

The composition determined from the Rutherford backscattering spectrum is an areal composition in atoms/cm^2. This composition is most accurately described as the average composition within the film. In order to determine the metallurgical phase(s) of the reacted material, other techniques must be used. Typically glancing angle X-ray diffraction is used for this separate determination of film composition. The X-ray diffraction measurement, to be described in the next section, will indicate whether the reacted film is single phase, multiple phases, or single phase with metal and/or silicon precipitates.

For the majority of the experiments to be discussed, glancing angle X-ray diffraction was employed to determine the phase composition of the films and Rutherford backscattering was used to indicate whether or not the entire film was reacted to form the X-ray measured phase.

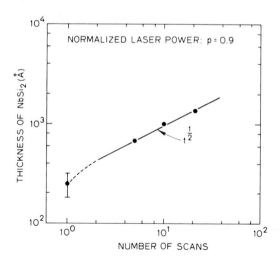

FIGURE 6.4. The thickness of NbSi$_2$ determined from backscattering spectra as a function of the number of laser scans. The data shows the parabolic growth for NbSi$_2$ formation since the number of scans is proportional to the annealing time. ©1981 Elsevier Science Publishing Co.

The uniformity of the reacted films can also be qualitatively understood from the Rutherford backscattering spectrum. Again if we refer to Fig. 6.3, we see that for the as-deposited niobium on silicon, a sharp edge exists at the niobium–silicon interface. This sharp edge is displayed as a rapid decrease in the niobium signal at an energy of about 1.70 MeV. However, a rapid rise in the silicon concentration in a nonuniform manner at the metal/silicon interface will cause these spectrum edges to become diffuse. As can be seen in the 21-scan reacted niobium silicon layer, the interface between the niobium metal and the $NbSi_2$ formed (solid dots near 1.75 MeV) is still quite steep, indicating a uniform interfacial layer between the $NbSi_2$ and metallic niobium.

6.2.4 · Analytical Methods: Read Camera

As we mentioned in the previous section, Rutherford backscattering measurements provide only information on the average composition of a reacted film. If we take a completely reacted $NbSi_2$ film as an example, RBS can verify that the Nb and Si are distributed throughout the film in exactly a 1:2 ratio, but it cannot determine whether the compound $NbSi_2$ has actually been formed, or whether the film is composed of a uniform mixture of other Nb_xSi_y compounds whose average composition is $NbSi_2$.

A unique determination of the phases present in the reacted film can be obtained from X-ray diffraction analysis. Several techniques are candidates for such experiments, including the recording X-ray diffractometer [6.22], the Seemann–Bohlin camera [6.23] and the Read camera [6.24]. Although contrast in the Read camera is limited and great accuracy is hard to achieve, exposure times are short (a few hours) and the instrument is therefore well suited to rapid compositional surveys of thin films.

The Read camera is a wide film cylindrical instrument in which the sample is held with its surface at a small angle (typically 13°) to the X-ray beam. The sample is both parallel to and along the cylindrical axis of the instrument. The camera geometry is illustrated in Fig. 6.5. Commercial instruments will accept samples up to approximately 1" x 1" with a suitable sample holder.

The principle of operation of the camera follows directly from the Bragg Law,

$$2d \sin \Theta = \lambda \qquad (6.16)$$

It is clear from this formula that the angle of reflection is related to the spacing between lattice planes d for a given X-ray wavelength λ (Cu-$K\alpha$ radiation is typically used). If a

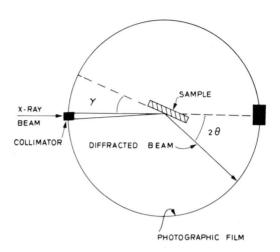

FIGURE 6.5 Schematic diagram of the Read camera illustrating the geometry used for the development in the text.

silicide film consists of small, randomly-oriented polycrystals of a given composition, then in general we expect to find reflections at the angles $\Theta(\langle a_1,b_1,c_1\rangle)$, $\theta(\langle a_2,b_2,c_2\rangle)$, etc., corresponding to the lattice plane spacings $d(\langle a_1,b_1,c_1\rangle)$, $d(\langle a_2,b_1,c_2\rangle)$. The reflections produce lines of film exposure as suggested in Fig. 6.5. Comparison of these lines with ASTM powder diffraction files [6.25] then permits positive identification of the compositional phase of the set of crystallites that are responsible for the separate reflections.

The accuracy of identification is improved dramatically by careful analysis of the exposure lines that correspond to large angle reflections. For this case, we have, by differentiating the Bragg Law,

$$\frac{\Delta d}{d} = -\frac{\Delta \Theta}{\tan \Theta} \tag{6.17}$$

For $\Theta \rightarrow 180°$, $\tan \Theta \rightarrow 0$ and then small changes in the spacing of the lattice planes will give rise to large changes in the reflection angle.

The Read camera image obtained for the Nb/Si film that was used as an example in the previous section on Rutherford backscattering is shown in Fig. 6.6. This pattern was obtained with the sample that had been given 21 scan frames of laser processing. The open circles in the pattern identify reflection lines (and therefore lattice parameter spacings) appropriate to crystallites of NbSi having a variety of orientations. The solid circles identify

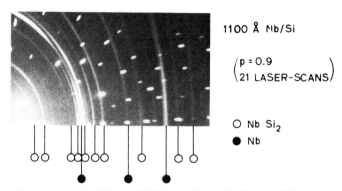

1100 Å Nb/Si

$\begin{pmatrix} p = 0.9 \\ 21 \text{ LASER-SCANS} \end{pmatrix}$

O Nb Si$_2$
● Nb

X-RAY DIFFRACTION PATTERN FROM READ CAMERA

FIGURE 6.6. Glancing angle X-ray diffraction pattern for the Nb/Si structure laser annealed at p = 0.9 for 21 frames. Formation of the NbSi$_2$ phase (o) is evident.

reflection lines arising from pure Nb metal. To within the accuracy of the Read camera these are the only two phases present in the film. (The spots in the pattern correspond to X-ray reflections from the underlying silicon single crystal). We conclude from data of this type and the corresponding RBS spectra that the reacted layer was exactly NbSi$_2$ for each of the laser scans employed.

6.3 EXPERIMENTAL RESULTS: LASER PROCESSING

In this section we will discuss the experimental results obtained on samples processed with scanning laser beams. For convenience we divide the discussion into the reaction of near-noble metals on silicon, including Pd and Pt, and then we consider the reaction of refractory metals such as Mo, W and Nb on silicon. In Section 6.4 we present data obtained with electron beam processing; and in Section 6.5 we discuss the formation of the superconducting compounds Nb$_3$Al and Nb$_3$Si using cw laser reaction. We summarize the beam process parameters and results of reactions that are described in detail in the following section in Table 6.3.

6.3.1 Near Noble Metals on Silicon: Pd/Si

In Fig. 6.7 we show an optical micrograph of a sample consisting of Si (200 Å)/Pd (1300 Å)/Si <100> that was laser beam scanned as a function of laser power level. It can be seen from this figure that the reaction of the Si/Pd/Si couple is initiated

TABLE 6.3 Summary of Beam Process Parameters and Resultant Reactions for the Silicides Discussed in this Work.

METAL FILM	REACTED PHASE	REACTED THICKNESS* (Å)	p (NORMALIZED LASER POWER)	POWER PARAMETER** (Watts/μcm)	T_{sub} (°C)	RESISTIVITY (μΩ·cm)
Pd	Pd_2Si	1930	0.71	0.154	50(±0.5)	37
	PdSi	2600	0.9 1.4	0.2 0.3		31
Pt	Pt_2Si⨍	--	0.69	0.148		--
	Pt_2Si⨍	--	0.86	0.185		--
Mo	$MoSi_2$	1450	0.88	0.093	350(±1.0)	190
W	WSi_2	1200	0.85 (10 scans)	0.087		110
Nb	$NbSi_2$	1350	0.9 (21 scans)	0.089		--
Pd	Pd_2Si	2280		0.112	50	39
	PdSi	2730		0.136		20
Pt	PtSi	1670		0.093		28
Nb	$NbSi_2$	1200≠		0.173		--

* Calculated from backscattering spectra using the near surface approximation for the energy loss

** Calculated values of absorbed power/beam radius: $(1-R)P/w$

⨍ A mixed phase compound including metastable Pt_2Si_3

≠ Thickness of a partially reacted film

at about 4 W of laser power. The laser-induced reaction is easily detected by the unaided eye through a surface color change. For the reaction initiated at a power level of 4 W, the surface color was observed to change from blue (unreacted film) to yellow (reacted film, low power phase). The low power phase was found to be stable up to a threshold power of about 7 W, above which a different high power phase appears in the center of the beam scan line. At this power level, the surface changes to a gray color that is characteristic of the high power phase. This high power phase was stable up to the power at which surface oblation was clearly evident. To obtain uniformly annealed areas for sub-sequent characterization, overlapped scans were performed with each scan line overlapped by at least 40%.

To simplify the discussion it is convenient to normalize the output power level of the laser (P) to the power at which the surface of a clean silicon wafer just begins to melt (P_O). The normalized power ($p = P/P_O$) is experimentally reproducible, thereby minimizing the dependence on the laser system and/or other experimental conditions. For typical experimental configurations, the values of P_O were 9.0 and 5.0 W for 50° and 350°C substrate temperatures, respectively. The normalized laser power is also related to a critical parameter of the cw beam process; i.e., the ratio of beam power to spot radius (P/w). This relation

100 μm

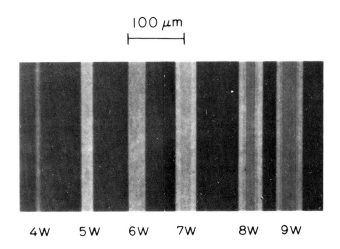

4W   5W   6W   7W   8W   9W

FIGURE 6.7.   Optical micrograph of a sample consisting of
Si(200 Å)/Pd(1300 Å)/Si⟨100⟩ after single laser scans at various
power levels.   The low power phase is formed between 4 and 7 W
and the high power phase at the above 8 W.

Reprinted by permission of the publisher,
the Electrochemical Society, Inc.

is given by P/w (watts/micron) = 0.337 p or 0.191 p for T_{sub} = 50°C
or 350°C respectively, using the formulation of Chapter 2. To summarize the argument presented in Chapter 2, we recall that the beam
radius, w, is defined by the 1/e point of the beam intensity assuming a Gaussian distribution. Since the actual power absorbed by
the sample is reduced by the reflection, the essential power parameter is (1 – R)P/w, where R denotes the reflectance of the sample.
This parameter is listed in Table 6.3 and has been calculated using
experimentally determined values for R.

Backscattering spectra for the Pd/Si samples which were laser
annealed at p = 0.71 (low power phase) or p = 1.1 (high power
phase) are shown in Fig. 6.8. The spectrum for the as deposited
sample is shown by the solid line. It is clearly seen that the entire metal layer was reacted with the Si after the laser irradiation. The average composition of the low power phase and high power
phase was determined to be Pd_2Si and PdSi, respectively, using the
near surface approximation for [S]. It should be noted that the
spectrum of PdSi did not change significantly with different laser
powers ranging from p = 0.89 up to p = 1.4. These results are
quite different from those obtained with pulsed laser annealing,
where the average composition of the silicide layer changes continuously with increasing laser power [6.14].

Figure 6.9 shows the X-ray diffraction pattern obtained from
the high power phase sample (p = 1.1). The diffraction lines from
PdSi are indicated by solid circles with the most intense line from

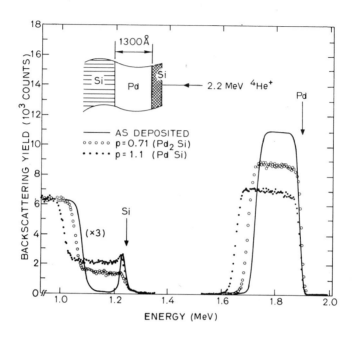

FIGURE 6.8. Backscattering spectra for Si/Pd/Si laser annealed samples for various normalized laser power levels. The
solid line (____) represents data taken on unreacted samples, the
circles (oooo) on samples reacted at low powers and the closed
circles (●●●) samples reacted at high power levels.

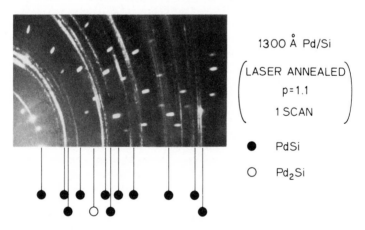

FIGURE 6.9. Glancing angle X-ray diffraction pattern from
a Read camera on the sample reacted at high power as shown in
Fig. 6.8. The line indicated by 0 is the most intense diffraction
line from Pd2Si.

Pd_2Si by an open circle. It can be concluded that the compound is essentially single phase PdSi with a trace amount of Pd_2Si. A similar result was obtained for the low power phase sample; however, here the compound formed was mainly Pd_2Si including a trace amount of PdSi.

6.3.2 Growth Mechanisms for $NbSi_2$ and Pd_2Si

To investigate the mechanisms by which $NbSi_2$ and Pd_2Si grow, it is useful to first consider what the growth rate would be if the cw beam process were equivalent to furnace annealing. To explore this possibility, we first note that the data for the formation of silicides in a furnace are well characterized by the relation:

$$\Delta x \text{ or } (\Delta x)^2 = At \exp\left(-\frac{E_a}{kT}\right) \tag{6.18}$$

Here, Δx is the thickness of the reacted layer, A a material constant, t the annealing time, E_a the activation energy, and T the annealing temperature. The selection of Δx or $(\Delta x)^2$ is made according to the formation kinetics; i.e., linear or parabolic growth, respectively. If the laser induced reaction is due to a solid phase mechanism, then the analytical treatment developed in Chapter 2, can be used to characterize the silicide formation. According to this formulation, the thickness of the reacted layer due to cw laser annealing can be obtained from Eq. (6.18) simply by replacing t and T by their effective values: t_{eff} and T_{eff}. T_{eff} is the peak value of the temperature distribution produced by the laser beam (T_{max}), which is calculated in terms of the absorbed power divided by beam radius: $(1 - R)P/w$. The beam radius, w, is defined by the $1/e$ point of the beam intensity assuming a Gaussian cross section, P the laser beam power, and R is the reflectivity. An example of the temperature calculations is shown in Fig. 6.10. The expressions for t_{eff} differ according to the laser scan mode, i.e., single scan or multiple overlapped scan, and are discussed in Chapter 2.

The data presented in Fig. 6.4 show that $NbSi_2$ formed by laser annealing is characterized by a parabolic growth rate. As detailed growth kinetics data does not exist in the parabolic growth regime for $NbSi_2$, we have used laser annealing data to determine A and E_a for $NbSi_2$.

The $Si(200 \text{ Å})/Nb(1100 \text{ Å})/Si\langle100\rangle$ samples were laser annealed at various power levels with multiple (20) scans with the substrate held at 350°C. The reacted layer thickness was calculated using the near surface ^4He energy loss factor for 2.2 MeV ^4He of $[S_o]_{Nb}^{NbSi_2} = 83.7$ eV/Å. The results for $(\Delta x)^2$ are plotted

362 T. SHIBATA ET AL.

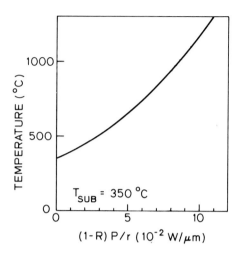

FIGURE 6.10. Example of reaction temperature calculation as a function of absorbed power/radius for a given (350°C) substrate temperature.

as a function of the calculated reciprocal temperature in Fig. 6.11. It can be seen that the data lie on a straight line. A least squares fit of the data results in the activation energy $E_a = 1.43$ eV for the process. Since t_{eff} is dependent upon both E_a and T_{max}, an iteration procedure was employed to obtain a more accurate value for E_a. For each point, t_{eff} was calculated using the multiple overlap scan model with $E_a = 1.43$ eV, then a least squares fit was performed to obtain $(\Delta x)^2/t_{eff}$ vs. $1/T$. This resulted in an activation energy $E_a = 1.47$ eV. The next iteration did not alter the value of E_a, giving the final results $\underline{E_a = 1.47}$ $\underline{eV}$ and $\underline{A = 1.08 \times 10^{14} \; A^2/sec}$. This activation energy is different from that ($E_a = 2.1 - 2.7$ eV) reported in Ref. [6.26]; however, in that work the linear, early stage of growth was used to determine the rates.

Figure 6.12 demonstrates the dependence of T_{max}, t_{eff} and the thickness of the NbSi$_2$ layer as a function of the normalized laser power. Figure 6.12 also shows that the experimental data are well described by a solid phase reaction model. This is consistent since the maximum temperature value T_{max} obtained in the experiment is about 1100°C (p = 0.96). This value is well below the published eutectic temperature (~ 1300°C) for the Nb-Si system [6.27].

Similar experiments were performed to study the reaction kinetics for the Si(200 Å)/Pd(1300 Å)/Si<100> system. The samples were laser annealed under conditions that form the Pd$_2$Si phase. For this case only one laser scan was used, and therefore the single scan model was used for the calculation of t_{eff}.

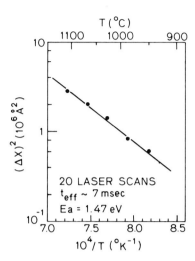

FIGURE 6.11.   Arrhenius plot of the square of the reacted thickness ($NbSi_2$) as a function of calculated temperature [6.10] for a constant number of laser scan frames (21) and velocity.

The thickness of the Pd_2Si layer determined from backscattering measurements is plotted as a function of the normalized laser power in Fig. 6.13.   Also plotted is the calculated Pd_2Si thickness vs laser power using the furnace anneal data of Ref. [6.28]. A systematic discrepancy between the experimental data and the calculation is apparent, the fit cannot be improved by adjustment of the constant A.   However, using the recently reported furnace annealing data for the activation energy for Pd_2Si formation from samples similar to ours (E_a = 0.93 eV) [6.29], the calculation was repeated, giving the results shown by the solid line in Fig. 6.13.   The constant A was calculated from the best fit of the curve to the data.   The agreement of this calculation with the experimental data indicates that the laser induced Pd_2Si formation occurs by a mechanism similar to that discussed in Ref. [6.29].

6.3.3 Formation Mechanism for PdSi

For the formation of PdSi, furnace annealing experiments have produced only inhomogeneous layers [6.30].   Recently, the formation of homogeneous single phase PdSi layers has been achieved by ion beam mixing [6.12] and cw-laser [6.7] or electron beam [6.8] processing.

The formation of PdSi by cw-laser annealing cannot be explained by a solid phase or grain boundary diffusion mechanism.

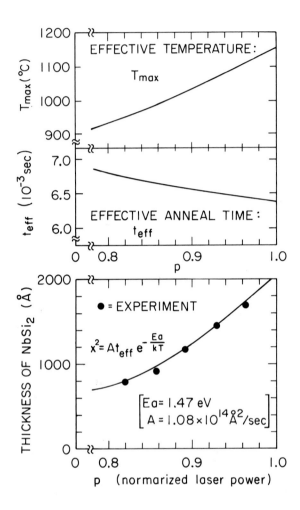

FIGURE 6.12.   Effective temperature, effective anneal time and NbSi$_2$ thickness calculated as a function of the normalized laser power (p).

Eutectic melting provides one possible mechanism for the process, supported by the following observations:

    (a)   the surface of the PdSi phase shows a laminar-like morphology which suggests that some sort of melting could be occurring;

    (b)   we have been unable to detect a gradual transition from Pd$_2$Si to PdSi; and

    (c)   the transition from Pd$_2$Si to PdSi occurs abruptly above a threshold power ($P_{th} \geqslant 0.8\ P_o$).

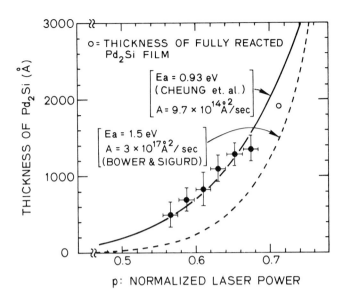

FIGURE 6.13.   Thickness of Pd_2Si as a function of normalized laser power. Calculated results are shown for two different activation energies.

The calculated temperature at this threshold power is about 830°C – 860°C, which is close to the eutectic temperature for 58% Si in the Pd–Si system. However, if this eutectic mixture is formed during laser annealing, the excess Si must be rejected during the formation of the PdSi. So far, we have been unable to observe the excess Si either by backscattering or X-ray diffraction. However, we cannot rule out the possibility that the excess Si is recrystallized epitaxially back onto the Si substrate.

The above observations can be alternatively interpreted in terms of the nucleation controlled reaction model proposed in Ref. [6.31]. We believe that this model is compatible with the concept of a moving hot spot produced by the scanning beam causing the propagation of a nucleation reaction from a nucleation site. Evidence for such a model is provided in Fig. 6.14, which is a Nomarsky interference micrograph of the Pd/Si sample after laser irradiation at p = 0.95. Here two single line scans were performed from two different directions. The center region of each scan line is seen to consist of light and dark regions. The darker region corresponds to the PdSi phase, while the lighter region consists of a nonuniform mixture of Pd and Si with no definite crystallite structure. It is especially interesting to note that the PdSi phase starts to form when the scanning beam passes over either the scratch (first scan) or the already existing

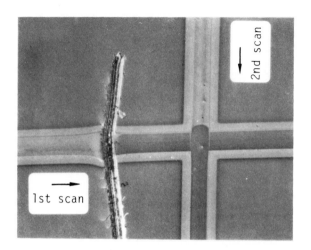

FIGURE 6.14 Nomarsky contrast optical micrograph of a Pd/Si sample laser scanned from two different directions. The PdSi phase is the darker center region which forms when the scanning beam passes over the scratch (1st scan) or the already existing PdSi phase (2nd scan). ©1981 Elsevier Science Publishing Co.

PdSi phase (second scan). If the process is begun at the scribed edge of the wafer, a uniform PdSi layer can be formed over the entire area of the wafer.

These observations strongly suggest that the formation of PdSi by laser process is a nucleation-controlled reaction, as suggested by Ref. [6.31]. However, the calculated temperature at the threshold power of the PdSi formation is close to the eutectic melting temperature, and we cannot rule out this mechanism.

6.3.4   Near Noble Metals on Si: Pt/Si

The cw laser annealing of $Si(200 \text{ Å})/Pt(1000 \text{ Å})/Si\langle100\rangle$ samples resulted in surface changes similar to those found in the Pd/Si sample (Fig. 6.7) when observed under an optical microscope. For this system the reaction started at a normalized power around $p \approx 0.5$, with a transition from a low power phase to a higher power phase occurring at $p \cong 0.74$. In contrast to this similarity however, the backscattering and X-ray diffraction results were quite different from those obtained for the Pd-silicide films.

The <u>average</u> composition of the low power phase calculated from the backscattering spectrum was close to Pt_2Si, but it was found from the X-ray diffraction analysis that the layer consisted mainly of Pt_3Si and $PtSi$ with a small amount of $Pt_{12}Si_5$ and Pt_2Si. The average composition found for the higher power phase was approximately $PtSi_2$, a phase which does not exist in the equilibrium phase diagram. The X-ray diffraction pattern was identified as a mixture of $PtSi$, $Pt_{12}Si_5$ and Pt_3Si. However there existed three definite lines which could not be fitted to any data for Pt–Si compounds listed in the Powder Diffraction File. It should be noted that no Si precipitation or unreacted Pt was found in either of the above samples. We will return to a discussion of this data in Sec. 6.4 where similar reactions promoted by an electron beam were found to produce substantially different results.

6.3.5 Formation of Pt_2Si_3

In Fig. 6.15, we show a typical X-ray diffraction pattern found for a $Si(200 Å)/Pt(1000 Å)/Si\langle100\rangle$ sample laser annealed at $p = 0.89$. The backscattering spectrum for this sample indicated an average composition of $PtSi_2$. However, X-ray analysis shows that a mixed phase silicide has been formed. Several unidentifiable lines are well explained by the metastable Pt_2Si_3 phase recently discussed in Ref. [6.32]. Further verification

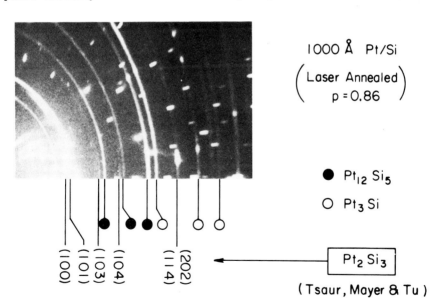

FIGURE 6.15.   Glancing angle X-ray diffraction pattern for the Pt/Si structure laser annealed at $p = 0.86$. Formation of superconducting Pt_2Si_3 phase is evident.

of this phase is provided by the fact that the layer exhibited superconducting behavior with a transition temperature around 4°K. The transition to this mixed phase Pt-silicide can be understood as arising from eutectic melting since it occurs at p $\geqslant$ 0.75 with the calculated temperature for this power being about 840°C. This temperature is slightly higher than the lowest eutectic temperature in the Pt-Si system (830°C). For this case the heating and cooling rates achieved by the scanning beam are of primary importance for rapid quenching of the metastable phase.

6.3.6 Refractory Metals on Si: (Mo, W, Nb)/Si

For pulsed laser annealing, it was reported that only a limited amount of reaction was observed in the Mo/Si system and no reaction whatsoever was observed in the Nb/Si system. For comparison, Fig. 6.16 shows the backscattering spectra obtained before and after cw laser annealing on Si(200 Å)/Mo(530 Å) Si<100> samples. It is clear from these data that the entire film of Mo was completely reacted after laser annealing at p = 0.88. The average composition of the silicide was calculated to be $MoSi_2$ from the spectrum (thickness = 1450 Å). The X-ray diffraction pattern shown in Fig. 6.17 indicates that the reacted layer is <u>single phase</u> $MoSi_2$, free of unreacted Mo or Si precipitation.

For the Si(200 Å)/W(440 Å)/Si(xtl) and Si(200 Å)/Nb(1100 Å)/ Si(xtl) systems, laser annealing at p = 0.85 and 0.9, respectively, produced only a limited amount of reaction at the metal-Si interfaces. Therefore, multiple frame laser scans were performed for these two systems. The results of the backscattering analysis for W/Si and Nb/Si are given in Fig. 6.18 and Fig. 6.19, respectively.

Figure 6.18 shows that the W layer of about 440 Å was completely reacted to form a WSi_2 film of about 1200 Å. That the film was single phase WSi_2 was also verified by X-ray diffraction analysis (Fig. 6.20). It should be pointed out that ten complete scan frames were required to fully react this film. From the backscattering data one can conclude that the reacted film is quite uniform with a well defined silicide/silicon interface.

In Fig. 6.19 the backscattering data clearly shows the advancement of the silicon niobium interface, occurring for a constant laser power, with the number of laser scan frames. The reacted film composition was calculated to be $NbSi_2$ and verified by X-ray diffraction analysis (Fig. 6.21). The analysis of this film has been discussed in the section on Growth Mechanisms (Sec. 6.3.2) and will not be discussed further in this section.

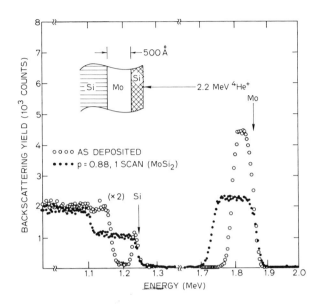

FIGURE 6.16. Backscattering spectra for Si(200 Å)/Mo(500 Å)/
Si⟨100⟩ structures before (ooo) and after (ooo) laser reaction.
A layer of MoSi$_2$ of about 1450 Å thickness was obtained after 1
laser scan at p = 0.88. Reprinted by permission of the publisher,
the Electrochemical Society, Inc.

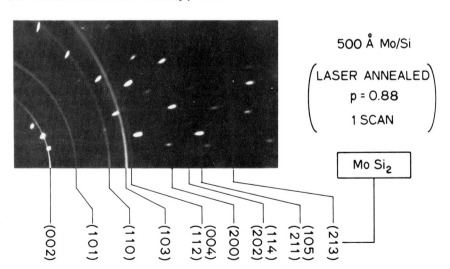

FIGURE 6.17.   X-ray diffraction pattern obtained from the
Mo/Si sample laser annealed at p = 0.88. Reprinted by permission
of the publisher, the Electrochemical Society, Inc.

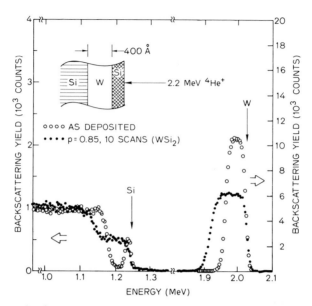

FIGURE 6.18.   Backscattering spectra for Si(200 Å)/W(440 Å/ Si⟨100⟩ structures.   About 400 Å of W (ooo) is completely reacted to form a WSi$_2$ film (ooo) of about 1200 Å thickness after 10 laser scans at p = 0.85. Reprinted by permission of the publisher, the Electrochemical Society, Inc.

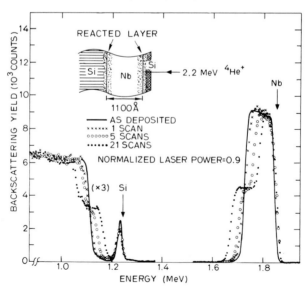

FIGURE 6.19   Backscattering spectra for Si(200 Å)/Nb(1100 Å)/ Si⟨100⟩ samples with multiple laser scans.   The thickness of the silicide (NbSi$_2$) increases with the number of scans.

6.4 EXPERIMENTAL RESULTS: ELECTRON BEAM PROCESSING

In this section we discuss the formation of silicides by the use of a scanned cw electron beam. Although all experimental results to be discussed utilized the electron beam system described in Sec. 6.2.2 to promote the reactions, it should be pointed out that similar results have been obtained with a modified SEM [6.33]. In Table 6.4 and Fig. 6.1a the materials used and the sample configuration for the electron-beam experiments are summarized. Table 6.4 lists the important experimental parameters and results achieved for these electron-beam reacted silicides.

6.4.1 Near Noble Metals on Si : Pd/Si

The formation of Pd_2Si and PdSi by scanning electron-beam reaction proceeds in a manner analogous to the scanned laser. Again the choice of the phase is determined by selection of the e-beam P/r value. Figure 6.22 shows backscattering data taken on a 1400 Å layer of Pd on ⟨100⟩ Si both before and after scanned e-beam processing. Two distinct phases are clearly discernible in this data. Glancing angle X-ray analysis confirms the phases calculated from the backscattering spectra. Again, similar to the scanned laser results, essentially single phases are found with only a trace amount of the other phase present.

TABLE 6.4. List of Properties of Scanned Electron-Beam Reacted Silicide Thin Films

METAL FILM	REACTED PHASE	UNREACTED THICKNESS	REACTED THICKNESS	P/r (E_B = 31 keV)	ESTIMATED SURFACE TEMP. (T_{sub} = 50°C)	RESISTIVITY ($\mu\Omega$-cm)	SURFACE TEXTURE
Pt	PtSi	~ 1100 Å	~ 1670 Å	0.93 kW/cm	240°C	27.9	Fine grain (1 µ) orange peel
Pd	Pd_2Si	~ 1400 Å	~ 2280 Å	1.12 kW/cm	290°C	38.5	Fine grain (2 µ) orange peel
	PdSi		~ 2730 Å	1.36 kW/cm	375°C	20.2	Laminar like
Nb	$NbSi_2$	~ 1250 Å	~ 510 Å	1.36 kW/cm	375°C	--	Unknown
			~ 780 Å	1.55 kW/cm	450°C	--	
			~ 1200 Å	1.73 kW/cm	525°C	--	

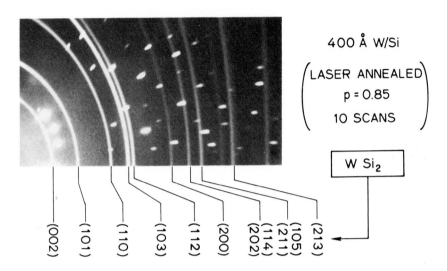

400 Å W/Si

$\left(\begin{array}{c}\text{LASER ANNEALED}\\ \text{p} = 0.85\\ \text{10 SCANS}\end{array}\right)$

$\boxed{\text{W Si}_2}$

(002) (101) (110) (103) (112) (200) (202) (114) (211) (105) (213)

FIGURE 6.20.   X-ray diffraction pattern obtained from the Si/W/Si structure laser annealed at p = 0.85 for 10 scans.

Reprinted by permission of the publisher, the Electrochemical Society, Inc.

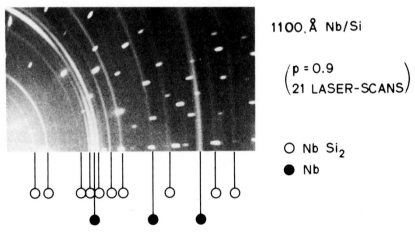

1100. Å Nb/Si

$\left(\begin{array}{c}\text{p} = 0.9\\ \text{21 LASER-SCANS}\end{array}\right)$

○ Nb Si$_2$
● Nb

X-RAY DIFFRACTION PATTERN FROM READ CAMERA

FIGURE 6.21.   X-ray diffraction pattern obtained from the Si/Nb/Si structure after laser annealing.

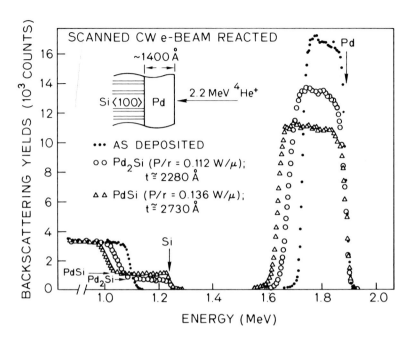

FIGURE 6.22   Backscattering spectra for Pd/Si before (●●●) and after reaction by a scanning electron beam; (ooo) Pd_2Si and (ΔΔΔ) PdSi.

The e-beam-reacted PdSi exhibits a laminar like surface structure along the direction of the scan. In Fig. 6.23 we show optical Nomarsky micrographs of both the Pd_2Si and PdSi films. A point of potential interest for device fabrication is that the PdSi film was found to have a very low resistivity (20 $\mu\Omega$-cm) as compared to the other films discussed in this chapter (see Table 6.4).

The theoretical development of Chapter 2 can be used to study scanned e-beam reactions by using the beam voltage and current to calculate the beam power (P). The reflectivity is set to zero and the beam diameter w is measured experimentally.

6.4.2   Near Noble Metals on Si : Pt/Si

In Fig. 6.24 we show backscattering data taken on electron-beam reacted Pt with ⟨100⟩ Si. The data clearly shows that a uniform reaction of the film has taken place at a P/r value of 0.093 W/μ. The average composition of the film calculated from the spectrum is PtSi. X-ray diffraction analysis verifies that the film is <u>single phase</u> PtSi.

e - BEAM REACTED Pd/Si

20 μm

Pd₂Si (P/r = 1.12 kW/cm)   PdSi (P/r = 1.36 kW/cm)

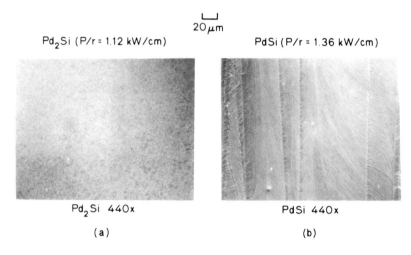

Pd₂Si 440x   PdSi 440x

(a)   (b)

FIGURE 6.23.   Normarski micrographs of electron beam reacted Pd/Si, (a) and Pd₂Si and (b) PdSi.

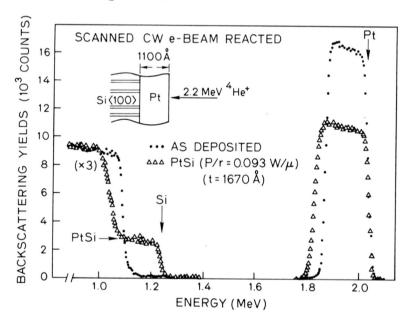

FIGURE 6.24.   Backscattering spectrum shown for Pt/Si film reacted by a scanning electron beam.   The closed circles (●●●) represent the spectrum from the As-deposited film while the triangles (ΔΔΔ) are for the e-beam reacted.

This result is in sharp contrast to that reported for the laser-beam reacted PtSi in Sec. 6.3.4, where for similar P/r values and scan speeds, a mixed phase silicide resulted. This difference can be caused by two phenomena: the basic physical differences between the laser and e-beam experiments are sample structure and beam energy deposition profile. Although we cannot rule out the possibility that interfacial effects such as bond breaking at the Pt/Si interface by the penetrating electrons may help this reaction proceed, recent experimental evidence suggests that the difference is due to the presence of the thin silicon overcoating present on the laser samples.

6.4.3 Refractory Metals on Si : Nb/Si

The formation of $NbSi_2$ from Nb/Si by use of a scanned e-beam has also been reported [6.18]. As in the case of the laser reacted films progression of the reaction interface has been observed by backscattering measurements. The existence of a single phase ($NbSi_2$) compound was verified by using glancing angle X-ray analysis.

6.5 THERMAL STABILITY AND OXIDATION PROPERTIES OF CW BEAM REACTED SILICIDES

The thermal stability and oxidation characteristics of silicide compounds are important properties for the application of these materials in semiconductor technology. Future integrated circuits will require both gate and interconnect metallizations that have low sheet resistivity and can withstand various wet and dry processing steps, including high temperature furnace processing and oxidation. The capability of oxidizing a silicide film without any phase change and degradation of its electrical properties will be especially important prerequisites for the use of these materials.

To date a number of refractory metal silicides have been proposed as alternatives to the polysilicon gate material and electrical interconnections, because of their lower bulk resistivities and compatability with present and future MOS process technologies. However, the extremely low bulk resistivity and unique properties afforded by such noble metal silicides as PdSi and $PtSi_2$ also makes these materials promising candidates for possible device application.

Thermal oxidation studies have been reported for several metal silicides, including $MoSi_2$, WSi_2, $TaSi_2$, $NiSi_2$, and $TiSi_2$ [6.34-6.43]. However, the reported oxidation properties have been found to vary widely, depending on the technique used to deposit the films. For tungsten-based films, a tungsten-rich silicide phase (W_5Si_3) is found after steam oxidation of sputtered WSi_2 films [6.38]. This is in contrast to the behavior of coevaporated WSi_2 films, where no change in the film stoichiometry during oxidation has been reported [6.37]. Also, it has been observed that dry oxidation of WSi_2 results in films of poor surface quality [6.38]. Furthermore we would expect that $MoSi_2$ films on silicon would behave similarly to WSi_2 because $MoSi_2$ is chemically and structurally alike to WSi_2. However, when these refractory silicides are deposited on oxide substrates, decomposition of the film to metal-rich phases are observed [6.34, 6.36-6.38]. Data on the oxidation properties or thermal stability of PdSi films have been reported because of the difficulty of forming these films by conventional furnace techniques. In the following sections we present studies of the thermal stability and oxidation properties of WSi_2, $MoSi_2$, and PdSi as formed by scanned cw laser beam reaction, and compare the results with those reported for both sputtered and coevaporated films.

6.5.1 Refractory Metal Silicides : WSi_2, $MoSi_2$

In Fig. 6.25 we show Rutherford backscattering data taken on oxidized, laser-reacted WSi_2 films as a function of oxidation time. The data shown are for samples oxidized in steam at 900°C. Also shown is the spectrum of an unoxidized WSi_2 structure. It can be seen from Fig. 6.25 that as the oxidation process proceeds, the tungsten peak is shifted to lower energies. This shift arises from the energy loss of the [4]He atoms in the SiO_2 layer being grown on top of the WSi_2 film. Conversion of this energy loss into a corresponding thickness allows measurement of the oxidation rate to be made.

At the silicon edge of the spectrum it is also seen that the interface between the silicon crystal and the WSi_2 (occurring at about 1.1 MeV for the unoxidized sample) shifts to lower energies following the thermal oxidation. A new peak at around 0.8 MeV, superimposed on this silicon spectrum, also appears due to [4]He particles backscattered from the oxygen atoms in the SiO_2 film. If one further examines the tungsten peaks in this figure, a decrease in the slope of the leading edges of the peaks is seen to occur with increasing oxidation time. This causes the peak height to decrease with increasing oxide thickness. This result indicates that the oxidation of the WSi_2 film is somewhat nonuniform, as the areas under these peaks remain constant. The constant area also indicates that no tungsten is lost from the film during the oxidation.

By using glancing angle X-ray diffraction analysis, the stoichiometry of the WSi_2 films after oxidation was studied. Diffraction lines corresponding only to single phase WSi_2 films in steam at 900°C for similar periods of time [6.38]. Although it has been

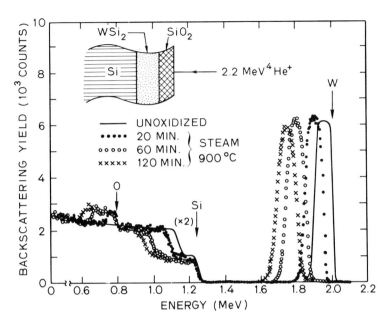

FIGURE 6.25.   2.2 MeV ^{4}He backscattering spectra for laser reacted and oxidized WSi_2. The solid line (———) represents the unoxidized control WSi_2 layer.   Spectra are shown for steam oxidation at 900°C for 20 min. (···), 60 min. (ooo), and 120 min (xxx) oxidation times.

reported that dry oxidation of sputter deposited WSi_2 films results in films of poor quality, laser-reacted WSi_2 films sub-jected to dry oxidation at 1000°C appear to have smooth surfaces, suggesting that the SiO_2 film grows on the WSi_2 without decompo-sition of the silicide.   Also the X-ray diffraction pattern obtained from laser reacted films subjected to a 60 min. dry oxidation at 1000°C shows no diffraction lines present other than those of the single phase WSi_2 compound.   Furthermore, no indication of WO_3 formation was found and surface morphology was good after the process.   One can conclude from these observations that the oxidation mechanism for the laser reacted films is simi-lar to that proposed for coevaporated WSi_2 films where it is believed the silicon atoms are supplied only from the silicon substrate by solid state diffusion through the silicide film.

Similar to WSi_2, the dry oxidation of laser-formed $MoSi_2$ on silicon proceeds without any change in phase nor any indication of MoO_3 formation.   However, as in sputtered $MoSi_2$ on oxides, $MoSi_2$ transforms to the metal rich silicide once the underlying silicon is completely consumed or silicon transport from the silicon substrate to the silicide surface is impeded by impurities at the silicon-silicide interface. By assuming local equilibrium at the growing oxide-silicide interface and favorable formation kinetics, the phases observed at this interface can be predicted from the ternary phase diagram for all oxidation conditions. A simplified ternary phase diagram was constructed from available thermodynamics data (Fig. 6.26). Only the stable tie lines between two phases have been determined. Solubility regions in the single and two phase regions, miscibility gaps, and ternary phases have been neglected. Details on the construction can be found in several references [6.44, 6.45]. Although the phase diagram has been oversimplified, the diagram allows us to explain the co-existence of certain phases and the absence of metal oxides.

For SiO_2 growth to continue and yet maintain the $MoSi_2$ phase, a silicon flux to the $MoSi_2$-SiO_2 interface is required. Therefore, an interface silicon concentration is established at this interface so that the overall composition at this interface lies

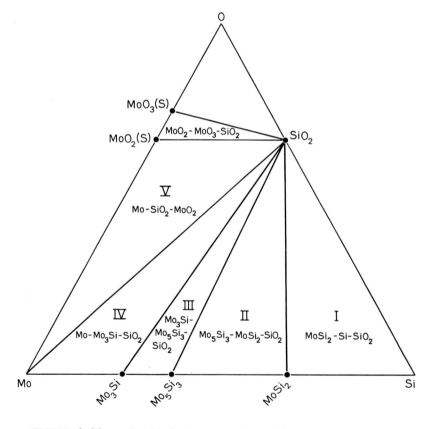

FIGURE 6.26.   Mo-Si-O ternary phase diagram.

along the $MoSi_2-SiO_2$ tie line. However, if silicon transport is impeded during oxidation by impurities, the silicon concentration decreases and the overall interface composition may lie in regions II, III, IV of the ternary phase diagram or along any of the tie lines, such as $Mo_5Si_3-SiO_2$ or $Mo_3Si-SiO_2$ (Fig. 6.26). From the thermodynamics we can predict that only SiO_2 will form as long as the overall interface composition lies in the silicon rich region of the phase diagram. Phases corresponding to regions II, III and IV have been observed when oxygen or SiO_2 is present at the $MoSi_2$-silicon interface prior to oxidation [6.46].

Figure 6.27 shows the backscattering spectra of $MoSi_2$ films on polysilicon oxidized at 909°C. Although after 15 hours of oxidation in dry O_2 the molybdenum peak has decreased and the ratio of the silicon to molybdenum peak has increased beyond 2, glancing angle X-ray diffraction indicates $MoSi_2$ as the only molybdenum phase present. Since less than 0.5% of molybdenum was detected at the $MoSi_2$-oxide interface and the molybdenum peak areas remain essentially constant during oxidation, these results

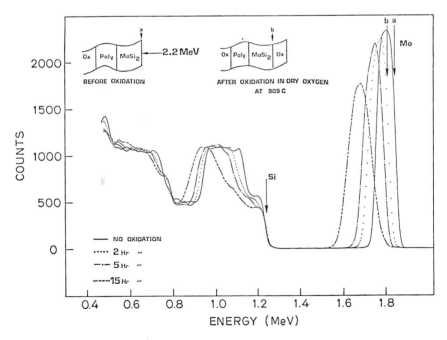

FIGURE 6.27. ^{4}He backscattering spectra for laser reacted $MoSi_2$ on polysilicon oxidized for 0 hrs. (solid), 2 hrs. (dots), 5 hrs. (dash–dots), and 15 hrs. (dash) at 909°C. The shift in the molybdenum peak is due to the growth of SiO_2 on the surface.

indicate very little loss of molybdenum from the film. Furthermore, the decrease in the slopes of the leading and trailing edges suggest that the oxidation is nonuniform. Therefore, the decrease in the molybdenum peak height possibly results from the nonuniformity of the oxidation and/or internal oxidation along grain boundaries. The TEM micrograph from the cross-sectional view of the oxidized film confirms that the interface has become nonuniform from long oxidations (Fig. 6.28). Grains of $MoSi_2$ suspended in the SiO_2 also suggest that oxidation along grain boundaries has occurred. For similar oxide thicknesses grown in wet O_2, but with shorter times, the molybdenum peak height remains constant and the interface nonuniformity is comparable to that observed for WSi_2 oxidized in steam.

The use of these silicides in integrated circuits also requires that the film retains its conductivity. Table 6.5 summarizes the resistivities of 100 nm of $MoSi_2$ oxidized for 0.5, 4, and 10 hours in dry O_2. Only after 10 hours of oxidation does the resistivity of the film degrade from 12 to 27 $\Omega/\square$. This degradation corresponds to the observed increase in the silicon to molybdenum ratio.

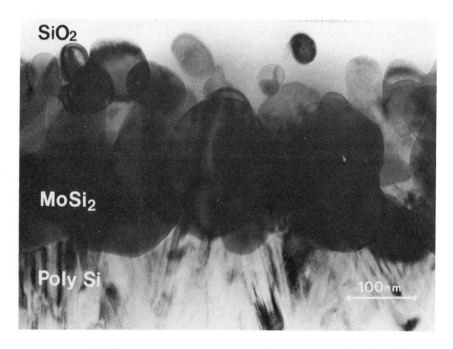

FIGURE 6.28. Cross-sectional TEM micrograph of $MoSi_2$ on polysilicon oxidized at 900°C for 15 hours.

TABLE 6.5.   Resistivities of 1000 Å of $MoSi_2$ on Polysilicon Oxidized at Various Times at 1007°C.

Dry Oxidation Condition	N	Resistivity $\Omega/\square$	Resistivity $\mu\Omega\text{-cm}$
0 hr.	2.1	12 $\Omega/\square$	120
0.5	2.2	10 $\Omega/\square$	200
4	3.0	12 $\Omega/\square$	120
10	3.9	27 $\Omega/\square$	270
1000 Å of Doped Polysilicon		~ 100 $\Omega/\square$	10^{-3} Ω cm

The oxide thicknesses were measured by Rutherford backscattering for films oxidized in dry O_2 at temperatures of 800 to 1100°C for times that ranged from 0.5 to 42 hours, and in wet O_2 at 800 to 1000°C for 0.5 to 2 hours. For both wet and dry oxidation, the growth of SiO_2 on the silicide surface is parabolic with time. By making an Arrhenius plot of

$$x_o^2 = Ae^{-E_a/kT}$$

where A is the pre-exponential factor, k is the Boltzmann constant, and E_a is the activation energy; the activation energies for the dry and wet oxidation processes are determined to be 1.6 and 1.3 ± 0.2 eV, respectively (Fig. 6.29).

6.5.2 Near Noble Metal Silicides: Pd/Si

Annealing and oxidation studies on PdSi films formed by scanned cw laser beam reaction have been performed over a temperature range from 750°C to 900°C. In Fig. 6.30 we show backscattering spectra taken using 2.2 MeV ^{4}He on PdSi films that were isochronally annealed in an N_2 ambient at temperatures from 750° to 900°C for one hour. Also shown for comparison on this figure is a spectrum of a laser reacted unannealed film. It is clear from Fig. 6.30 that the PdSi film has completely decomposed during the 900°C one hour anneal, with deep diffusion of Pd into the silicon substrate readily observable. The spectrum is shown into the silicon edge (~ 1.24 MeV). The difference in the height of the silicon spectrum for the case of the 900°C anneal compared to that of the unannealed sample is due to dilution of the silicon substrate by the Pd and the resultant difference in the stopping

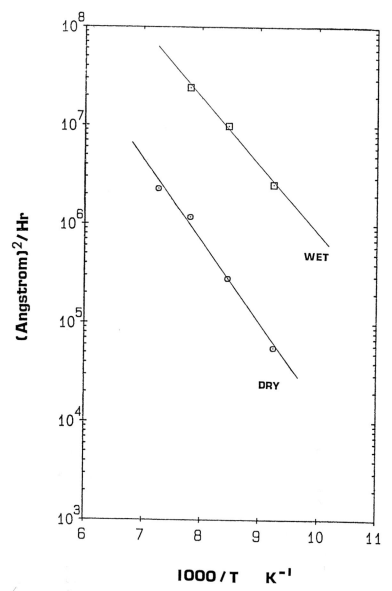

FIGURE 6.29. Determination of the activation energy from the Arrhenius plot of the oxide growth on MoSi$_2$.

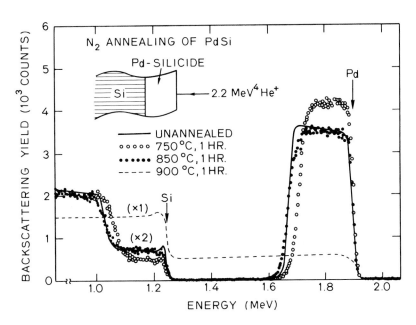

FIGURE 6.30.   Backscattering spectra for N_2 annealing of laser reacted PdSi as a function of temperature. Also shown is the as-reacted sample for reference.

power. Nomarsky microscopy indicates that the surface morphology of the 900°C annealed sample is bad with visible surface damage present.

In contrast to the sample annealed at 900°C the backscattering spectrum for the sample annealed at 850°C for one hour shows very little change compared to the spectrum of the unannealed sample. A very slight change is observed at the Si/PdSi interface (~ 1.65 MeV), and a trace amount of Pd_2Si is found to appear in the X-ray diffraction pattern of this sample. However, a complete phase change is observed for the sample annealed at 750°C for one hour, as seen in Fig. 6.30. Both the backscattering and X-ray diffraction analysis confirm that the entire film decomposed from PdSi into Pd_2Si with no trace of PdSi or polycrystalline silicon evident. The data thus suggest that the excess silicon in the PdSi film has deposited at the silicon Pd_2Si interface epitaxially. In Table 6.6 a summary of the film properties obtained during one hour anneal in flowing N_2 are listed.

TABLE 6.6.   Phase Changes of PdSi Under Various Heat Treatments.

Temperature (°C)	Time	Oxide Thickness (A)	Results of Film Characterization	
			X-ray Diffraction	RBS
750	1 hour	---	Pd_2Si (No PdSi)	Well-defined silicide-silicon interface
850		---	PdSi (Trace of Pd_2Si)	"
900		---	Pd_2Si (Trace of PdSi)	Deep metal penetration into Si substrate
750	1 hour	560	Pd_2Si (No PdSi)	Well-defined silicide-silicon interface
850		440	PdSi (Trace of Pd_2Si)	"
750	1 hour	1750	Pd_2Si (No PdSi)	Gradual interface
850		2410	PdSi > Pd_2Si (Mixed Phase)	"
900		1610	Pd_2Si (Trace of PdSi)	Deep metal penetration

In Table 6.6 we also summarize the effect of dry oxidation on laser-reacted PdSi films. From the data listed in this table it can be seen that the behavior of the PdSi at 750 and 850°C in the oxygen ambient is similar to that observed during the nitrogen ambient anneals, except for the growth of an SiO_2 overlayer on the film surface.

Since the oxides formed during dry oxidation were found to be quite thin (400-600 Å), steam oxidation of the PdSi film was investigated as a technique for increasing the oxide film thickness without increasing the time-temperature product. We show the results of the steam oxidation at 750 and 850°C for one hour in Fig. 6.31 the resultant data is summarized again in Table 6.6.

The backscattering spectrum in Fig. 6.31 clearly shows that the PdSi phase remains stable during the 850°C oxidation but changes to Pd_2Si for the 750°C oxidation. We can also see that the interface between the PdSi film and the silicon substrate is no longer well defined after the oxidation but is spread out in energy indicating intermixing and possible deep metal penetration into the substrate. This result is in contrast to that observed for these temperatures for both the N_2 and dry O_2 anneals. X-ray diffraction analysis on the sample subjected to wet oxidation at 850°C shows the existence of essentially the two phases Pd_2Si and PdSi.

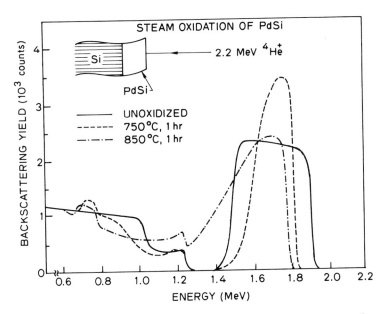

FIGURE 6.31.   Backscattering spectra for PdSi samples subjected to steam oxidation at 750°C (---) and 850°C (-•-•-) for
one hour. Also shown is the as-reacted sample for reference.

6.6   LASER PROCESSING OF Nb₃X SUPERCONDUCTORS*

6.6   LASER PROCESSING OF Nb_3X SUPERCONDUCTORS*

The synthesis of metastable superconducting compounds with
the A15 crystal structure, such as Nb_3Ge, Nb_3Al, and Nb_3Si, has
received considerable attention because of the well known potential of these materials for high temperature superconductivity
($T_c > 20°K$). The difficulties that have been encountered in the
synthesis are well illustrated by Nb_3Al, which is the least unstable member of the group [6.47]. Nb_3Al is known to exist as
a niobium rich A15 compound over a large temperature range in
the equilibrium phase diagram [6.48-6.49]. However, the A15
phase is believed to approach stoichiometry only at relatively
high temperatures (1700–1960°C) [6.48-6.49]. Hence, even though
the exact details of the phase diagram at high temperatures
remain unsettled, it is to be expected that a nearly stoichiometric A15 phase will be disproportionate at low temperatures unless it is cooled very rapidly. Rapid cooling, on the other

---

*The work reported in this section was done in collaboration with
J. Kwo, R. D. Feldman, and T. H. Geballe and R. H. Hammond [6.47,
6.51].

hand, tends to freeze in atomic disorder and structural defects. Much of the effort in the synthesis of a high-T_c Nb$_3$Al has therefore been directed toward finding a suitable compromise between stoichiometry and ordering.

Even greater difficulty is encountered in the synthesis of Nb$_3$Si, since a stoichiometric A15 phase apparently does not even exist in the equilibrium phase diagram [6.50]. As a result, materials synthesis procedures that are known to be capable of forming metastable phases are very essential to the synthesis of a high-T_c form of Nb$_3$Si [6.51].

As indicated in the previous section, deposition of concentrated energy with the cw laser beam has been successfully applied to the modification of electronic materials, including the formation of metastable alloys [6.7, 6.18]. In particular, metastable silicides, which have so far proven difficult or impossible to form by either conventional furnace annealing or pulsed beam annealing, can readily be obtained with cw beam annealing. In effect, the annealing process using a scanning cw beam can be described as a "high temperature, short-time furnace". With such a process, the annealing temperature can easily be raised to above 2000°C by a proper choice of the laser power; the annealing time can be increased by employing multiple laser scans; and heating and cooling rates can be controlled by changing the beam scan speed (up to a maximum cooling rate of ~ 10^6°C/sec).

These features are quite attractive for the synthesis of Nb-Al and Nb-Si A15 compounds. For example, it is reasonable to expect that with cw beam annealing, nearly stoichiometric Nb$_3$Al can be quenched from a high temperature A15 state to room temperature without decomposition into low temperature phases. In addition, the formation of the metastable A15 Nb$_3$Si phase might be favored by this process. In this section we describe recent attempts to apply cw laser annealing to the synthesis of Nb$_3$Al and Nb$_3$Si, and we discuss the preliminary results and the potential application of cw beam annealing for forming metastable phases of these materials.

6.6.1 Sample Preparation

Thin films of Nb-Al (5000 Å) and Nb-Si (1000 Å) were prepared by the electron-beam coevaporation pressure of (1.0 → 3.0) x 10^{-7} Torr. Typical deposition conditions are ~ 30 Å/sec deposition rate, ~ 300°C substrate temperature for the Nb$_3$Al, and ~ 50 Å/sec deposition rate, ~ 500°C substrate temperature for the Nb$_3$Si. Further details on the thin-film deposition are given in Refs. [6.47, 6.51, 6.52]. These films were deposited onto a single crystal Si substrate coated with a thin layer of Si$_3$N$_4$

or SiO_2 as a buffer. This substrate is used because (a) single crystal silicon is a good heat conductor at high temperatures and (b) analytical calculations for the laser-induced temperature rise [10] are well established for a silicon substrate.

6.6.2 Results and Discussion

Figure 6.32 shows a typical x-ray diffraction pattern obtained from as-deposited samples of Nb-Al. The principal lines shown were identified as arising from the bcc structure. Some weak lines from the A15 phase were also observed. A variety of laser annealing experiments were performed with calculated laser-annealing temperatures (T_{max}) in the range 1300°C to 1800°C. Figure 6.32 also shows the x-ray diffraction pattern of a sample (of 24.6 at. % Al) which was laser-annealed at 1400°C with 10 laser scan frames. Scan lines were overlapped by ~ 30% in each frame. Formation of the A15 structure after laser annealing is clearly observed, with some weak diffraction lines from the bcc phase. All laser annealed Nb_3Al samples examined exhibited

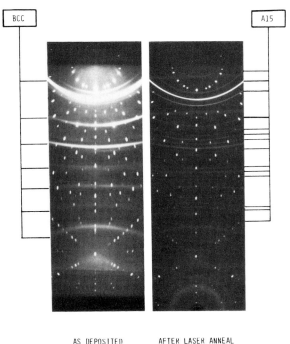

AS DEPOSITED   AFTER LASER ANNEAL
Nb-Al   1400°C, 10 SCANS

FIGURE 6.32.   Read-Camera diffraction pattern for a Nb-Al sample (24.6 at.% Al); (a) as-deposited (b) after laser anneal 10.

similar diffraction patterns, where both the A15 and bcc phases exist. The relative abundance of the A15 to the bcc phase varied substantially with the composition of the films and laser anneal-ing conditions (T_{max} and number of laser scan frames). One impor-tant point is that diffraction lines from the tetragonal Nb_2Al phase, which is the second phase typically obtained at lower temperatures were not observed. This is particularly important since the phase boundary between the A15 and Nb_2Al phase is less than 22 at. % Al for temperatures below about 1500° C [6.47-6.48]. This result therefore shows that the cw laser processing succeeded in quenching the high temperature phase to room temperature with-out decomposition into lower temperature phases that accompanies the conventional thermal-quench method.

In Fig. 6.33 T_c is plotted as a function of the number of laser scan frames. The laser annealing temperature is about 1400°C for all samples. Data are shown for two sets of Nb-Al samples with different compositions (21.0 and 24.6 at. % Al). T_c is seen to increase linearly with the number of scan frames up to 10 frames, after which it remains essentially constant. X-ray diffraction analysis showed that this T_c enhancement was cor-related with the growth of A15 phase relative to the bcc phase. Therefore, ten laser scans are probably sufficient to carry the

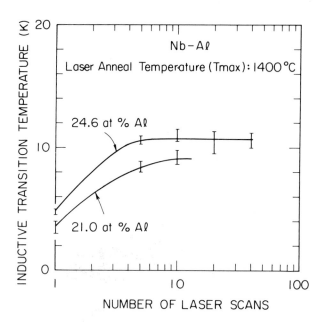

FIGURE 6.33. Superconducting transition temperatures (T_c) as a function of high temperature laser scans at 1400°C, for two sets of NbAl samples (of 2.10 and 24.6 at. % Al). ©1981 Elsevier Science Publishing Co.

reaction to completion. It is also shown in Fig. 6.33 that T_c is consistently higher for films with higher Al composition.

Figure 6.34 shows the dependence of T_c on laser annealing temperature for the same two sets of NbAl samples discussed above. Ten laser scans were employed at each temperature. It is seen that higher annealing temperature reduces T_c, particular for the low Al composition (21.0 at. % Al) samples. X-ray diffraction analysis showed that the samples annealed at high temperatures had less A15 phase (and more bcc phase) than the samples annealed at lower temperatures for both Al compositions. These observations are qualitatively consistent with the equilibrium phase diagrams of Nb-Al [6.38-6.39], considering the uncertainty in temperature determinations. Briefly, the compositional range of the bcc solid solution is more Al-rich ($\geqslant$ 21 at. % Al) at high temperatures ($\geqslant$ 1800°C). Hence the increased formation of the bcc phase at high temperatures, particularly for the sample of 21.0 at. % Al, leads to the lower values of T_c.

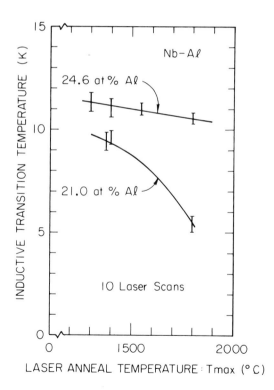

FIGURE 6.34. Superconducting transition temperature (T_c) as a function of low temperature laser scans at 800°C.

Low Temperature Anneal

From the discussion above one can conclude that the high temperature state in the phase diagram can be quenched to room temperature by scanning cw laser annealing, without a corresponding phase change. However, atomic disorder is usually observed following a rapid quench to room temperature. Low temperature furnace annealing at 700-800°C can of course improve the ordering, but such annealing usually leads to decomposition of the quenched in phase to the unwanted low-temperature phases.

As discussed earlier, a single laser beam scan is equivalent to furnace annealing for very short times (~ msec) with very short heating and cooling cycles. By repeated scanning, one can therefore extend the effective annealing time without changing cooling and heating rates. This is exactly what is required for annealing Nb-Al samples. Using such a procedure, three different samples were subjected to 800°C laser annealing for 100 scan frames.

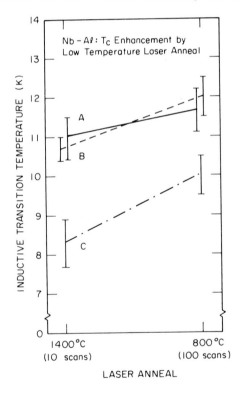

FIGURE 6.35. Superconducting transition temperatures (T_c) as a function of laser annealing temperature for the same two sets of Nb-Al samples.

Values of T_c before and after this process are shown in Fig. 6.35. A T_c enhancement of 0.5–1.5°K is observed for all samples. There is no significant change in the relative amounts of the A15 to the bcc phase, and the second phase Nb_2Al was not observed in any of these cases. We tentatively attribute this T_c enhancement to the atomic ordering obtained by low-temperature laser annealing.

6.6.3 Experimental Results for Nb–Si

X-ray diffraction analysis showed that the as-deposited films of Nb–Si (1000 Å) were highly disordered or amorphous. Laser annealing was performed with a scan speed of ~ 50 cm/sec. The crystal structures obtained after a single laser scan are summarized in Table 6.7.

TABLE 6.7.   Summary of Various Crystalline Structures of NbSi Films Produced by Laser Annealing.

	Composition of NbSi Film		
	19.9 at % Si	20.5 at % Si	22.8 at % Si
9 W	A15 (single phase)	Ti_3P type $\gg$ A15	Tetragonal (Cr_5Si_3 type) + Ti_3P type (no A15)
11 W	Ti_3P type + Tetragonal (Cr_5B_3 type)	------	--------

Figure 6.36 shows the x-ray diffraction pattern obtained from the 19.9 at. % Si sample after laser annealing at 9 W. Formation of the single phase A15 compound is clearly observed. T_c measurements performed on this sample, both resistively and inductively, gave values of ~ 4°K.

From Table 6.7 it is seen that the Ti_3P structure was obtained at higher laser annealing temperatures (11 W). Table 6.7 also shows that the A15 structure becomes more difficult to form when the composition is close to stoichiometric Nb_3Si.

It should be noted that laser annealing can crystallize 19.9 at. % Si Nb–Si samples to the single phase A15 structure without melting, while melting is clearly involved in case of the splat growth of a stoichiometric A15 phase. An amorphous Nb–Si film

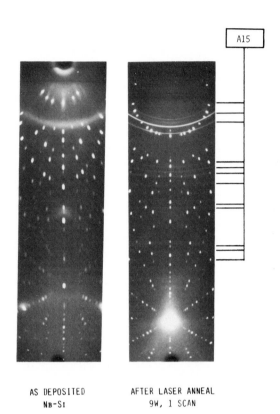

AS DEPOSITED   AFTER LASER ANNEAL
Nв-Si   9W, 1 SCAN

FIGURE 6.36.   Read–Camera diffraction pattern for a Nb–Si sample (19.9 at. % Si) (a) as deposited (b) after laser anneal.

consisting of layers in which the composition changes gradually from 19.9 to 25 at. % would be interesting for laser annealing, since a single laser scan could possibly crystallize the 19.9 at. % Si layer into the A15 phase, which might then propagate into the layer of 25 at. % Si.

REFERENCES

6.1   Crowder, B. L., and Zirinsky, S., IEEE Trans. on Elec. Dev. ED-26, 369 (1979).

6.2   Shah, P. L., IEEE Trans. on Elec. Dev. ED-26, 631 (1979).

6.3   Yu, H. N., Reisman, A., Osburn, C. M., and Critchlow, D. L., (1979). IEEE Trans. on Elec. Dev. ED-26, 318.

6.4   "Thin Films Interdiffusion and Reactions" (J. M. Poate, K. N. Tu and J. W. Mayer, eds.) Wiley Interscience, New York (1978).

6.5   Poate, J. M., Leamy, H. J., Sheng, T. T., and Celler, G. K., Appl. Phys. Lett. 33, 918 (1978).

6.6   Liau, Z. L., Tsau, B. Y., and Mayer, J. W., Appl. Phys. Lett. 34, 221 (1979).

6.7   Shibata, T., Gibbons, J. F., and Sigmon, T. W., Appl. Phys. Lett. 36, 566 (1980).

6.8   Sigmon, T. W., Regolini, R. L. and Gibbons, J. F., "Symposium on Laser and Electron Beam Processing of Electronic Materials," Los Angeles, p. 350 (1979).

6.9   Tsau, B. Y., Liau, Z. L., and Mayer, J. W., Appl. Phys. Lett. 34, 1968 (1979).

6.10   Tsau, B. Y., Liau, Z. L., and Mayer, J. W., Phys. Lett. 71A, 270 (1979).

6.11   Kanayama, T., Tanoue, H., and Tsurushima, T., Appl. Phys. Lett. 35, 222 (1979).

6.12   Tsau, B. Y., Lau, S. S., and Mayer, J. W., Appl. Phys. Lett. 35, 225 (1979).

6.13   "Laser and Electron Beam Processing of Semiconductor Structures" (J. W. Mayer and J. M. Poate, eds.) to be published.

6.14   von Allmen, M., and Wittmer, M., Appl. Phys. Lett. 34, 68 (1979).

6.15   Wittmer, M., and von Allmen, M., J. Appl. Phys. 50, 4786 (1979).

6.16   van Gurp, G. J., Eggermont, G. E. J., Tamminga, Y., Stacy, W. T., and Gijsbers, J. R. M., Appl. Phys. Lett. 35, 273 (1979).

6.17   Lau, S. S., Mayer, J. W., Tsau, B. Y., and von Allmen, M., in "Laser and Electron Beam Processing of Materials" (C. W. White and P. S. Peercy, eds.) p. 511, Academic Press, New York (1980).

6.18   Shibata, T., Sigmon, T. W., and Gibbons, J. F. in "Proceedings of the Symposium on Thin Film Interfaces and Interactions" (J. E. E. Baglin and J. M. Poate, eds.) Electrochem. Soc., p. 458, Princeton (1980).

6.19   Shibata, T., Sigmon, T. W., and Gibbons, J. F., in "Laser and Electron Beam Processing of Materials" (C. W. White and P. S. Peercy, eds.) p. 530, Academic Press, New York (1980).

6.20   Lietoila, A., Ph.D. thesis, Stanford University, Dept. of Applied Physics (1981). See also Sec. 2.3.6 of this volume.

6.21 Chu, W. K., Mayer, J. W., and Nicolet, M. A., "Backscattering Spectrometry", Academic Press, New York (1978).

6.22 Azaroff, L. V., "Elements of X-Ray Crystallography," 383, McGraw-Hill, New York (1968).

6.23 Feder, R., and Berry, B. S., J. Appl. Cryst. 3, 372 (1970).

6.24 Read, M. H., and Hansler, D. H., Thin Solid Films 10, 123 (1972).

6.25 American Society for Tresting and Materials, Powder Diffraction Data Files, Philadelphia, PA.

6.26 Wagner, R. J., Lau, S. S., and Mayer, J. W., "Proceedings of the Symposium on Thin Film Phenomena – Interfaces and Interactions", 59 (1978).

6.27 Hansen, M., "Constitution of Binary Alloys," McGraw-Hill, New York, 1016 (1959).

6.28 Bower, R. W., Sigurd, D., and Scott, R. E., Solid State Electron. 16, 1461 (1973).

6.29 Cheung, N., Lau, S. S., Nicolet, M. A., Mayer, J. W., and Sheng, T. T. in "Proceedings of the Symposium on Thin Film Interfaces and Interactions" (J. E. E. Baglin and J. M. Poate, eds.) 80-2, p. 494, Electrochem. Soc., Princeton (1980).

6.30 Hutchins, G. A. and Shepola, A., Thin Solid Films 18, 343 (1973).

6.31 Anderson, R., Baglin, J., Dempsey, J., Hammer, W., d'Heurle, F., and Peterson, S., Appl. Phys. Lett. 35, 285 (1979).

6.32 Tsaur, B. Y., Mayer, J. W., and Tu, K. W., in "Proceedings of the Symposium on Thin Film Interfaces and Interactions" (J. E. E. Baglin and J. M. Poate, eds.) 80-2, p. 264, Electrochem. Soc., Princeton (1980).

6.33 Johnson, N., Sigmon, T., Nemanich, R., Mayer, D., and Lau, S. S., in "Proceedings of Materials Research Society Symposium on Laser and Electron Beam Processing", Boston (1980).

6.34 Inoue, T., and Koike, K., Appl. Phys. Lett. 33, 826 (1978).

6.35 Mochizuki, T., Shibata, K., Inoue, T., and Ohuchi, K., Jpn. J. Appl. Phys. Suppl. 17-1, 37 (1977).

6.36 Majni, G., Nava, T., and Ottaviani, G., in "Proceedings of Materials Research Society Symposium on Thin Film Interfaces and Interactions," Boston (1979).

6.37 Zirinsky, S., Hammer, W., d'Heurle, F., and Baglin, J., Appl. Phys. Lett. 33, 76 (1968).

6.38 Mohammadi, F., Saraswat, K. C., and Meindl, J. D., Appl. Phys. Lett. 35, 529 (1979).

6.39 Shibata, T., Wakita, A., Sigmon, T. W., and Gibbons, J. F., Appl. Phys. Lett. 40, 77 (1982).

6.40 Muraka, S. P., Fraser, D. B., Lindenberger, W. S., and Sinha, A. K., J. Appl. Phys. 51, 3241 (1980).

6.41 Razouk, R., Thomas, M., and Pressacco, S., J. Appl. Phys. 53, 5342 (1982).

6.42 Bartur, M., and Nicolet, M-A., Appl. Phys. Lett. 40, 175 (1982).

6.43   Chen, J., Houng, M-P., Hsiung, S-K., and Liu, Y-C., Appl. Phys. Lett. 37, 824 (1980).

6.44   Schwartz, G.P., Sunder, W. A., Griffiths, J. E. and Gualtieri, G. J., Thin Solid Films, 94, 205–212 (1982).

6.45   Beyers, R., submitted for publication.

6.46   Wakita, A. S., Sigmon, T. W., and Gibbons, J. F., submitted for publication.

6.47   Kwo, J., Hammond, R. H., Geballe, T. H., J. Appl. Phys. 51(3), 1726 (1980).

6.48   Lundin, C. E., and Yamamoto, A. S., Trans. Am. Inst. Mech. Eng. 236, 863 (1966).

6.49   Svechinkov, V. N., Pan, V. M., and Latzshev, V. I., Metallafigik 32, 28 (1970).

6.50   Pan, V. M., Pet'kov, V. V., Kulik, O. G., in "Physics and Metallurgy of Superconductors", Moscow, USSR, 1965–1966, E. H. Savitskii and V. V. Baron, eds. (Consultants Bureau, New York, 1970).

6.51   Feldman, R. D., Hammond, R. H., and Geballe, T. H., Appl. Phys. Lett. 35, 818 (1979), and references therein.

6.52   Gat, A. and Gibbons, J. F., Appl. Phys. Lett. 32(3), 142 February 1, 1978).

CHAPTER 7

CW Beam Processing of Gallium Arsenide

Yves I. Nissim* and James F. Gibbons

STANFORD ELECTRONICS LABORATORIES
STANFORD UNIVERSITY
STANFORD, CALIFORNIA

7.1 INTRODUCTION

As is suggested in the previous chapters, the use of cw beams to process silicon can provide significant advantages in both material and devices properties. Compound semiconductors, and in particular GaAs, have from the beginning been potential candidates for the application of the beam annealing technology. In particular the conventional thermal annealing of GaAs often gives unsatisfactory results in both activation of dopants and removal of damage induced by ion implantation. Furthermore, encapsulation or controlled arsenic overpressure are required to prevent the substrate from surface decomposition (As evaporation) at elevated temperatures.

A short, localized heat treatment under beam irradiation can be expected to suppress decomposition and thereby possibly improve the electrical properties of the annealed layers. However, the results to date on cw beam annealing of implanted layers in GaAs have not shown substantial improvements as compared to thermal annealing. Major difficulties arising from the fragility and dis-sociation of the material under laser irradiation are responsible for the limited success achieved in early attempts to anneal implanted GaAs with a cw laser system. However, the temperature calculations presented in Chapter 2 demonstrate that a laser beam with an elliptical cross section produces a more gradually dis-tributed temperature gradient at the surface of a semiconductor than a beam with a circular cross section. The combination of

---

*Present address: CNET, 92220 Bagneaux, France.

the elliptical geometry and an appropriate set of annealing parameters in different environments does result in efficient electrical activation (Section 7.2). Multiply scanned electron beam can also be used as an alternative to laser, giving results that also will be presented in the same section.

The mechanism that could produce annealing of ion implanted amorphous layers in GaAs at low temperatures is solid phase epitaxial regrowth. While this process is well characterized in silicon, it is much less well understood in GaAs. Recently, however, solid phase epitaxial recrystallization in GaAs has been observed at different laboratories using a well controlled, precisely defined ion implantation schedule. These results are presented in Section 7.2 with their limitations.

Cw laser processing of GaAs has made a more significant contribution as a means of reacting deposited thin films with the underlying substrate. Laser alloying of thin metallic film to n-GaAs has resulted in alloyed ohmic contacts that are significantly better than conventional thermally alloyed contacts. It has been also demonstrated that a scanned cw laser can assist the diffusion and activation of tin in semi-insulating GaAs from an SnO_2/SiO_2 thin film source. This process produces thin n^+ layers that display excellent ohmic behavior directly after metal evaporation, (i.e., without an alloying step). This nonalloyed ohmic contact technology is well suited to device technology since the required laser treatment operates at incident power levels that are well below the laser induced damage threshold of the GaAs substrate. This subject is discussed in Section 7.3.

More recently, rapid thermal processing has been used to anneal implanted layers in GaAs, InP and HgCdTe; and to provide a technique for studying impurity diffusion. A novel process employing controlled evaporation of plated Zn doping source has also been used to produce significant improvements in Schottky barrier heights. We will discuss these topics in Sections 7.4-7.6 below.

7.2   CW LASER ANNEALING OF ION IMPLANTED GaAs

Among the different doping techniques for GaAs, ion implantation is used extensively to control both the doping levels and junction depth. A reproducible annealing sequence that is capable of activating the implanted dopants (on substitutional sites) and removing the ion implantation induced damage is required. In what follows we discuss attempts to anneal ion implanted GaAs with a scanning beam.

7.2.1   The Elliptical Beam Approach

The temperature induced in a semiconductor by a scanning elliptical beam has been calculated in Chapter 2. As can be seen from Fig. 2.7, a laser beam with an elliptical cross section narrow in the direction of the scan and large in the direction perpendicular to the scan, produces a more uniformly distributed temperature gradient at the surface of the irradiated semiconductor. This property is of interest for processing brittle materials such as GaAs because it reduces the thermal shock that a circular beam would create. This beam geometry also permits a large area to be annealed in a single scan, with substantially less power than a circular beam of large radius would require. The size of the annealed area is nevertheless limited by the amount of power available, as can be seen in Fig. 7.1. This figure compares the maximum temperature obtained in GaAs for different value of the aspect ratio β as a function of the absorbed power per unit radius. For instance, the maximum temperature of 1000°C is obtained in GaAs for β = ω_y/ω_x = 20 with seven times the power required in the case where β = ω_x/ω_x = 1. Since the thermal conductivity of GaAs is small compared to the thermal conductivity of silicon, this technique will be more applicable in GaAs using a cw laser of moderate output power (~ 10 watts).

Significant electrical activity after cw laser annealing of ion implanted layers in GaAs was first reported by Fan et al., [7.2]. A cw Nd:YAG laser (operating at 1.06 μm) and a set of unconventional annealing parameters as compared to those utilized for silicon were used. The thermal shock induced by the laser was reduced by a combination of a high substrate temperature (580°C) and a laser beam with an elliptical cross section. Scanning speeds between one and two orders of magnitude lower than those reported for silicon were used. Typically, speeds between 1 cm/sec and 0.5 mm/sec resulted in reasonable electrical activation. The substrate was in a forming gas environment during irradiation. The samples studied were implanted with 1×10^{14} Se$^+$/cm^2 at an

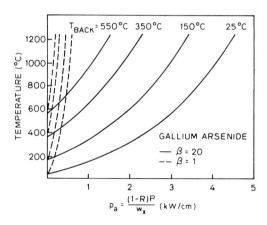

FIGURE 7.1. The "true" maximum temperature induced in GaAs for different values of the aspect ratio β = 1 and β = 20).

energy of 400 keV. The implant temperature was 300°C in the case of highest activity. Values of 27% for electrical activity with a Hall mobility of 2100 cm^2/V.sec for a single scan and 59% with a mobility of 1400 cm^2/V.sec for ten consecutive scans were the best results reported. It was observed that surface deterioration (slip lines) occurred when the incident power and/or the scanning speed were increased. The formation of slip lines implies a limit on the maximum temperature at the surface of the irradiated sample. The results presented in Ref. 7.2 demonstrate that the highest electrical activity is obtained when slip lines just start to appear; however, the slip lines also lead to a lower surface mobility. It is important to note that encapsulation of the implanted samples prior to laser annealing enhances surface deterioration, probably due to increased stress at the semiconductor-encapsulant interface. These results, obtained for medium implanted doses, become poorer for higher doses; and the technique was found to be unsuccessful for low implant doses [7.3] (typically 10^{13} ions/cm^2).

If the incident power is increased above the threshold for slip line formation, the GaAs substrate decomposes due to As evaporation. When the substrate is held in a forming gas environment as described in the previous paragraph, the substrate decomposition is observed by an excess of Ga in the form of a grey powder at the surface of the irradiated sample. However if the substrate is held in a laboratory environment or an oxygen environment, an oxide is formed when decomposition occurs [7.4]. This oxide has been identified, using X-ray diffraction, as the gallium oxide $\beta-Ga_2O_3$. Sputter Auger profiling has been used to study the composition of this oxide. The resulting profile is shown in Fig. 7.2. It can be seen that the gallium oxide is completely free of As (within the sensitivity of the measurements) and is about 1200 Å thick, corresponding to an observed blue color. The interface with the underlying substrate is relatively sharp ($\simeq$100 Å). In this experiment an argon ion laser was focused by a cylindrical lens onto a GaAs substrate heated to 550°C. A scanning speed of 0.5 mm/sec was used. The resulting temperature profiles and the different working areas are shown in Fig. 7.3. The electrical activity obtained after oxide growth on low dose implanted layers (1×10^{13} Si^+/cm^2) in <100> semi-insulating GaAs was studied. The annealing area is characterized by a low electrical activity (10 to 13%) and high mobility (2500 cm^2/V.sec) in the lower power regime and higher activation (30%) with reduced mobility ($\simeq$100 cm^2/V.sec) accompanied by the formation of slip lines in the higher power regime. When the oxide is formed, higher temperatures can be induced, resulting in very high dopant activation. The electrical activity as judged by sheet electrical measurements was found to be 100% when the impurity content in the oxide layer is subtracted. Hall mobilities in the range of 1800 cm^2/V.sec (for 1×10^{13} e^-/cm^2) were reported

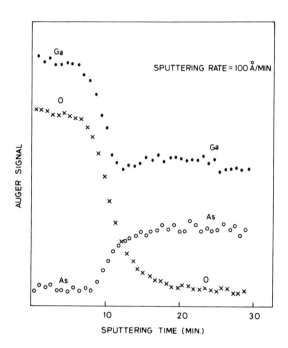

FIGURE 7.2. Sputtering Auger profile of the laser grown oxide. ©1981 Elsevier Science Publishing Co.

when the laser parameters were chosen to obtain a uniform oxide layer. The overlap between the annealing and oxide growth areas shown in Fig. 7.3 indicates that the oxide can grow at lower powers if the growth is initiated by an enhanced absorption (edge or scratch on the wafer). While scanning over an already grown layer, the oxide growth will continue. The decomposition of the surface of the GaAs and the formation of $\beta-Ga_2O_3$ acts as a stress release and thus prevents the slip lines from forming.

An alternate method to obtain a uniform heating at the surface of a semiconductor has been reported using a multiply-scanned electron beam [7.5]. The sample is thermally isolated and exposed to multiple randomly interlaced raster scans of the beam. This results in an uniform exposure over the scanned area. Exposure times on the order of 5 sec have resulted in high induced temperatures and successful annealing of low dose Si implants ($6 \times 10^{12}/$ cm^2) in GaAs [7.6]. The resulting annealed profile is presented in Fig. 7.4 and compared to an as implanted LSS profile. Activation of 50 to 60% of the implanted dopant were reported with a mobility of 3800 $cm^2/V.sec$. This value of mobility for an active dopant concentration of $10^{17}/cm^3$ is the highest obtained by beam annealing techniques so far and comparable to the best thermal annealing.

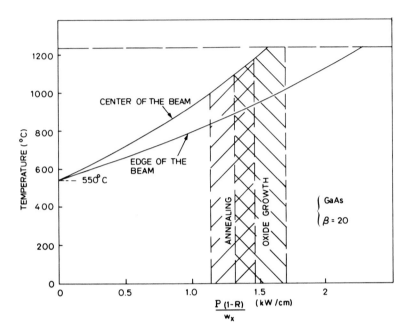

FIGURE 7.3.   Temperature induced in GaAs by a laser focused on an elliptical spot of aspect ratio β = 20. The maximum temperature at the center, and the temperature at the edge of the beam are plotted. ©1981 Elsevier Science Publishing Co.

7.2.2   Solid Phase Epitaxy in GaAs: Low Temperature Annealing

Solid phase epitaxial recrystallization of ion implanted amorphous layers has been observed to occur in silicon at temperatures on the order of 500°C [7.7].   In this process, the crystallinity of the material is recovered and most of the dopant impurities are incorporated on substitutional sites.   As presented in Chapter 3, cw laser annealing of amorphous layers in Si proceed via the same mechanism.   The difficulties presented in the previous section to obtain good crystal recovery and electrical activation following ion implantation in GaAs would be overcome if a complete epitaxial recrystallization process in solid phase could be induced by laser irradiation.   Unfortunately this process is more complex in GaAs and has only recently been demonstrated.   We begin our discussion of this topic by reviewing furnace annealing studies.

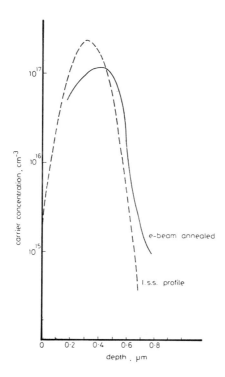

FIGURE 7.4. Carrier concentration profile of electron beam annealed GaAs implanted with silicon ($6x10^{12}$ cm^{-2} at 360 keV).

7.2.2.1   Furnace Annealing

Gamo et al. [7.8] reported experiments performed on Zn implanted GaAs substrates for a range of annealing temperatures between 200 and 600°C. Incomplete or highly defective crystal recovery was observed for temperatures up to 500°C and complete epitaxial regrowth was obtained at 600°C. Local stoichiometric imbalance produced during the implantation cycle was proposed as a limiting factor. More recently, Williams and Austin [7.9] have shown that crystallinity recovery could occur at temperatures as low as 180°C. The isochronal annealing of an amorphous layer obtained by an implant of $5x10^{13}$ Ar^{+}/cm^{2} at 100 keV with the substrate held at liquid nitrogen during the implant is shown in Fig. 7.5a. The implantation schedule is just sufficient to create an amorphous layer continuous to the surface. The epitaxial growth is started at 130°C and is almost complete at 180°C. If the incident dose is increased to $2x10^{14}$ Ar^{+}/cm^{2} it is then necessary to reach 600°C before the epitaxy is complete as seen in Fig. 7.5b. As the implant dose increases, the quality of the

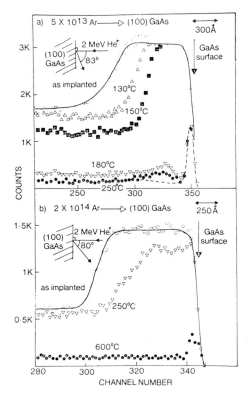

FIGURE 7.5.   Isochronal (10 min) annealing of (a) 5×10^{13} Ar^+/cm^2 and (b) 2×10^{14} Ar^+/cm^2 implanted in GaAs. The dotted curve in (a) is for unimplanted GaAs. The GaAs was encapsulated with AP metal for the 600°C anneal in (b).

regrown material degrades and the regrowth rate decreases at constant annealing temperature. This illustrates the strong dose dependence of this mechanism and precludes any meaningful regrowth kinetic calculations. Further studies by Williams and Harrison [7.10] have shown that the furnace annealing of ion implanted amorphous layers in GaAs at doses well above amorphization occurs in stages. This is illustrated in Fig. 7.6 where the normalized disorder as measured by channeling for 8×10^{13} Ar^+/cm^2 implant at 100 keV in <100> GaAs is plotted as a function of annealing temperature. It can be seen that the crystalline to amorphous transition occurs at very low temperatures (125-230°C) but 50% disorder is still observed after the transition. Only in a second stage (400-600°C) does the annealing of extended defects occur and the crystalline quality is recovered. TEM analysis by Kular et al. [7.11] has shown that the annealed layers are a highly twinned single crystal in the first stage of annealing (low temperature)

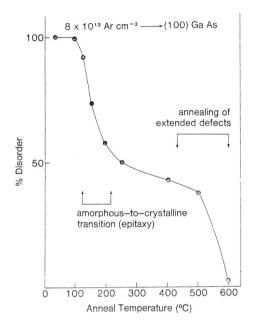

FIGURE 7.6.   Normalized disorder as measured by channeling for $8x10^{13}$ Ar^+/cm^2 (100 keV) implanted in <100> GaAs plotted as a function of furnace anneal temperature (15 min. anneals) indicating two annealing stages. ©1981 Elsevier Science Publishing Co.

and a single crystal with dislocation loops at higher temperature. An even higher temperature results in the reduction of the density of loops.

The process of solid phase epitaxial regrowth in GaAs has been extensively studied by Grimaldi et al. [7.12]. In contrast to the work reported above, the regrowth has been observed to depend strongly on the initial thickness of the amorphous layer. Complete recovery after furnace annealing was obtained for thicknesses < 400 Å. A model in which the growing crystal front accumulates local defects has been proposed. The thickness of 400 Å would then be the maximum distance that the interface can move before the defects begin to precipitate in clusters. These clusters in turn could then be removed only if the annealing temperture is elevated to a range between 700 and 800°C. Below this critical thickness the incident dose, type of incident atoms and substrate temperature during implantation become less critical for the regrowth.

In the light of these results, a number of different experimental choices were made by Nissim et al. [7.13] to study this mechanism. The work reported in Ref. 7.9 shows that a very fast regrowth rate is observed along the <100> direction of GaAs. In an attempt to control the process, the <511> orientation in GaAs was chosen (it has been observed in silicon [7.14] that the <511> orientation had a slower rate than the <100> direction). The implantations were carried out with the substrate held either at ice water (for the highest doses) or liquid nitrogen temperature, to obtain a sharp crystalline amorphous interface and reduce the necessary incident dose of amorphization. The contribution of dopant atoms to the regrowth rate was minimized by the implantation of As^+ ions. Annealing was carried out at a range of temperatures between 400 and 500°C. When high doses are used to amorphize the GaAs (typically above 10^{15} As^+/cm^2), epitaxial growth from the crystalline amorphous interface is observed, but it terminates before reaching the surface (40% of the initial amorphous layer consistently recrystallized). The recrystallized region has a high concentration of defects as revealed by a high backscattering yield. When the annealing temperature reaches 500°C, a slow annealing of defects accompanies the regrowth. This is characterized by a reduction in the backscattering yield with an increase in the annealing time without evolution of the damaged layer. These results have been supported by TEM analysis as shown in Fig. 7.7. After annealing, polycrystallites of GaAs are detected within the implanted region as shown in Fig. 7.7(b). Careful examination of the diffraction pattern inset indicates the possible existence of residual dislocation line structure beneath the polycrystalline layer. Using controlled chemical stripping, 750 Å of material was etched from the surface of the sample and specimens were again prepared for TEM examination. Figure 7.7(c) shows that all the polycrystalline layers have been removed, and a complex array of dislocation nesting is observed. The diffraction pattern indicates the presence of an imperfectly regrown single crystal layer. The formation of this polycrystalline region is interpreted as a competitive mechanism to the regrowth that proceeds from the surface down to the crystalline-amorphous interface and meets the recrystallizing front. In order to understand the large defect concentration in the regrown material, damage distribution and net stoichiometric imbalance resulting from the implantation were calculated using a Boltzmann transport approach [7.15,7.16]. The result of this calculation is shown in Fig. 7.8 for an implant of 1.1×10^{15} As^+/cm^2 at 145 keV. The measured thickness of the amorphous layer and recrystallized layer after annealing are indicated in this plot. A minimum damage density of 2.8×10^{21} keV/cm^3 is required to obtain an amorphous layer, and the damage density at which the regrowth stopped is 9×10^{21} keV/cm^3. This upper limit is not well established since the recrystallized region was observed to be highly damaged. A series of As^+ implants (3×10^{14} As^+/cm^2 at 145 keV followed by 1×10^{14} As^+/cm^2 at 80 keV), whose doses and energies were selected to achieve a damage density of $\simeq 3 \times 10^{21}$ keV/cm^3, and to ensure that this value was maintained in the near surface region, led

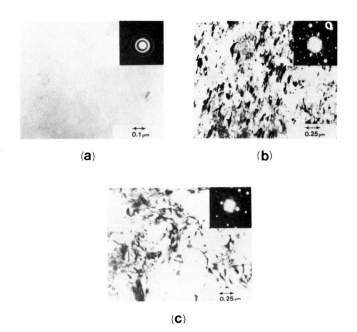

(a)   (b)

(c)

FIGURE 7.7. Representative bright field transmission electron micrograph obtained from samples implanted with 180 keV As ions to a dose of 4.6×10^{15} cm^{-2} in ⟨511⟩ GaAs. (a) As implanted; (b) implanted and annealed at 475°C for 10 min; (c) implanted, annealed and 750 Å stripped from the surface. Insets show selected area diffraction patterns from regions.

to complete recrystallization. This is illustrated in the channeling spectra of Fig. 7.9. The nature of the microstructure in the as-implanted and annealed samples in this experiment was studied by TEM (Fig. 7.10). After implantation [Fig. 7.10(a)] an absence of microstructure and the presence of an amorphous layer is observed. After annealing complete recrystallization is obtained and well defined single crystal diffraction patterns are detected. Within the implanted region, residual damage is observed in the form of dislocation loops of 150–200 Å average diameter [Fig. 7.10(b)] at a concentration of $\simeq 10^{11}$/cm^2. It should be noted that the same implant schedule performed in ⟨100⟩ GaAs led to complete recrystallization with some discontinuous areas at the surface of amorphous material.

From the results obtained in these different experiments, sufficient conditions to obtain solid phase epitaxial regrowth of ion implanted layers in GaAs can be summarized in the following way:

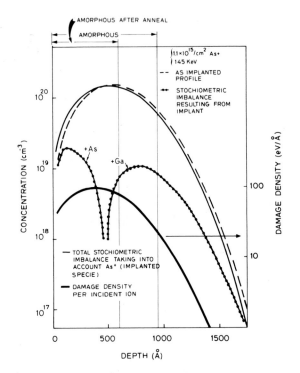

FIGURE 7.8. Net stoichiometric imbalance and damage distribution in GaAs produced by an implantation of 1.1×10^{15} As$^+$/cm^2 at 145 keV (calculation).

(1)   The damage density induced by the implantation must be below a critical threshold but above the level of amorphization. Such windows can eventually be found for different incident ions by holding the substrate at liquid nitrogen temperature during the implantation.

(2)   Polycrystalline growth from the surface must be suppressed with a continuous damage density through the implanted layer.

(3)   An alternative to these conditions can be obtained by keeping the amorphous layer thickness below 400 Å.

7.2.2.2 CW Laser Annealing

As shown in the previous section, the mechanism of solid phase epitaxial growth in GaAs induced by furnace annealing is not fully understood. For this reason very little work on CW

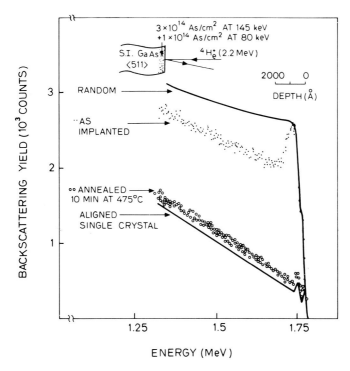

FIGURE 7.9. Channeling spectra illustrating the complete epitaxial regrowth at 475°C for 3×10^{14} As$^+$/cm^2, 145 keV plus 1×10^{14} As$^+$/cm^2, 80 keV implanted into <511> GaAs.

beam annealing has been reported. Since the regrowth can be initiated at low temperature, backsurface substrate heating cannot be used here. Therefore, formation of slip lines will occur at low temperatures, preventing sufficient power increase to achieve high temperatures. Williams et al. [7.17] have compared the behavior of the regrowth when induced by a furnace or a beam irradiation. A cw argon laser was used and annealing parameters were set to be just below the threshold of slip line formation with a dwell time of 15 msec. At doses below amorphization the cw laser induces a complete recrystallization similar to the 180°C furnace annealing. Above the amorphization level, only partial regrowth is observed. These results indicate that the two types of annealing give essentially similar results at low temperatures. The annealing of extended defects observed at higher temperature (600–900°C) furnace annealing is prevented by slip line formation in the case of laser annealing.

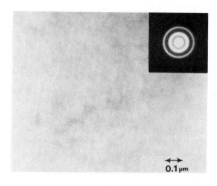

(a)

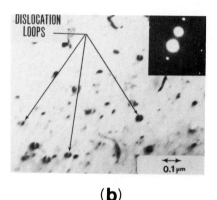

(b)

FIGURE 7.10. Representative bright field transmission elec-
tron micrographs obtained from samples implanted with 3×10^{14} As$^+$/
cm^2 at 145 keV plus 1×10^{14} As$^+$/cm^2 at 80 keV. (a) as implanted;
(b) implanted and annealed at 475°C. Selected area diffraction
patterns are shown in insets.

7.2.2.3   Electrical Activation

During the recrystallization of the lattice, one would expect
to be able to incorporate dopant atoms onto lattice sites that
would result in electrical activity. Unfortunately, different
laboratories have reported unsuccessful dopant activation during
the epitaxial growth for ranges of temperatures between 400-600°C.
The best implantation schedules as determined in Section 7.2.2.1
were carried out with the dopants either introduced at low dose
within the amorphous layer or with the dopants being the amorphiz-
ing species, but in both cases very little or no dopant activation
was observed after the low temperature anneal. A further anneal
at a higher temperature with an encapsulated sample is required

to obtain reasonable electrical activity. Compensation of the dopants after low temperature anneals can be speculated to come from the remaining dislocation loops or the redistribution of background impurities during epitaxial regrowth.

7.3 CW LASER PROCESSING OF THIN FILMS DEPOSITED ON GaAs

In this section we consider the doping of GaAs that may be obtained when a cw beam is used to react a thin deposited film with the underlying substrate. Both alloyed and nonalloyed ohmic contacts to n.gaAs have been obtained from irradiated thin films deposited on GaAs. The performance and reliability of a number of GaAs microwave, logic and optoelectronic devices are determined in large part by the properties of their ohmic contacts. In microwave MESFET's, for example, parasitic source resistance due to the ohmic contact is a major contributor to noise in state-of-the-art devices. The electrical performance of a typical FET (1 μm gate-length) degrades significantly for values of specific contact resistance ρ_c above the mid10^{-6} $\Omega\text{-cm}^2$ range. As device geometries become smaller, the demands will be even more stringent. The two different approaches reported here can result in the formation of low specific contact resistance ohmic contacts well suited to the small geometry device processing requirements.

7.3.1 CW Laser Alloying of Au-Ge/GaAs

The conventional technique for the formation of ohmic contacts to n-GaAs consists of the evaporation and subsequent thermal alloying of a layer of eutectic-composition Au-Ge [7.18]. When heated to a temperature typically 100 to 150°C above the Au-Ge eutectic point of 356°C, the contact metal melts and dissolves a portion of the GaAs. Upon cooling, the epitaxially regrown GaAs layer incorporates Ge as a substitutional dopant. Provided the Ge is located primarily on Ga sites, it acts as a donor, and a degenerate n$^+$ layer is formed. The lowered energy barrier at the metal-semiconductor interface then permits ohmic conduction [7.19]. There are many limitations to this technique. Although a Ni or Pt overlayer is almost always included to improve the wetting of the molten Au-Ge, the alloyed contacts are characterized by significant surface roughness and poor edge definition, by microprecipitates of nonuniform composition, and by uneven penetration into the underlying GaAs [7.20, 7.21]. The alloy temperature and time, moreover are constrained by the amphoteric nature of Ge and the resulting need to prevent the loss of the extremely volatile As dissolved in the Au-Ge melt. It has been shown that rapid heating rates and short times lead to

improved contact resistance and morphology [7.22], but there is
a practical limit (on the order of 50°C/sec) to the rate which can
be achieved with conventional furnace processing.

The ability of cw beam irradiation to produce high surface
temperatures for short periods of time has the potential to
eliminate many of the problems just discussed. The first attempt
to form laser-alloyed contacts was reported by Pounds et al.,
[7.23] in 1974 using a Q-switched laser. Although the process
was not well controlled (melting of few microns of substrate
resulted from irradiation) and the resulting specific contact
resistance were relatively high ($>10^{-4}$ Ω–cm^2), the basic feasi-
bility of the technique was demonstrated. Gold et al., [7.24]
reported the use of lasers (free running ruby laser and cw argon
laser) to alloy Au–Ge/structures resulting in improved surface
morphology as compared with furnace alloying. This is illustrated
in Fig. 7.11 where the micrographs of furnace, pulsed and cw
laser, alloyed contacts morphologies are compared. To further
quantify the improvement a stylus profiler was used to characterize the surfaces of both furnace and laser alloyed samples [7.25].
These profile are shown in Fig. 7.12; the laser-alloyed surface
(pulse laser shown here) is seen to be significantly smoother.
A similar improvement was observed on alloyed samples whose contact metal had been stripped to expose the metal–semiconductor
interface. The metallization system reported by Gold [7.25] was
designed specifically to be compatible with laser processing. The
deposition sequence was Au (200 Å) followed by Ge (100 Å); some
samples also had a thin layer of Ni, nominally 20 Å, at the Au/GaAs
interface. The presence of Ge as the top layer, rather than Au
or eutectic composition Au–Ge resulted in high light absorption
(reflectivity at λ =5145 Å was measured to be only 31%) and
allowed alloying to take place at laser powers well below that
at which damage occurred. This basic technique of self limiting

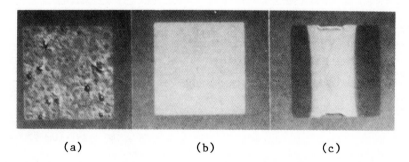

(a)   (b)   (c)

FIGURE 7.11. Microscopic appearance of alloyed 115 μm–square
Au–Ge contacts to n–GaAs. The alloyed metal are Au (200 Å) followed by Ge (100 Å) (a) thermal anneal 450°C, 60 sec; (b) ruby
laser 15 J/cm^2, 1 msec; (c) scanned argon laser anneal 2.5 W.

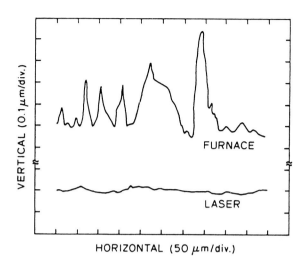

FIGURE 7.12. Surface profilometer characterization of pulsed alloyed Ge–Au/GaAs ohmic contacts. Laser energy was 15 J/cm^2; furnace alloy was 450°C, 10 sec.

absorption has also been successfully applied to the cw laser processing of metal–silicides (Chapter 6). The scanning cw argon ion laser system described in Chapter 1 was focused on a 40 μm spot and scanned at a speed of 12.5 cm/sec. An incident power of 2.5 W resulted in the surface morphology shown in Fig. 7.13(a). When Au–Ge layers are used, the metal tends to pull back from the edge of the contact, due most likely to the surface tension of molten Au–Ge. However if a Ni underlayer is utilized this behavior is not seen [Fig. 7.13(b)] which can be explained by the

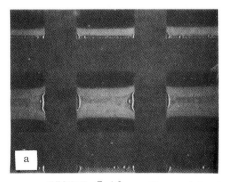
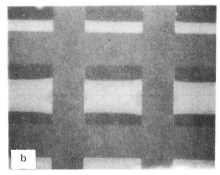

FIGURE 7.13. Nomarski photomicrographs of cw argon laser alloyed contacts on GaAs. (a) Ge–Au (b) Ge–Au–Ni. Size of contact squares is 115 μm.

improved adhesion of the metal. Excellent adhesion was observed
in any configuration with a free running ruby laser irradiation.
In this case energies in the vicinity of 15 J/cm^2 were used
leading to a specific contact resistance of 2×10^{-6} Ω-cm^2 in the
best case, but with a poor reproducibility.

An extensive comparison of different beam irradiation technique to alloy Au:Ge based contacts on GaAs was reported by
Eckhardt [7.26]. The best results (lowest ρ_c, firmest adhesion
between contact and GaAs and highest reproducibility) were obtained with the single mode (TEM$_\infty$) cw Ar-ion laser. The summary
of the results, and experimental conditions is presented in Table
7.1. All the samples used in this study consisted of semi-
insulating GaAs implanted with 4×10^{12} Si$^+$/cm^2 at 100 keV (n $\simeq 10^{17}$
cm^{-3} at the sample surface). The laser beam was always focused
onto the sample in a 66 μm spot size.

A number of 1 μm gate GaAs MESFET's were prepared with a
laser alloyed Au:Ge based contact and compared with furnace alloyed contacts from the same wafer. The laser annealed contacts
were found to be superior or at least as good in every respect.
Their resistance was lower by 10-20% and the dc characteristics
were excellent [7.27]. Average values of gain, and noise measured
at 14 GHz were somewhat better for the laser annealed devices. In
all cases the surface morphology and edge definition of the laser
annealed contacts was far superior to that of their thermally
annealed counterparts. Furthermore, because of the shortness of
the laser-annealing period (1-100 msec), interdiffusion of semiconductor and metal constituents is reduced (verified by Auger
spectroscopy studies [7.26]) and microscopic phase separation
due to metal flow is prevented.

A number of other laboratories have reported the formation
of low contact resistance alloyed ohmic contacts using pulsed
laser (Q-switched or free running) or pulsed electron beam. But

TABLE 7.1.   Best Ohmic Contacts Formed on GaAs by CW Ar-ion
Laser.

Contact Metal	Specific Contact Resistance (Ω-cm)	Laser Characteristics	
		Power (W)	Scan Velocity (cm/s)
Au:Ge-Ni-Au	4.8×10^{-6}	4.0	0.43
Au:Ge-Pt-Au	1.5×10^{-5}	3.8	0.43
Au:Ge-Ag-Au	2.0×10^{-4}	4.1	0.43
Au:Ge-Ti-Au	1.8×10^{-5}	3.8	0.2
In-Au:Ge	1.3×10^{-6}	3.5	0.43

in all cases reported, no reliability or lifetime data was obtained. If the beam alloying technique has brought an important improvement in the formation of ohmic contacts to GaAs, the degradation of these contact at elevated temperatures continues to be a limiting factor to the lifetime of GaAs devices. The next section will describe a different technique to form ohmic contact to nGaAs that are stable at elevated temperatures.

7.3.2 Cw Laser Assisted Diffusion and Activation of Sn in GaAs

Among the different doping techniques available, ion implantation and epitaxy (liquid, vapor phase or molecular beams) are preferred for GaAs. Diffusion processes in GaAs are difficult due to the high vapor pressure of arsenic at diffusion temperatures. They usually require sophisticated equipment such as sealed ampoules or controlled environments. An open tube diffusion technique is presented here using thin film solid sources. It represents a low-cost, reproducible method to obtain thin heavily doped n layers on semi-insulating GaAs. Thermal treatments or a combination of thermal and cw laser processing can be used to control the diffusion process. The thin film source is a mixture of SnO_2 and SiO_2 in solution deposited using a conventional spinner. In addition to the diffusion, an interface reaction is observed to occur. A detailed analysis of this reaction and the role of the different atoms involved in the complete process is presented. The combination of the interface layer and Sn doped GaAs layers will result in the formation of low contact resistance non-alloyed ohmic contacts to n-type material.

7.3.2.1 Source and Sample Preparation

The sample preparation steps are schematically described in Fig. 7.14. Semi-insulating chrome-doped GaAs substrates, oriented along the <100> direction are used. The source consists of a tin silica spin-on film obtained from Emulsitone [7.28]. The solution is a mixture of SnO_2 and SiO_2 in ethyl alcohol. A photoresist spinner is used to spin the film on. Rotation speeds of 3000 to 5000 rmp are selected depending on the size of the substrate processed. The films obtained are typically 0.3 μm thick. A thermal treatment of 200°C for 15 min. removes the binder (ethyl alcohol) leaving behind a layer of SnO_2/SiO_2 with a molecular concentration of about 1:10, according to the vendor.

As an encapsulant for further thermal treatments, a layer of CVD SiO_2 (Silox) is deposited on the SnO_2/SiO_2 film. The temperature of deposition is 450°C. The thickness of this second layer has been found to be critical to avoid the thermal

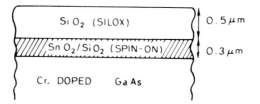

1 – SPIN. ON A SnO_2/SiO_2/ETHYL ALCOHOL SOLUTION

2 – BAKE 200°C /15min TO REMOVE BINDER

3 – DEPOSIT 0.5 μm OF SiO_2 (SILOX)

FIGURE 7.14.   Schematic of the sample preparation for tin diffusion in GaAs.

stress occurring at high temperature caused by the difference in thermal expansion between the substrate and the deposited layers. It was found that the oxide thickness should be no more than 0.8 µm. A thickness of 0.5 µm is chosen to be the standard condition.

7.3.2.2   Thermal Diffusion

Sn is an element in column IV of the periodic table, and therefore is a possible amphoteric dopant for the III–V compound GaAs. In this system, Sn has a strong tendency to occupy Ga sites and then is frequently used as a n-type dopant. The diffusion of this impurity in GaAs from Sn doped silicon dioxide has been reported from various types of thin film deposition [7.29–7.32]. The study of the thermal diffusion from the spin-on solution presented here shows significantly different chemical and electrical behavior of Sn compared to the doped silicon dioxide sources reported before.

Any abrupt changes in temperature of the double layer "source-cap" (SnO_2/SiO_2-SiO_2) structure described in the sample preparation produce stresses that cause the films to peel. For this reason, a slow thermal ramp is required before any subsequent thermal treatment. Since this procedure initiates the diffusion of Sn from the source, a very precise ramping cycle is performed to insure reproducibility of the experiments. The total cycle is done in 15 min in a N_2 ambient. As an example, a detailed temperature ramp from room temperature to 900°C in 15 min is described below:

(a) Room temperature to 700°C in 7 min ($\simeq$ 100°C per min)
(b) 700°C to 850°C in 3 min (50°C per min)
(c) 850°C to 875°C in 1 min
(d) 875°C to 890°C in 1 min
(e) 890°C to 900°C in 1 min
(f) 900°C for 2 min

If the final temperature is not 900°C then the temperature reached in step (a) is changed in order to follow the same ramp from step (b) to (f).

A Van der Pauw technique is used to measure sheet resistivity, Hall mobility and sheet carrier concentration after the ramping cycle. The results are presented in Table 7.2. These data show that this sequence is responsible for both diffusion and electrical activation of Sn. Measurable activity starts when the final temperature of 800°C is reached with an active dose of 4.7×10^{12} atoms/cm^2. This value can reach 4.4×10^{13} atoms/cm^2 for a ramp to 950°C. The low value of the mobility obtained at 800°C indicates that the diffusion has barely started and is very nonuniform. Expected values of Hall mobility are obtained as the final temperature is increased.

TABLE 7.2   Electrical Characterization of the Diffused Layers Obtained After a Ramping Cycle Only.

T_{ramp}(°C)	ρ ($\Omega/\square$)	μ_H(cm^2/V.sec)	N_s(cm^{-2})
750	--	--	--
800	974	1353	4.7×10^{12}
850	224	2255	1.2×10^{13}
900	122	2098	2.4×10^{13}
950	64	2034	4.4×10^{13}

A comparison of the chemical profile and the electrically active one is obtained by a combination of SIMS and differential Van der Pauw analysis. Data for the two measurements are presented in Fig. 7.15 for a sample ramped to 900°C in 15 min. It can be seen in this figure that only 3 to 4% of the total concentration of Sn (7×10^{14} cm^{-2} as measured by RBS) is electrically active. The total diffusion depth is $\simeq$ 1600 Å. The large difference between the two profiles is located in the first 300 Å of the surface where the total amount of Sn reaches a value above 10^{20}/cm^3. Following this near surface region, most of the impurities are electrically active. Further studies, reported later, indicate the formation of a Sn–As compound in this region, explaining the high Sn concentration. Consequently the yield of

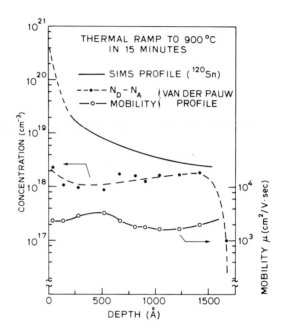

Fig. 7.15.   SIMS and Van der Pauw stripping profiles for a sample ramped to 900°C in 15 min. The Hall mobility is also shown.

Sn in this matrix does not correspond exactly to the scale shown in Fig. 7.14 since the calibration was performed for Sn in GaAs. A dotted SIMS profile in the near surface region is plotted for this reason. The variation of mobility in the diffused layer is also shown in Fig. 7.15. These measured values agree to within 10% of those expected from Irvin-Sze [7.33]. Extensive studies of the thermal diffusion following the initial thermal ramp was reported by Nissim et al., [7.34]. They have shown that if further diffusion is carried out with the "source-cap" (SnO_2/SiO_2: SiO_2), a carrier concentration of 2×10^{14} cm^{-2} with sheet resistivity as low as 20 $\Omega/\square$ can be achieved. Due to the large amount of Sn introduced into the substrate during the ramp, this cycle can be viewed as a conventional predeposition step. The equivalent of a drive in diffusion was performed by removal of the "source-cap", deposition of a new encapsulant (As-doped CVD SiO_2), and thermal annealing. The sheet electrical characteristics of the diffused layer thermally annealed at 800, 850 and 900°C for 30 min are shown in Fig. 7.16 as a function of ramping (predeposition) temperature. It can be seen that here again very low sheet resistivity ($20/\square$) can be achieved. Electrical depth profiles were performed for the 850°C/30 min anneal using a differential Van der Pauw technique and chemical stripping. The results are shown in Fig. 7.17.

(a)

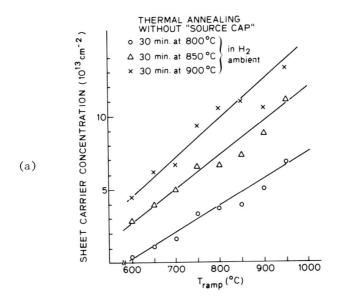

(b)

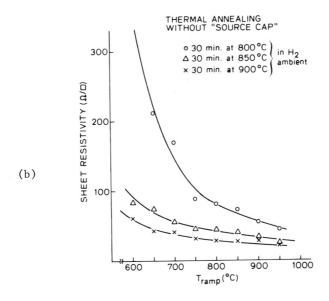

FIGURE 7.16. Evolution for different ramping (predeposition) conditions of the sheet electrical characteristics of the diffused layer when thermally annealed without the "source-cap" of different temperatures (a) sheet carrier concentration (b) sheet resistivity.

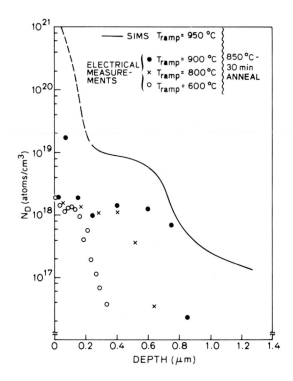

FIGURE 7.17.   Electrical profiles for ramping cycles to 600, 800 and 900°C followed by an anneal at 850°C/30 min without the "source-cap". A SIMS profile for the same anneal and a ramping temperature of 950°C is also shown.

7.3.2.3   Cw-Laser Assisted Diffusion and Activation

Following an initial thermal ramp, the diffusion of Sn impurities can also be assisted using a scanning cw laser beam [7.35]. The characteristic features of this process are presented here. The ramping cycle prior to laser treatment is necessary to prevent the peeling of the double layer "source-cap" during irradiation and apparently to break down an interfacial barrier to start the diffusion process. It has been impossible to start the diffusion of Sn with the scanning laser alone.

The thermal ramp gives an electrically active Sn concentration of 2 to 3 $\times 10^{18}$/cm^3 to a depth of about 1600 Å in the GaAs substrate as seen in the previous section. The wafers were then laser scanned using a cw argon ion laser. The beam was focused by a 136 mm lens, leading to a 50 μm beam diameter at the focal plane. The spot was then scanned with 15 μm spacing between adjacent lines at a speed of 12 cm/sec. In order to reduce the thermal stress created by the incident beam at the surface of the wafer, the substrate was held at 350°C during irradiation. These parameters were found to be optimum and were kept constant. The investigation of different temperature regimes was accomplished by changing only the incident power level. The reflectivity of the substrate with its double layer "source-cap" was directly measured and found to be 24%. Using the calculated curves of laser induced temperature in GaAs [7.1] and the above parameters, a maximum surface temperature was associated with each incident laser power.

The diffusion of Sn was first studied in a regime where the laser power was kept below the level required to melt the substrate. This working window was determined by observing the laser power required to just produce visible thermal etching and then reducing the setting by 5%. A power level of P = 0.61W, leading to a maximum induced temperature of about 800°C, was obtained. A series of 1, 3 and 5 scan frames were performed with the double layer " source cap" remaining on the substrate. The Van der Pauw technique was used to characterize the resulting sheet resistivity, Hall mobility and sheet carrier concentration. The results are summarized in Table 7.3. The increase of the sheet carrier concentration from 2.44 to 3.01×10^{13} cm^{-2}, accompanied by a decrease of sheet resistivity with the larger number of scans, illustrates either diffusion from the source and/or activation of Sn impurities induced during the thermal cycle.

SIMS analysis was performed on a sample that was thermally ramped only, and on a sample scanned five times after the ramp. The resulting profiles shown in Fig. 7.18 illustrate an increase in Sn concentration and an in-diffusion of about 150 Å. To investigate the contribution of the irradiation to the diffusion from the "source-cap", compared to the activation of the impurity present in the substrate after the ramp, the "source-cap" was removed before laser scanning. The reflectivity of the bare

TABLE 7.3   Sheet Electrical Characterization of the Laser-Diffused Layers With the "Source-Cap" On.

Thermal ramp: 900°C in 15 min.	$\rho\,(\Omega/\square)$	$\mu_H(cm^2/V.sec)$	$N_s(cm^{-2})$
+ 0 scan	122	2098	2.44×10^{13}
+ 1 scan	118	2017	2.63×10^{13}
+ 3 scans	117	1838	2.91×10^{13}
+ 5 scans	107	1940	3.01×10^{13}

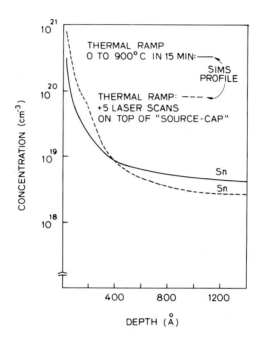

FIGURE 7.18. SIMS profiles of Sn obtained on ramped and ramped + laser scanned samples.

substrate (after the thermal ramp) was measured and the incident laser power adjusted to reach the same maximum temperature ($\simeq 800°C$) as before. The wafer was scanned five times and the sheet electrical measurements were carried out. The results, when compared with those obtained with the "source-cap" on, suggest that the total increase in active impurities due to laser irradiation is 77% from the source and 23% from the activation of Sn introduced during the ramp.

The action of the laser is found to be more significant if the sample is left "at temperature" for a short time after the ramp. For this reason the above experiment was repeated after thermally ramping the substrate and its source to 900°C and leaving it for 5 min longer at 900°C. A series of 5 scans inducing a maximum temperature of about 800°C was then performed. A net increase in carrier concentration from 3 to 8 x 10^{18} cm^{-3} is observed after the scans. The total Sn profile was obtained using RBS. The analysis was done with 2.2 MeV helium atoms after removal of the source. The backscattered particle spectra obtained for random orientation are presented in Fig. 7.19. A sample having received a thermal treatment only, another one receiving additional laser scans ($T_{max} \simeq 800°C$) and third scanned at higher laser power ($T_{max} \simeq 850°C$), were analyzed. The Sn

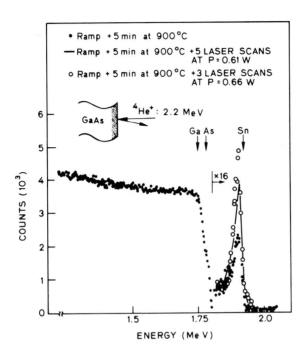

FIGURE 7.19.   RBS spectrum of the Sn profile in the laser diffused layers for different incident laser power.

profile is characterized by the large peak in concentration at the surface which increased after laser scan, showing the diffusion of Sn from the source. An increase in the depth of the profile after irradiation is also observed. This illustrates the diffusion of Sn in GaAs.

The temperature regime where the incident power is above the melting threshold of the GaAs substrate has also been investigated. Melting occurs at a power of about P = 1W. Different experiments were tried in the range of 1 to 2W. When melting occurs, the surface of the substrate becomes very smooth, and depending on the incident power level, different colors appear in the scanned area. This is illustrated in Fig. 7.20 for incident powers of 1.2 and 1.9W. The behavior of this system is very similar to the silicon–metal (silicide) reactions induced by a scanning laser (Chapter 6). The formation of a series of compounds involving Sn and As is believed to occur. However those formed during the high power scanning could not be identified since the substrate cracks during an overlapping scan (P > 1.4W) as shown in Fig. 7.20(b). The compound formed by the lower power scan, has been identified and will be discussed in the following section. Electrical characterization of these diffused

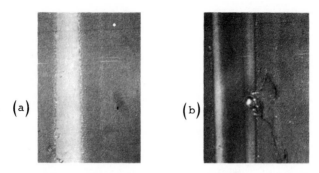

FIGURE 7.20.   Nomarski optical micrograph of single laser scan at power above the melting threshold of the substrate (a) 1.2 W; (b) 1.9 W.

layers showed high carrier concentrations (3 x $10^{14}/cm^2$) but anomalously low mobilities as compared to those obtained in the solid phase regime.

7.3.2.4 Surface Reaction Analysis

The chemical profiles obtained using the RBS or SIMS techniques show an anomalously high concentration of Sn close to the surface in all the diffused layers analyzed. However there is no corresponding high surface concentration of electrons (see Fig. 7.15 for example). This observation suggests that a chemical reaction takes place in addition to the diffusion process. To study this possibility, a sample ramped to 900°C in 15 min. and a sample ramped and laser scanned five times at a power inducing a maximum temperature of 800°C were prepared for transmission electron microscopy and diffraction analysis using conventional jet thinning techniques. Bright field (Fig. 7.21) and dark field transmission electron micrographs indicated precipitation after both the thermal ramp and laser processing. Selected area electron diffraction patterns revealed that the precipitates were composed of a tin-arsenide compound (As_2Sn_3). Additionally, an amorphous gallium oxide was detected in the sample ($\beta-Ga_2O_3$). After laser irradiation, an increase of the amount of surface coverage by $\beta-Ga_2O_3$ crystallites was observed. In all cases the surface oxide film appeared as a laterally discontinuous film.

After a thermal ramp to 900°C in 15 min, the compound layer is 300 Å thick (from SIMS profile). This indicates that there is a large excess of Ga atoms free or involved in the formation of the oxide. To investigate the possible diffusion of the species out of the interface region, RBS analysis was performed before and after the thermal ramp with the double layer "source-cap" remaining on the substrate. The results of this experiment are shown in

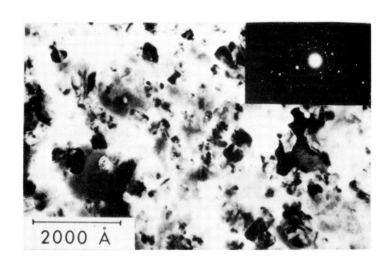

FIGURE 7.21.   Bright field transmission electron micrograph showing formation of As_2Sn_3 plates. In the inset are the rings associated with As_2Sn_3 and $\beta-Ga_2O_3$.

Fig. 7.22. After the ramp, diffusion of Sn in the GaAs substrate and in the SiO_2 is observed. A slight shift of the GaAs edge indicates that part of the encapsulant was lost during the cycle. But the most striking feature is the formation of an extra peak after the ramp. The energy of this peak corresponds to a gallium signal at the surface. This demonstrates that Ga is dissolved from the GaAs by the SiO_2 layers and diffuses toward the outer-most surface.

These analyses have shown that the different components in this system have a precise behavior resulting in the formation of the doped layers: Sn diffuses into the GaAs substrate to create an n^+-type layer, and reacts at the interface with As to form As_2Sn_3. The extra Ga diffuses through the "source-cap" and forms $\beta-Ga_2O_3$ oxide.

7.3.2.4   Nonalloyed Ohmic Contact Formation

Relatively high carrier concentration can be achieved in GaAs when Sn impurities are diffused from a spin-on SnO_2/SiO_2 source. The thermal cycle initiating the diffusion of impurities has been optimized for the best values of contact resistance with two main criteria: a high value of active carrier concentration and short diffusion length. These conditions are met after a thermal ramp from room temperature to 900°C in 15 minutes. The

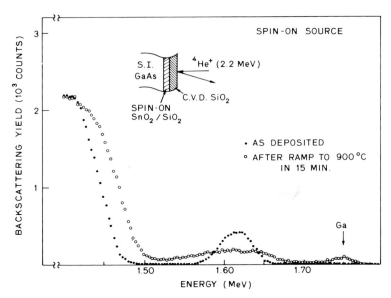

FIGURE 7.22.   RBS spectra of a GaAs sample with its "source-cap" on before and after a thermal ramp.

resulting profile studied in the previous section is shown in Fig. 7.15. It corresponds to a relatively flat active carrier concentration profile at $2\text{--}3 \times 10^{18}$ atoms/cm^3 to a depth for about 1600 Å. In addition to the diffusion, an interface reaction occurs, leading to the formation of a tin-arsenide compound (Sn_3As_2). This reaction takes place in the first 300 Å of the diffused layer. Subsequent metallization of these layers shows ohmic behavior without an alloying cycle.

It has been seen that following the thermal ramp, cw laser scans can assist the diffusion and activation of Sn. Nonalloyed ohmic contact formation has therefore been investigated on the thermally ramped and laser scanned layers. The laser irradiation parameters described previously were kept constant, and contact resistance of these layers was determined as a function of incident power [7.36]. Values of specific contact resistance were measured using the Transmission Line Model (TLM) [7.37-7.38]. The pattern is isolated by a mesa etch and the metal is patterned using a standard photolithographic lift-off. The spacing between metal pads was 6, 4, 2, 4 and 6 μm. The symmetry of this structure is intended to eliminate any systematic errors due to lateral gradients in sheet resistance. Following diffusion, the "source-cap" is removed and the TLM patterns are prepared. The values of specific contact resistance and sheet resistivity obtained from these measurements are plotted as a function of incident laser power in Fig. 7.23. All the samples were ramped to 900°C in 15

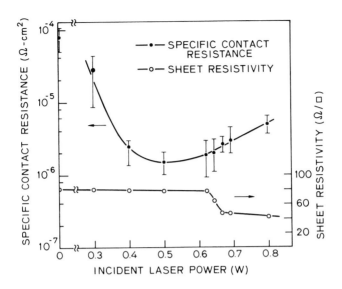

FIGURE 7.23. Effect of laser power on the properties of the diffused n^+ layers and nonalloyed Tc-Pt-Au contacts.

minutes before laser irradiation. Immediately after the ramp, the specific contact resistance (ρ_c) of the nonalloyed contacts is about 10^{-4} Ω-cm^2. A dramatic decrease of two orders of magnitude is observed when the wafer is scanned at incident laser powers between 0.4W and 0.6W. Specific contact resistance between 1 and 2×10^{-6} Ω-cm^2 are obtained for this processing window while the GaAs surface remains smooth and free of visible damage. Above 0.65W a slight increase of specific contact resistance is observed, probably due to deterioration of the surface. The bars on the ρ_c data represent the range of values obtained for various runs. The temperature calculation developed in Chapter 2 has been applied in this experiment to define the temperature limits of the different regions. The results are shown in Fig. 7.24. The three distinct regions that appear on this figure are:

(i) The processing window, that leaves a mirror finish on the surface and corresponds to induced temperature between 600 and 800°C. This window is large enough to insure very reproducible specific contact resistance in the low 10^{-6} Ω-cm^2 range. It has been experimentally observed that five consecutive laser scans of the sample are optimum in achieving the lowest value of ρ_c.

(ii) The region where the induced temperature is between 800°C and the melting point of the GaAs substrate. Slip lines and surface decomposition occur in this window and the specific contact resistance of the resulting layers is increased.

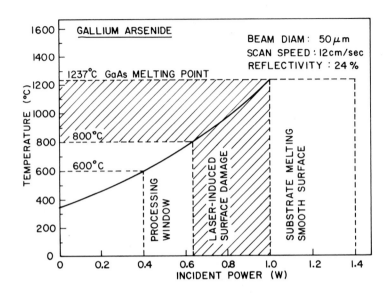

FIGURE 7.24. Temperature induced in the GaAs during laser irradiation as a function of incident power. The different regimes for ohmic contact formation are shown.

(iii) Above the melting point of GaAs, the surface becomes shiny again. This regime has been discussed in the previous section. The Hall mobility of the diffused layer is affected but values of ρ_c as low as 2.6×10^{-7} Ω-cm^2 have been measured.

In the remainder of this section the properties and obtained within the processing window are considered. A standard non-alloyed ohmic contact fabrication is established by a ramp to 900°C in 15 minutes followed by laser scans at P = 0.54W.

For a given metal semiconductor barrier height (0.7 eV in this system) theoretical calculations of specific contact resistance as a function of carrier concentration are reported in Ref. 7.19. From these calculations it can be seen that the values of $\rho_c = 10^{-4}$ Ω-cm^2 and $\rho_c = 10^{-6}$ Ω-cm^2 are obtained for carrier concentration of 8×10^{18} and 5×10^{19}/cm^3, respectively. The carrier concentration in the diffused layer obtained after a thermal ramp to 900°C (3×10^{18}cm^{-3}) would result in values of specific contact resistance in the vicinity of 10^{-4} Ω-cm^2. But it has been impossible to measure any value of carrier concentration that would result in $\rho_c = 10^{-6}$ Ω-cm^2 as observed. Furthermore, Fig. 7.23 shows that a decrease of two orders of magnitude of ρ_c is obtained at incident laser powers (0.4 to 0.6 W) where the sheet resistivity of the diffused layers remains constant at 80 $\Omega/\square$. Only above 0.6 W is there a significant laser-assisted diffusion and/or

activation of tin. These two observations suggest that a contribution other than the doping level is responsible for low values
of ρ_c. A very distinct characteristic of the diffused profile
is the presence of the tin-arsenide compound (Sn_3As_2) in the first
300 Å. The phase diagram of the tin-arsenic system [7.39] is
plotted in Fig. 7.25. The melting point of Sn_3As_2 is shown to be
596°C. It is striking to notice that this temperature corresponds
to the laser-induced threshold temperature at which the specific
contact resistance drops by 2 orders of magnitude (from the temperature calculation of Fig. 7.24). The action of the laser scan
can then be interpreted as causing the melting of isolated Sn_3As_2
plates nucleated during the thermal ramp, after which they regrow
into a more uniform film. This argument is supported by the fact
that when the sample was scanned at laser powers inducing temperatures very close to the melting point of the compound, multiple
scans were necessary (typically 5) to obtain the lowest ρ_c. This
compound would then form an interface layer that reduces the barrier between the metal and the semiconductor. In view of these
results, the low value of specific contact resistance is believed
to arise from two contributions. The high active carrier concentration ($3 \times 10^{18}/cm^3$) in the diffused layer provides values of ρ_c
in the 10^{-4} Ω-cm^2 range. The second contribution comes from the
interface layer (Sn_3As_2) that reduces the metal–semiconductor barrier and brings ρ_c into the 10^{-6} Ω-cm^2 range.

To check the stability of these nonalloyed contacts, aging
experiments were carried out at elevated temperatures. The
degradation of contact resistance, normalized to 1 mm of contact

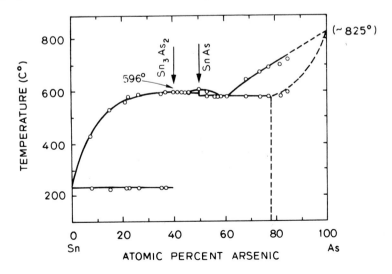

FIGURE 7.25.   Phase diagram of the Sn-As system.

width, is plotted for two different temperatures: 320 and 400°C
in Fig. 7.26. One thousand hours at 320°C produces an increase
in resistance of less than 50%. Even at 400°C the slope of the
resistance increase is still small. These results are compared
to the behavior at elevated temperature of an alloyed Au-Ge ohmic
contact [7.40-7.41]. As can be seen in Fig. 7.26 the contact
resistance is three times smaller for the case of the nonalloyed
diffused contacts and they sustain temperatures that would destroy
a conventional ohmic contact. These characteristics may make this
diffused contact technology very attractive in the improvement of
the reliability of GaAs MESFETS. Application of these techniques
to the fabrication of such devices is described in Ref. 7.42.

7.4   RAPID THERMAL PROCESSING

Sealy and Surridge [7.43] first showed that rapid thermal
processing in the 1 to 100 second time scale could result in good
electrical activation and high mobility for ion implanted n-type
layers in GaAs. Since then a number of studies have been performed
in an effort to establish the annealing mechanisms and the range
of system operating parameters required to obtain results equivalent to or better than those obtained in furnace annealing. At
present the results are mixed, but there is clear promise for
the future. In what follows we review the main results of the
research to date.

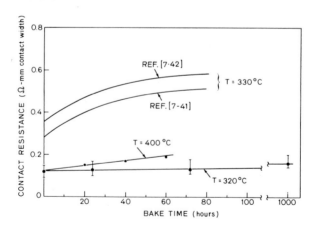

FIGURE 7.26.   High temperature stability of planar nonalloyed
contacts.

7.4.1   Resume of Furnace Recrystallization

Williams [7.44] has characterized the furnace annealing of ion implanted GaAs as occurring in three stages. A basic amorphous-to-crystalline transition, labelled as stage I, occurs at temperatures in the range of 150–200°C. This transition is characterized by a major reduction in damage as measured by He$^+$ ion channeling and/or TEM. However, unlike silicon, this transition leads to a highly defective surface layer containing a high density of twins and other, more complex, polycrystalline structures.

A second, broad annealing stage, labelled as stage II, occurs in the temperature range from ~ 400°C to 800°C and is characterized by the removal of extended crystalline defects. However, removal of these defects does not guarantee electrical activity. Instead, for n-type dopants especially, a third annealig stage leading to dopant activation (stage III) occurs at temperatures in the range 850°C–1500°C. As suggested earlier, it is important to protect the surface against dissociation and arsenic loss, by using either encapsulating layers or As overpressure, at annealing temperatures in excess of about 450°C.

7.4.2   Annealing of Si-implanted GaAs and its Potential for MESFETS

Arai et al. [7.45] first reported using halogen lamps to anneal GaAs. For "encapsulation", the GaAs was placed face down on a Si wafer and annealed for 5 seconds at a temperature of 950°C. It is believed that under these circumstances only that amount of As (~ one monolayer) will evaporate that is required to establish an equilibrium pressure of As over GaAs in the small space between the wafers.

Arai et al. obtained nearly 100% activation of a $3 \times 10^{12}/cm^2$ Si implant at 70 keV. Kuzuhara et al. [7.46], using a similar system, obtained ~ 75% activity for a 150 keV, $5 \times 10^{12}/cm^2$ Si implant at a peak temperature of 950°C and an annealing time of 2 seconds. Longer annealing times resulted in arsenic loss and the formation of gallium pits on the surface, consistent with the data of Rose et al. [7.47].

Davies et al. [7.48] have used filament lamps focussed by elliptical mirrors to anneal GaAs implanted with Si to a dose of $4 \times 10^{14}/cm^2$ at an energy of 200 keV. The samples were capped with a 900 Å layer of Si_3N_4 deposited by CVD at 400°C. They obtained a maximum of 50% electrical activity following an anneal at a peak temperature estimated to be 1000°C. The peak temperature is reached in about 1 second, with the lamps being shut off 2–3 seconds after turn on. The impurity profile obtained for a total elapsed time of 2-1/2 seconds is shown in Fig. 7.27. The maximum carrier concentration of $\sim 6 \times 10^{18}$/cm^3 represents high electrical activity. The flat top is attributed by the authors to solid solubility effects. However, the complete temperature history includes a long fall time due to the nearly lossless optical cavity used, and this may also have an important bearing on the carrier concentration profiles they obtain.

Finally, Tabatabaie-Alavi et al. [7.49] have used an arc lamp annealing system to activate a GaAs sample implanted with Si to a dose of 4×10^{14}/cm^2 at an energy of 200 keV. These authors comment that their system is capable of uniform illumination over a 100 mm diameter wafer and as such is appropriate for IC processing.

The data they obtain for a 3 second exposure at 1160°C and a 10 second exposure at 1100°C are shown in Fig. 7.28. The

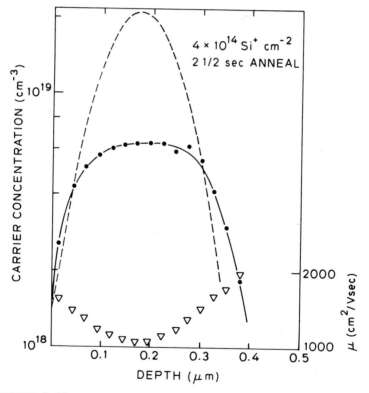

FIGURE 7.27.   Electron and Si impurity distributions in ion implanted, filament lamp annealed GaAs (Davies et al. [7.48].

©1982 IEEE

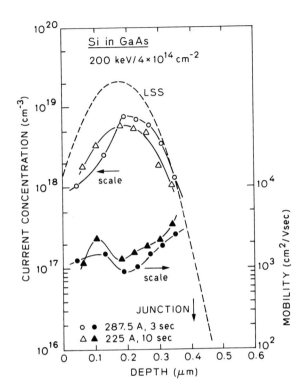

FIGURE 7.28. Electron and Si impurity distributions in ion implanted GaAs, annealed with argon ac lamp (Tabatabaie-Alavi et al. [7.49].

significant difference in the profile shapes is apparent. It is interesting to compare these profile shapes with calculations of Christel and Gibbons [7.50] on the stoichiometry imbalance produced by recoil effects in such an implant. The result of the calculation is given in Fig. 7.29; when laid over the 3 second data of Tabatabaie-Alavi et al., it is apparent that for short enough anneal times the stoichiometric imbalance is a critical factor in the electrical activity obtained. This imbalance is evidently modified upon further annealing, though the use of an SiO$_2$ cap and the dissolution of Ga from the sample must also be included in a thorough analysis of the problem.

7.4.3   FET Fabrication Using Rapid Annealing

Kuzuhara et al. [7.46] have reported on the properties of FETs fabricated using Si-implanted channels with annealing carried out in both furnace and halogen lamp systems. The test devices

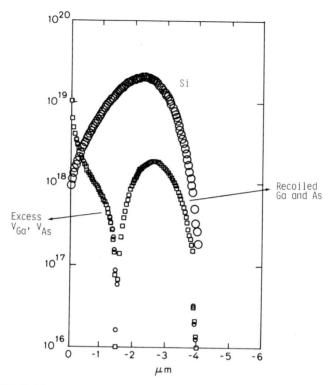

FIGURE 7.29.   Stoichiometric disturbances produced by recoil effects in Si-implanted GaAs.

had gate lengths of 1 μm and gate widths of 300 μm. The g_m obtained for the devices annealed in the halogen lamp system was 100 mS per mm of gate width compared to 80 mS per mm of gate width for devices subjected to a furnace annealing cycle. There thus appears to be an improvement arising from the use of rapid thermal annealing. However, the transconductance obtained with rapid thermal annealing is not as good as in state-of-the-art devices, where a similar geometry device made with an implanted channel and a capless furnace anneal in a gaseous environment containing Ga and As at 850°C for 20 minutes gives g_m of 140–160 mS per mm. Devices made with epitaxially grown channels also have gm of 160 mS/mm. Even more dramatic improvements are possible by using molecular beam epitaxy to produce an undoped GaAs layer for the FET channel and a doped GaAlAs layer to supply electrons to it (so-called High Electron Mobility Transistor, or HEMT). Hence, as with the microwave Si bipolar transistor, the current state-of-the-art in GaAs still represents something of a challenge for rapid annealing, and the direction of the technology toward MBE might also represent a challenge for any ion implantation process.

7.4.4   Rapid Thermal Annealing of Be and Zn Implants in GaAs

Be and Zn are generally implanted into GaAs to form p regions in pn junction devices, including both diodes and JFETs. Early work on rapid thermal annealing of these dopants [7.51–7.54] showed that halogen lamps, graphite strip heaters, incandescent lamps and dc Ar$^+$ arc lamps could all be used to obtain reasonable majority carrier properties in the annealed layer. In addition, Tabatabaie-Alavi et al. [7.55] have extended their work to the fabrication and characterization of pin diodes made by Be implants followed by annealing with a water-walled argon arc lamp.

The hole density and mobility profiles obtained by Tabatabai-Alavi et al. [7.55] following a double implant (4.4×10^{14}/cm^2 at 50 keV followed by 5.1×10^{14}/cm^2 at 150 keV) and anneal at 950°C for 10 seconds are shown in Fig. 7.30 (approximately 5 seconds are used to reach the final temperature, after which annealing is

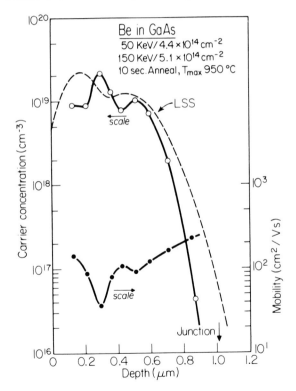

FIGURE 7.30. Room-temperature carrier concentration and mobility profiles measured by differential Hall technique on Be-implanted (4.4×10^{14} cm^2, 50 keV and 5.1×10^{14} cm^{-2}, 150 keV) GaAs samples arc lamp annealed for 10 s to 950°C.

carried out for an additional 5 seconds). The implant was performed into a 14 μm thick, undoped ($N_D-N_A \sim 10^{15}/cm^3$) layer grown
on an n^+ substrate by vapor phase epitaxial techniques.

Residual damage in such structures usually manifests itself
as excess leakage current, poor diode ideality factor (n) and low
electric field at breakdown. The lamp annealed diodes exhibited
an electric field at breakdown that was within about 20% of the
theoretical value, which is superior to results obtained with
furnace annealing. Ideality factors n as low as 1.6 (1 being
ideal) were measured, with leakage current densities of 1.4×10^{11}
A/cm^2, which is difficult to achieve with furnace annealing. The
process therefore appears to offer some advantage over furnace
processing with respect to residual defects following the annealing step.

7.4.5   Effects Due to Extended Hold Time in Annealing of Zn-Implanted GaAs

Suzuki et al. [7.56] have discovered a previously unreported
effect in the annealing of moderately high dose Zn implants into
Cr:GaAs. Zn was implanted to a dose of 5×10^{14} ions/cm^2 at 150 keV
and annealed at a variety of annealing temperatures, temperature
hold times and heating rates. With no encapsulant and no As vapor
pressure, an activation of over 90% was achieved by annealing at
800°C with zero hold time. Prolonging the hold time and/or
decreasing the heating rate reduces the activation to values in
the range of 70%. Suzuki et al. find from He backscattering
measurements that the degree of crystallogaphic recovery of
implantation-induced damage is similar to the dependence of
electrical activation on heating rate. In prolonged hold time
experiments, the electrical profiles show a double peak, implying
a depletion of Zn atoms around R_p (or the point of maximum damage).
Suzuki et al. attribute this effect to gettering of Zn atoms in
the implantation-induced damage layer. Further measurements are
required to clarify this process.

7.5   NOVEL APPLICATIONS OF RAPID THERMAL PROCESSING IN GaAs

To conclude the discussion of GaAs, we describe two novel
applications that are presently being explored. The applications
represent attempts to dope GaAs and control its surface properties
without using ion implantation, thus avoiding the stoichiometric
disturbances associated with the implantation process.

7.5.1 Controlled Si Diffusion

Greiner and Gibbons [7.57] have observed that Si diffusion from a thin elemental source can be controlled by the use of a thin SiO_2 encapsulant deposited by CVD at 400°C. The experimental data suggest that Si diffuses rapidly when two Si atoms are located on nearest neighbor Ga-As sites, making a neutral donor-acceptor pair.

The samples used for these experiments were <100> Cr-doped GaAs wafers that were cleaned and etched in standard solutions [7.57]. Immediately following the cleaning cycle, the samples were loaded into an electron-beam evaporator where they received 100 Å of Si at a base pressure of $2x10^{-8}$ Torr. After the Si deposition, the samples were encapsulated with either Si_3N_4 or SiO_2 or left uncapped. SiO_2 was deposited by chemical vapor deposition at 420°C to a thickness of 0.5μm. 1000 Å of silicon nitride was deposited by plasma enhanced chemical vapor deposition at 200°C. During the anneal, each GaAs sample was placed face up on a Si sample holder and another Si cover wafer was placed on top in contact with the sample to increase the thermal mass of the sample for feedback stabilization. Silicon was diffused at temperatures of 850°C to 1050°C for times from 3 to 300 sec. Rise times from 200°C were 7-8 sec. After the anneal, samples were etched in HF to remove the encapsulant. Residual Si from the elemental source was removed in a CF_4 plasma, leaving a smooth GaAs surface.

Secondary Ion Mass Spectroscopy (SIMS) was used to measure the chemical profiles of the diffused Si. The electrical profiles were determined using the differential van der Pauw technique and chemical stripping.

Annealing of both Si proximity and Si_3N_4-encapsulated samples resulted in no measurable electrical conductivity. However, samples encapsulated with SiO_2 showed considerable conductivity with sheet resistances as low as 50 ohms/square. These results are summarized in Table 7.3. A similar encapsulant effect has been observed for low dose Si implants in GaAs [7.58]. Electrical profiling of the SiO_2 encapsulated samples showed an abrupt junction and a maximum electron concentration of $5-6x10^{18}/cm^3$. The surface concentration of Si as measured by SIMS was $2x10^{20}/cm^3$ and a sharp diffusion front was observed. The SIMS and electrical profiles for a 3 sec anneal at 1050°C are shown in Fig. 7.31. These measurements indicate nearly complete compensation of Si. We will use this amphoteric character of Si in the diffusion model.

TABLE 7.3. Summary of Electrical Measurements For SiO_2 Encapsulated Si Diffusion Using Rapid Thermal Processing

Diffusion			van der Pauw	
$T(°C)$	$t(sec)$	$\rho_s (\Omega/\square)$	μ_H $(cm^2/Vsec)$	C_s $(\times 10^{13}/cm^2)$
850	300	60.4	1119	9.3
900	20.5	169.3	954	3.9
950	15.5	97.1	884	7.3
1000	10.5	70.7	848	10.4
1050	10.0	51.3	777	15.7

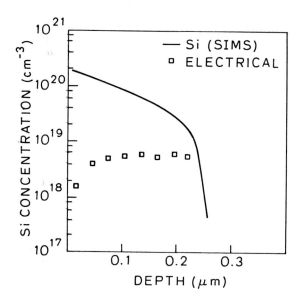

FIGURE 7.31. SIMS and electrical (van der Pauw) profiles for Si diffused into GaAs at 1050°C for 3 sec under 0.5 μm SiO_2 encapsulation.

Mechanisms for Si diffusion that involve motion of isolated silicon atoms through Ga and As vacancies require changes in vacancy type and charge state that would presumably produce very low diffusion coefficients. However, the diffusion of a nearest neighbor (Si-Si) pair could be much simpler. Paired Si atoms can move substitutionally by exchanging sites with either a Ga or As vacancy. This is expressed by

$$(Si_{Ga}^+ - Si_{As}^-) + V_{Ga} \qquad V_{Ga} + (Si_{As}^- - Si_{Ga}^+)$$

with a similar expression for the exchange with an As vacancy. This mechanism conserves both charge and vacancy type.

Greiner and Gibbons model the diffusion of Si in GaAs by assuming that Si-Si nearest neighbor pairs (referred to as Si complexes) are the main diffusing species in heavily Si-doped GaAs. If equilibrium exists between single and complexed Si, the fraction of Si existing as complexes can be obtained from the law of mass-action. The mass-action relation between paired and single Si atoms is

$$[Si_{Ga}^+] \cdot [Si_{As}^-] = K \cdot [Si_{Ga}^+ - Si_{As}^-]$$

where the subscript indicates site location and [ ] denotes concentration.

Electrical and SIMS measurements indicate that heavy Si doping ($10^{20}/cm^3$) results in almost complete compensation (n = $5 \times 10^{18}/cm^3$). Using the approximation that $[Si_{Ga}^+] = [Si_{As}^-]$, the concentration of Si complexes in terms of the total Si concentration, [Si] can be expressed as

$$[Si_{Ga}^+ - Si_{As}^-] = 1/2 \qquad \{[Si] + K(1 - \sqrt{1 + 2[Si]/K})\}$$

Here we have assumed that only single and complexed substitutional Si exists. For [Si] = K, 27% of the Si is predicted to be complexed.

The equilibrium constant for Si complex formation, K, can be estimated by calculating the probability of finding Si atoms residing on adjacent lattice sites with an added term containing the energy of formation of the complex [7.59]. The energy of coulombic interaction from an ionized Si donor and acceptor on adjacent sites is taken as the energy of complex formation. A nearest neighbor spacing of 2.45 Å was used [7.59]. The result is

$$K = N/Z \exp(-.56eV/kT)$$

where N is the density of the Ga or As sublattice and Z is the number of configurations of the pair (Z = 4). At 1000°C, K = $8.3 \times 10^{19}/cm^3$ [7.57].

Assuming that Si complexes are the main diffusing species, the equation for the total Si flux simplifies to

$$Flux_{Si} = 2 \cdot D_{complex} \cdot (\partial[Si-Si]/\partial[Si]) \cdot (\partial[Si]/\partial t)$$

The equation for the flux explicitly in terms of the total Si concentration is

$$Flux_{Si} = D_{eff} \cdot (\partial [Si]/\partial t)$$

where

$$D_{eff} = D_{complex} \left[1 - (1+2[Si]/K)\right]^{-1/2}$$

This effective diffusion coefficient is proportional to [Si] for [Si]<<K and constant for [Si]>>K, representing the high and low concentration regimes.

The diffusion equation was solved numerically assuming that (a) solid solubility of Si at the surface is maintained, and (b) the diffusion coefficient for the Si complexes, $D_{complex}$, is constant. The results of the calculations are shown in Fig. 7.32 along with the SIMS data for a 1050°C, 3 sec diffusion.

To demonstrate the diffusion of amphoteric pairs in GaAs, silicon and germanium were co-diffused from a thin film Si–Ge alloy. A $Si_{.6}Ge_{.4}$ alloy was substituted for the thin Si layer used previously and diffused under identical conditions.

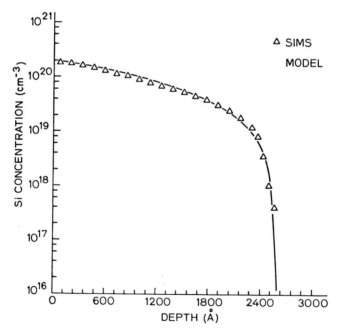

FIGURE 7.32. Calculated Si profile using the complexed Si diffusion model and fitted to the SIMS data.

The SIMS profiles for Si and Ge co-diffused at 1050°C for 100 sec are shown in Fig. 7.33. The separate profiles are seen to be the same within the resolution of SIMS. This is the expected result based on the formation and diffusion of Si–Ge pairs. For comparison, the co-diffusion of Si and Ge indicates a diffusion coefficient ten times lower than that for Si diffusion alone.

7.5.2   Schottky Barrier Modifications

Rapid thermal processing can also be used to modify the properties of Schottky barrier contacts in a way that is of potential importance in GaAs devices and integrated circuits. The experimental procedure builds on an idea that has been used to improve GaAs liquid junction solar cells.

Parkinsen et al. [7.60] have shown that hole lifetime at a GaAs surface can be improved by dipping the sample into a Ru Cl₃ solution. The Ru is thought to bond preferentially to As surface atoms atoms leaving Ga surface atoms open to further surface chemistry. This then provides the possibility of plating Zn onto the GaAs with the Zn atoms on the surface bonding to Ga

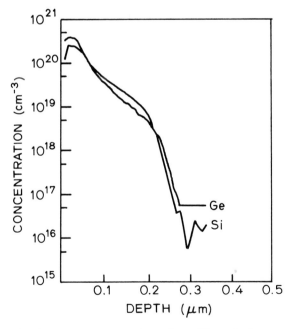

FIGURE 7.33.   SIMS profiles for Si and Ge co-diffused at 1050°C for 10 sec.

atoms. Experimentally this technique is found to provide a means of plating a uniform 100 Å layer of Zn onto the GaAs. Highly nonuniform plating results if the Ru treatment is omitted.

If this sample is then heated to 500–600°C for 25–50 seconds in an evacuated chamber, all of the Zn appears to evaporate except for about one monolayer bonded to the surface. A subsequent diffusion of the residual Zn into the GaAs permits carrier density profiling experiments which show the total residual Zn doping to be constant at a level of $4-8 \times 10^{14}/cm^2$, corresponding approximately to 1/2 monolayer of Zn.

If the samples are removed from the annealer prior to the diffusion cycle, Schottky barriers can be made with a standard Al evaporation. The properties of such barriers are of course modified by the presence of the thin, Zn-doped surface layer.

Barrier height measurements and diode ideality factors are shown in Table 7.4 for barriers made in this way. The results show that the Zn layer increases the Schottky barrier height by nearly 0.3 volt. FETs made with such improved barrier heights would provide significant improvement of the noise margins in GaAs digital integrated circuits.

7.5.3 Summary

Rapid thermal processing can provide efficient annealing of ion implanted layers in GaAs with properties that are at least promising for device applications. The technique can also be used to provide controlled diffusion from oxide-encapsulated, surface-deposited dopants and, for elements with high vapor pressure, a technique for monolayer-thickness doping. The properties of such surfaces may be of importance in device fabrication in situations where modified Schottky barrier properties are required or implantation doping is to be avoided.

TABLE 7.4.   Schottky Barrier Properties for Zn-treated GaAs Surfaces

Plating Time (sec)	Evaporation Time (sec)	Evap. Temp.°C	ϕ,V (C-V)	ϕ,V (I-V)	n
0	0	0	0.66	0.72	1.0
120	25	580	0.97	0.92	1.17
20	25	500	0.95	0.89	1.2
20	45	500	1.08	0.97	1.2

REFERENCES

7.1 Immorlica, A. A., and Eisen, F. H., Appl. Phys. Lett. 29, 94 (1976).

7.2 Fan, J. C. C., Donnelly, J. P., Bozler, C. O., and Chapman, R. L., Proc. 7th Intern. Symposium on GaAs and Related Compounds, p. 472. Institute of Physics, London (1979).

7.3 Olson, G. L., Anderson, C. L., Dunlap, H. L., Hess, L. D., McFarlane, R. A., and Vaidyanathan, K. V., in "Laser and Electron Beam Processing of Electronic Materials", (C. L. Anderson, G. K. Celler and G. A. Rozgoni, eds.) p. 467. ECS, Princeton (1980).

7.4 Nissim, Y. I., and Gibbons, J. F., in "Laser and Electron Beam Solid Interactions and Materials Processing" (J. F. Gibbons, L. D. Hess and T. W. Sigmon, eds.) p. 275. North Holland (1981).

7.5 MacMahon, R. A., Ahmed, H., Dobson, R. M., and Speight, J. D., Elect. Lett. 16, 295 (1980).

7.6 Shah, N. J., Ahmed, H., Sanders, I. R., and Singleton, J. F., Elect. Lett. 16, 433 (1980).

7.7 Csepregi, L., Mayer, J. W., and Sigmon, T. W., Phys. Lett. 54A, No. 2, 157 (1975).

7.8 Gamo, K., Ineda, T., Mayer, J. W., Eisen, F. H., and Rhodes, C. G., Rad. Effects 33, 85 (1977).

7.9 Williams, J. S., and Austin, M. W., Appl. Phys. Lett. 36, 994 (1980).

7.10 Williams, J. S. and Harrison, H. B., in "Laser and Electron Beam Solid Interaction and Materials Processing" (J. F. Gibbons, L. D. Hess and T. W. Sigmon, eds.) p. 209. North Holland (1981).

7.11 Kular, S. S., Sealy, B. J., Stephens, K. G., Sadana, D. K., and Pooker, G. R., Solid State Elec. 23, 831 (1980).

7.12 Grimaldi, M. G., Paine, B. M., Nicolet, M. A., and Sadana, D. K., J. Appl. Phys. 52, 4038 (1981).

7.13 Nissim, Y. I., Christel, L. A., Sigmon, T. W., Gibbons, J. F., Magee, T. J., and Ormond, R., Appl. Phys. Lett. 39, 598 (1981).

7.14 Csepregi, L., Kennedy, E. F., Mayer, J. W., Sigmon, T. W., J. Appl. Phys. 49, 3906 (1978).

7.15 Christel, L. A., Gibbons, J. F. and Mylroie, S., J. Appl. Phys. 51, 6176 (1980).

7.16 Christel, L. A. and Gibbons, J. F., J. Appl. Phys. 52, 5050 (1980).

7.17 Williams, J. S., Austin, M. W., and Harrison, H. B., in "Thin Film Interfaces and Interactions" (J. E. E. Baglin and J. M. Poate, eds.) p. 187. ECS, Princeton (1980).

7.18 Braslaw, N., Gunn, J. B., and Staples, J. L., Solid State Electron. 10, 381 (1967).

7.19   Chang, C. Y., Fang, Y. K., and Sze, S. M., Solid State Electron. 14, 541 (1971).

7.20   Robinson, G. Y., Solid State Electron. 18, 331 (1975).

7.21   Christou, A., Solid State Electron. 22, 141 (1979).

7.22   Yokoyama, N., Ohkawa, S., and Ishikawa, H., Jap. J. Appl. Phys. 14, 1071 (1975).

7.23   Pounds, R. S., Saifi, M. A., and Hahn, W. C., Jr., Solid St. Electron. 17, 245 (1974).

7.24   Gold, R. B., Powell, R. A., and Gibbons, J. F., in "Laser-Solid Interactions and Laser Processing" (S. D. Ferris, H. J. Leamy, and J. M. Poate, eds.) p. 635. AIP, New York, (1979).

7.25   Gold, R. B., Ph.D. Thesis, Stanford University, Department of Electrical Engineering (1980).

7.26   Eckhardt, G., in "Laser and Electron Beam Processing of Materials" (C. W. White and P. S. Peercy, eds.) p. 467. Academic Press (1980).

7.27   Eckhardt, G., Anderson, C. L., Hess, L.D., and Krumm, C. F., in "Laser-Solid Interactions and Laser Processing" (S. D. Ferris, H. J. Leamy, and J. M. Poate, eds.) p. 641. AIP, New York (1979).

7.28   Emulsitone Co., 19 Leslie Court, Whippany, N. J.

7.29   Van Much, W., IBM J. Res. Dev. 10, 438 (1966).

7.30   Gibbon, C. F., and Ketchow, D. R., J. Electrochem. Soc. 118, 915 (1971).

7.31   Yamazaki, H., Kowasaki, Y., Fujimoto, M., and Kudo, K., Jap. J. Appl. Phys. 14, 717 (1975).

7.32   Arnold, N. Daemkes, H., and Heime, K., Jap. J. Appl. Phys. 19, Supplement 19.1, 361 (1980).

7.33   Sze, S. M., "Physics of Semiconductors Devices", p. 38. Wiley, New York (1969).

7.34   Nissim, Y. I., Gibbons, J. F., Evans, C. A., Jr., Deline, V. R., and Norberg, J. C., Appl. Phys. Lett. 37, 89 (1980).

7.35   Nissim, Y. I., Gibbons, J. F., Magee, T. J., and Ormond, R., J. Appl. Phys. 52, 227, (1981).

7.36   Nissim, Y. I., Gibbons, J. F., and Gold, R. B., IEEE Trans. Electron. Dev. ED-28, 607 (1981).

7.37   Murrmann, H. and Widmann, D., IEEE Trans. Electron. Dev. ED-16, 1022 (1969).

7.38   Berger, H. H., Solid St. Electron. 15, 145 (1972).

7.39   Hansen, H., "Constitution of Binary Alloys", p. 181. McGraw-Hill (1958).

7.40   Christou, A., Sleger, K., Proc. 6th Biennial Cornell, Elec. Engr. Conf., 169 (1977).

7.41   Macksey, H. M., in "Gallium Arsenide and Related Compounds" (L. F. Eastman, ed.) p. 254, Inst. of Phys. Conf. Ser. 33b, London (1977).

7.42   Nissim, Y. I., Gibbons, J. F., Gold, R. B., and Dobkin, D. M., ECS Meeting Hollywood, Florida (1980).

7.43   Sealy, B. J. and Surridge, R. K., IBBM-78, Conference
       Proceedings (J. Gyulai, T. Lohner, and E. Pásztor, eds.)
       pp. 487-495. Budapest (1979).

7.44   Williams, J. S., "Laser-Solid Interactions and Transient
       Thermal Processing of Materials," (J. Narayan, W. L. Brown
       and R. A. Lemons, eds.), p. 621. North-Holland (1983).

7.45   Arai, M., Nisibiyama, K., and Watanabe, N., Jap. Int. Appl.
       Phys. 20, L 124 (1981).

7.46   Kuzuhara, M., Kohzu, H., and Takayama, Y., Appl. Phys.
       Lett. 41, 755 (1982).

7.47   Rose, A., Pollak, J. T. A., Scott, M. D., Adams, F. M.,
       Williams, J. S., and Lawson, E. M., ""Laser-Solid Inter-
       actions and Transient Thermal Processing of Materials,"
       (J. Narayan, W. L. Brown, and R. A. Lemons, eds.), pp.
       633-639. North-Holland (1983).

7.48   Davies, D. E., McNally, P. J., Lorenzo, J. P., and Julian,
       M., IEEE Elec. Dev. Lett. EDL-3, 102 (1982).

7.49   Tabatabaie-Alavi, K., Masum Choudhury, A. N. M., Fonstad,
       C. G., and Gelpey, J. C., Appl. Phys. Lett. 43, 5 (1983).

7.50   Christel, L. A., and Gibbons, J. F., J. Appl. Phys. 52,
       8 (1981).

7.51   Kuzuhara, M., Kohzu, H., and Takayama, Y., Appl. Phys.
       Lett. 41, 755 (1982).

7.52   Davies, D. E., McNally, P. J., Lorenzo, J. P., and Julian,
       M., IEEE Elec. Dev. Lett. EDL-3, 102 (1982).

7.53   Chapman, R. L., Fan, J. C. C., Donnelly, J. P., and Tsaur,
       B-Y., Appl. Phys. Lett. 40, 805 (1982).

7.54   Tabatabaie-Alavi, K., Masum Choudhury, A. N. M., Fonstad,
       C. G., and Gelpey, J. C., Appl. Phys. Lett. 43, 505 (1983).

7.55   Tabatabaie-Alavi, K., Masum Choudhury, A. N. M., and Kanbe,
       H., Fonstad, and Gelpey, J. C., Appl. Phys. Lett. 43, 647
       (1983).

7.56   Suzuki, T., Sakurai, H., and Arai, M., Appl. Phys. Lett.
       43, 951 (1983).

7.57   Greiner, M. E., and Gibbons, J. F., Appl. Phys. Lett., 15
       April 1983 (to be published).

7.58   Onuma, T., Hirao, T., and Sugawa, T., J. Elec. Soc. 129, 4,
       837 (1982).

7.59   Kroger, F. A., "Chemistry of Imperfect Crystals," p. 275.
       North Holland (1964).

7.60   Parkinsen, B., Heller, A., and Miller, B., Appl. Phys.
       Lett. 33, 6, 521 (1978).

Index

447

Contents of Previous Volumes

453